Organic and Bio-organic Mechanisms

Organic and Bio-organic Mechanisms

Michael Page
The University of Huddersfield

and

Andrew Williams
The University of Kent

LONGMAN

Addison Wesley Longman

Addison Wesley Longman Limited,
Edinburgh Gate, Harlow,
Essex CM20 2JE, England
and associated companies throughout the world

First published 1997

British Library Cataloguing in Publication Data

A catalogue entry for this title is available from the British Library

ISBN 0-582-07484-3

Library of Congress Cataloging-in-Publication Data

A catalog entry for this title is available from the Library of Congress

Set by 16 in 10/12pt Times
Produced by Longman Singapore Publishers (Pte) Ltd.
Printed in Singapore

Contents

Viewing the stereo-diagrams

This book includes stereoscopic images which consist of two non-intersecting half-pictures. With a little practice these can yield stereoscopic views with the naked eye. Stereoscopic viewing devices such as that designed by Vögtle and Schüss provide an easier method of visualization: these are commercially available.

Preface

The fascinating chemical logic of the sequence of the reactions which occur in living systems is matched by the fascination of the reaction mechanisms of the individual steps involved. An understanding of how bonds are made and broken is essential to the understanding of life. There have been enormous advances in the application of instrumentation to elucidating chemical structures—from the smallest molecule to the largest biopolymer. These static structures have made an invaluable contribution to chemistry and biology. However, it is the knowledge of the dynamic interconversion of these structures which remains an intriguing challenge. How are the bonds between atoms rearranged? What sort of structural changes take place to cause bond fission and formation? How do catalysts lower the activation energies of reactions?

To some extent chemistry and biology are still dominated by the consideration of static structures. The three-dimensional structures of enzymes and the identification of active sites, although necessary to understand mechanisms, are too often used as the only vehicle on which to base mechanistic speculation. An appreciation of the dynamic processes involved and a deeper understanding of the *assumptions* involved in many models and descriptions will advance our understanding of the processes involved.

It is now clear that an understanding of reaction mechanisms is essential for the application of enzymes to organic synthesis and their use as biosensors and as targets for drug design. The pharmaceutical industry has been extremely successful in realizing that the development of drugs as enzyme inhibitors is strongly dependent upon an understanding of reaction mechanisms. The chemical industry, interested in clean technology, high product yield and purity, wants to know how bonds are made and broken and uses this information to show how unwanted side reactions can be prevented.

The assumptions involved in elucidating standard reaction mechanisms, and indeed the basis of the models which are used in defining the structure of states along a reaction path, are often forgotten in advanced chemical and biochemical studies. The nature of a 'state' itself is often not made clear, but such omissions are natural considering the time between the student's learning the fundamentals and entering advanced work. This text aims to redress this failing and is an introduction to the diagnosis of mechanism particularly in its application to bio-organic chemistry; we hope it will provide a handbook for the student starting research into mechanisms and reactivity. Continued advancement and development in biochemistry and chemistry require an understanding of chemical reaction mechanisms and how they are elucidated.

We discuss bonding in terms of the line formalism which remains the best working model for most chemists and biochemists; although it has limitations the model has both the 'feel' of chemical intuition and graphic utility. The book is intended to help the specialist and the non-specialist come to some meaningful conclusions about mechanism and its elucidation. There are chapters on the fundamental assumptions involved in describing reactions and structures, and there are descriptions of the methods used to elucidate mechanisms as well as examples of biologically important reactions, catalysis and enzymes. Suggestions for reading 'in depth' are given at the end of most chapters; and readers can judge for themselves from the titles which references are useful for general reading.

M. I. Page and A. Williams. October 1995
Huddersfield and Kent

1 The transition state

1.1 Mechanism as a progression of states

The mechanism of a reaction is often described as the structure and energy of a molecule through its progress from reactants to products. Such a definition is obviously suspect because the properties of a single, static molecule cannot be measured and because structure itself requires a definition. Energy is required to transform reactants to products and the energy barrier in a single-step reaction arises from a transitional structure which has an existence of only 10^{-13} to 10^{-14} s. The assembly of transitional structures 'in passage' from reactant to product states is known as the *transition state*. The collection of reactant molecules is converted into the collection of product molecules through a transition state which embraces a collection of transient species which effectively has a 'normal' thermodynamic distribution of energies even though these structures/energies are not interconvertible within their lifetimes. Thus the mechanism of a reaction could be defined as the structures of states on progression from reactant through transition states to product. A definition of mechanism which can be fulfilled experimentally, at least in principle, is a description of any intermediates and all the transition-state structures connecting these intermediates, reactants and products.

Theory and gas-phase work have provided information on the energy of single entities as they go through to products for a limited number of reactions. A high-level definition of mechanism is the energy surface of such a progression as a function of all degrees of freedom. This definition is not attainable for reactions in solution.

1.2 Structure and its interpretation

In order to discuss the mechanism of any reaction and transition-state structures it is necessary to know precisely what the term 'structure' means. The chemist visualizes a pure compound as an assembly of molecules, each atom of which has identical topology relative to its neighbour. The relative position of each atom varies with time and the *average* positions of the atoms are measured. Even the most explicit method of structure determination, namely X-ray crystallography, cites the atomic coordinates with a degree of uncertainty, partly because the method depends on X-rays being diffracted by the electrons and not the nuclei. Moreover, analytical methods provide results for assemblies rather than for an individual molecule. A single crystal used for an X-ray structure determination is likely to contain up to 10^{20} molecules.

It is important to remember that, when chemists write structures with bonds represented by lines (Lewis bonds), these are hypothetical *models*. These structures (which are commonly called Kekulé structures) are often taken for reality but they are simply representations of hypotheses which fit experimental knowledge of compounds. There are many ways available to represent molecules, each with its own advantages. Most have the disadvantage that they refer to a single molecule and assume that the constituent atoms have time-stable spatial coordinates relative to each other. There is nothing superior to the line bonding model, which readily graphs an assembly with a facility that is readily comprehensible to all chemists. Most reactions are carried out in solution and there is a great, but unfulfilled, need for a simple, graphical, model of an assembly of molecules in solution.

Descriptions of mechanisms in this book are couched in a language devised for structural studies and can therefore be misleading if the assumptions are forgotten. For convenience, and following precedent, solvent is often omitted from descriptions of state in this text; moreover the term 'bond' is invariably used to mean the summation of electronic bonds (in its 'Lewis' sense) and solvation.

Structure requires a description of the relative positions of nuclei and the electron density distribution, and how these vary with energy. Even at absolute zero, the exact positions of the atoms in a molecule are uncertain, as reflected in zero-point energies. As the temperature is increased, higher quantum states are occupied for each degree of freedom so that fluctuations around the mean positions of the atoms increase. Most vibrational motions are decoupled from each other so that the apparently static pictures which are drawn of the relative positions of atoms in a molecule can occasionally be very misleading. For example, at room temperature, most covalent C–C bond lengths are $1.54\,\text{Å}$ but individual ones fluctuate with time by $\pm0.05\,\text{Å}$; bond angles at saturated carbon are, on average, $111°$ but vary between $106°$ and $116°$. The mean square amplitude of vibration is inversely proportional to the reduced mass and force constant, so that either a small mass or a low force constant gives rise to a large vibrational amplitude. For example, the classical turning points for the bending mode of water are at HOH angles of $83°$ and $127°$. Non-covalently bonded atoms move even more with respect to each other than normal vibrational motions; this means that $O-H\cdots O$ bond lengths and angles may be $\pm0.15\,\text{Å}$ and $\pm25°$, respectively.

Intrinsic uncertainty in nuclear positions is inversely matched by an equivalent ambiguity about electron density. The distribution of electrons is important because nuclear motion is much slower than that of the lighter electrons. It is often considered that nuclei can be imagined to move within a constant force field generated by the electrons. Crudely speaking, electron transfer occurs when a suitable nuclear configuration has been achieved. This simple fact explains the chemist's pre-occupation with electrons and 'curved arrows'. Most tools used to elucidate transition-state structures give a measure of the apparent electron density/charge around an atom and the geometrical arrangement of the atoms. However, the relationship between charge and structure is not straightforward, even in stable systems. For example, the resonance structures of amides include negative and positive charges on oxygen and nitrogen respectively. However, even if the absolute partial charges on these atoms were known, they would not necessarily be informative about the relative single/double bond character and bond lengths of the amide. Furthermore, the charge will vary on transfer of an amide hydrogen-bonded in water to a 'free-molecule' in a non-polar solvent.

Knowledge only of the structure of ground, intermediate and product states does not enable us to calculate the appropriate rate constants for an enzyme-catalysed reaction, or any other reaction. This problem is neatly summarized in an amusing analogy, attributed to Jeremy Knowles: *knowing the structure of a reactant such as an enzyme and possessing the picture of a horse tells us neither about the catalytic activity of the enzyme nor about the Derby winning propensities of the horse!*

The development of the idea of mechanism since the first studies at the beginning of the 20th century (Lapworth, 1903, 1904, 1907) went hand in hand with progress in methods for its determination. The concepts of mechanism sought by Lapworth were not very different from the descriptions pursued today in that they were couched in Lewis-type language. The most important advance since Lapworth's era is the development of the concept of the transition state. Our improved understanding of molecular structure demands an ability to think in terms of multidimensional space if this is extended to descriptions of mechanism.

The maximum of the potential energy along the reaction coordinate between reactants and products corresponds to the *transition-state structure,* or activated complex. The *transition state* is a quasi-thermodynamic *state* and is at the maximum of the Gibbs' free energy along the reaction coordinate. This free energy represents the pseudo-collection of molecules of the transition state distributed among the available quantum states of the various degrees of freedom as reflected in the entropy. The maximum in the potential energy along the reaction coordinate is temperature-independent, whereas the transition-state structure may be temperature-dependent because of entropy effects. 'Structure' usually refers to potential energy and strictly we should always refer to the transition structure or 'transition-state structure' but common usage abbreviates this simply to the 'transition-state'. In this text we adopt this rather casual approach but it is important to remember that it is a simplification of phraseology.

1.3   Interconversion of states—reaction and encounter complexes

A bimolecular reaction in solution occurs via the following series of events. Two reactant molecules diffuse through the assembly of solute and solvent molecules and collide to form an *encounter complex* within the same solvent cage. If the molecules are charged, then the ionic atmosphere adjusts to any changes in the combined charge. Reaction may still not be possible until any necessary changes in solvation occur (such as desolvation of lone pairs) to form a *reaction complex*, in which bonding changes take place. The encounter complex remains essentially intact for the time period of several collisions because of the protecting effect of the solvent surrounding molecules once they have collided. The products of the subsequent reaction could either be converted back to reactants or diffuse into the bulk solvent.

A similar description applies to a unimolecular reaction except that the transition state, formed from a single reactant molecule, is initiated by energy accumulation in the solvated reactant by collision between reactant and solvent molecules.

Scheme 1.1 gives typical half-lives for reactant molecules *destined* to react. Many encounters do not lead to reaction and only a small fraction of the complexes will have the appropriate transition-state solvation in place for reaction to take place.

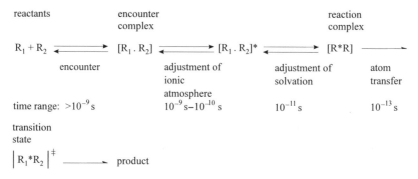

Scheme 1.1 Bimolecular mechanism in solution.

Reaction complexes in enzyme-catalysed reactions are more ordered than those in reactions of simple molecules and often constitute relatively thermodynamically favourable species. The enzyme active site provides a special microsolvation for the reaction compared with that for the uncatalysed reaction in bulk solution; time must elapse after the first encounter of substrate with enzyme molecule before the active site is occupied.

1.4   Methods of representing reaction mechanisms

The mechanism of nucleophilic aliphatic substitution was the first to be studied in depth. It is exemplified by reactions of alkyl halides (Eqn [1.1] and structure **1**). The reaction type is the paradigm for many bio-organic reactions, including biological methylation and the transfer of the glycosyl moiety between nucleophiles.

$$Nu^- \quad C-Cl \longrightarrow Nu-C \quad + \quad Cl^- \qquad [1.1]$$

$$\left[\begin{array}{c} \overset{\delta-}{Nu} - - \overset{|}{C} - - \overset{\delta-}{Cl} \\ \end{array} \right]^{\ddagger}$$

1

Equation [1.1] illustrates a 'mechanism' as represented in most studies of organic reactions. A knowledge of reactant and product structure indicates that only two bond changes are involved. The nucleophile donates electrons and the leaving group attracts electrons; the passage of pairs of electrons is represented by a sequence of curved arrows. This description does not indicate the structure of the transition state although some structure similar to the trigonal bipyramidal arrangement (**1**) must be traversed in the reaction.

The transition state is effectively an assembly of molecular structures which exists for less than 10^{-13} s; the measurement of its properties by conventional means is not possible because the measuring devices have relaxation times larger than this. Recent gas-phase work employing femtosecond (10^{-15} s) light pulses as probes can glimpse the molecule at various stages along the reaction coordinate (Pilling & Smith, 1987; Zewail, 1988; Baggott, 1989; Smith, 1990; Polanyi & Zewail, 1995). Since the velocity of separation of atoms constituting a bond is about 0.01 ångstroms per femtosecond (1 Å per 10^{-13} s) the time-scale for separation is a few hundred femtoseconds and resolution is therefore possible with light pulses of a few femtoseconds duration. Such studies are limited by the *uncertainty principle*, especially as it cannot be assumed that the interaction of the light pulses with the molecule is 'innocent'. It is only possible to study simple gas-phase reactions by this technique; nevertheless the results are very useful as models for more complicated systems.

In general it is not possible to determine the structure of species in a transition state in the same way that we can measure structure for a regular assembly of molecules. The structural information that we require involves the positions of the atoms and bond order (in particular that for the bonds undergoing major changes) and knowledge of the electronic structure such as the electronic charge at atoms.

The description represented by structure **1** is commonly called a mechanism, but the real reaction is between *assemblies* of molecules and in bio-organic chemistry the molecules are in solution. For example, in aqueous solution there will be a dramatic change in the solvation around the chlorine as it is converted from a relatively neutral entity to an anion. The solvent contributes to the activation energy and is fundamental in determining the relative charge and atom distribution drawn to represent the transition-state structure. It is the general convention to neglect the effect of the solvent molecules in the description as shown, because of the difficulties of graphical representation.

The only measurable property of the transition state is its energy relative to reactant or product states and this is obtained by kinetics, or indirectly, by product distribution studies. *All* experimental knowledge of transition-state structures for solution reactions comes from such measurements and includes stereochemical, trapping, isotopic labelling and product isolation techniques. These techniques will be discussed in Chapters 2, 3 and 4.

1.5   General considerations concerning reaction mechanisms

Single- and double-electron transfer

Ingold (1953) divided mechanisms broadly into those proceeding by bond fission through two-electron transfer (heterolytic) and those by single-electron transfer (homolytic) (Scheme 1.2).

Electron pair transfer
Heterolysis $A:B \longrightarrow A^+ + B:^-$

Single electron transfer
Homolysis A $A:B \longrightarrow A^{\cdot} + B^{\cdot}$

Scheme 1.2 Types of mechanism.

These divisions are still relevant except that there is now considerable overlap between the two types, and radical cations and radical anions can also be involved in many solution reactions. For example, nucleophilic substitution at an aromatic centre can involve a radical anion in an S_{RN} process (Kim & Bunnett, 1970). Reactions in polar solution are often heterolytic because of the massive solvation stability afforded to ions, whereas gas-phase reactions are often homolytic. The distinction between electron-pair and single-electron transfer (SET) may not be clear-cut if the apparent heterolytic reactions involve 'inner-sphere' single-electron transfers which do not express themselves as free radicals or even radicals caged in encounter complexes (Pross, 1985; Shaik, 1990; Savéant, 1990, 1993).

The observation of free radicals or radicals caged in encounter complexes by use of CIDNP (chemically induced dynamic nuclear polarization) experiments in NMR is incontrovertible evidence for SET processes; however, the absence of evidence for radicals is not sufficient to disprove the existence of 'inner-sphere' SET mechanisms (Ashby & Pham, 1987; Newcomb & Curran, 1988). 'Inner-sphere' and 'outer-sphere' single-electron transfer mechanisms for nucleophilic displacement (Rossi *et al.*, 1989) are illustrated in Scheme 1.3.

Inner sphere:

$$R-X + Nu{:}^- \longrightarrow [Nu \cdot \cdot R-X]^- \longrightarrow [Nu-R \cdot \cdot X]^- \longrightarrow Nu-R + X{:}^-$$

Outer sphere:

$$R-X + Nu{:}^- \longrightarrow R-X^{\cdot-} + Nu^{\cdot}$$
$$R-X^{\cdot-} \longrightarrow R^{\cdot} + X{:}^-$$
$$R^{\cdot} + Nu{:}^- \longrightarrow R-Nu^{\cdot-}$$

or

$$R^{\cdot} + Nu^{\cdot-} \longrightarrow R-Nu$$
$$R-Nu^{\cdot-} + R-X \longrightarrow R-Nu + R-X^{\cdot-}$$

Sum: $R-X + Nu{:}^- \longrightarrow R-Nu + X{:}^-$

Scheme 1.3 Inner- and outer-sphere single-electron transfer (SET) mechanisms.

The distinction between electron-pair and single-electron transfer involves identifying processes in which electron transfer, bond breaking and bond making are stepwise, and processes where they are concerted. Although the 'outer sphere'/'inner sphere' terminology was used originally for electron-transfer reactions involving metal complexes it can be applied to organic reactions (Lexa *et al.*, 1987). Bond making and bond breaking in outer-sphere reactions occur in separate steps distinct from electron transfer. If all the steps are concerted the reaction occurs by inner-sphere electron transfer mechanisms (Scheme 1.3) which is difficult to distinguish from a classical S_N2 mechanism involving electron-pair transfer.

The transition state of a single-electron transfer from $Nu{:}^-$ to X can be represented by the resonance hybrids represented by structure **2**.

$$\left[Nu{:}^- \quad R \quad X \longleftrightarrow Nu \quad R \quad {:}X^- \right]^{\ddagger}$$

2

The extent of bond breaking and the extent of electron transfer cannot readily be separated (Perrin, 1984). The length of the R–X bond is itself an essential coordinate in controlling the occurrence of electron transfer, subject to Franck–Condon restrictions, in the concerted electron transfer–bond breaking pathway and in the outer-sphere electron transfer in the framework of the Born–Oppenheimer approximation (see Section 1.9). The electronic 'reshuffle' from reactant to product configurations takes place 'instantaneously' when the nuclei, which move more slowly than electrons, adopt the appropriate intermediate configuration between that of reactants and that of products. The occurrence of electron transfer depends on solvent reorganization and vibrational modes other than R–X stretching, but it does not seem appropriate, in general, to regard electron transfer and bond breaking as two independent phenomena.

It is difficult to define whether two, one or a non-integral quantity of electrons transfer in a regular S_N2 reaction. The conventional hypothesis, until recently, has been that two electrons always 'go together'. Since the mid-1980s there has been considerable discussion (Shaik, 1985; Bordwell, 1987) indicating that electrons may move one at a time (Scheme 1.3), particularly with nucleophiles which are members of one-electron reversible redox couples. The factors governing whether radical reactions occur (i.e. whether the radical escapes from the reaction complex) are those which govern the coupling of the spin paired electrons following the electron shift. This enables us to understand why reactions sometimes involve radicals and sometimes involve straightforward heterolytic processes.

In the above discussion we have exemplified the problem of electron transfer with nucleophilic aliphatic substitution. Proton transfer between bases could also be considered as an SET process (Eqn [1.2]), as can electrophilic substitution in benzene (Eqn [1.3]).

$$B:^- + H\text{--}A \longrightarrow [B:H \cdots A]^- \longrightarrow [B \cdots H:A]^- \longrightarrow B\text{--}H + A:^- \qquad [1.2]$$

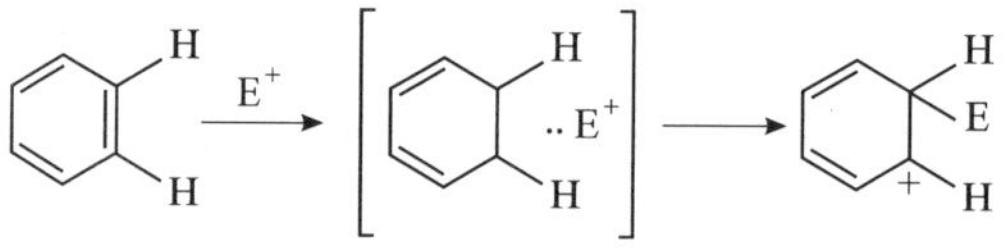

$$[1.3]$$

Classification of reactions

So far we have discussed the way in which individual bond changes can occur. Most reactions involve at least two major bonding changes and Ingold classified reactions into four main types—substitution, addition, elimination and rearrangement (Scheme 1.4). The classification is based on the stoichiometry and *not* on the mechanism.

Substitution (S)	$Y + B\text{--}X \longrightarrow B\text{--}Y + X$
Addition (Ad)	$A + B \longrightarrow A\text{--}B$
Elimination (E)	$A\text{--}B \longrightarrow A + B$
Rearrangement (not designated)	$X\text{--}A\text{--}B\text{--}Y \longrightarrow Y\text{--}A\text{--}B\text{--}X$

Scheme 1.4 Ingold's classification of reactions.

It is important to recall that this classification records observations about the structure of *reactants* and *products*. Since mechanism is strongly connected with classification, the symbols that Ingold and later workers used to denote reaction types have come to be used as symbols for mechanistic types. A IUPAC group has proposed a new symbolism to refer to *mechanisms*; its application is hotly disputed so we gather the most important current symbols together with a brief description of the IUPAC scheme in Appendix A.1 (Guthrie & Jencks, 1989).

When mechanisms are studied it is surprising how the basic types described above suffice, often in combination, to describe the overall reaction. For example, the mechanism for the reaction of hydroxide ion with esters is simply a combination of an addition and an elimination process (eqn [1.4]).

$$R-\overset{\overset{O}{\|}}{\underset{\underset{HO^-}{}}{C}}-OR \quad \xrightarrow{(Ad)} \quad R-\overset{\overset{O^-}{}}{\underset{\underset{HO}{}}{C}}-OR \quad \xrightarrow{(E)} \quad R-\overset{\overset{O}{\|}}{\underset{\underset{OH}{}}{C}} + RO^- \qquad [1.4]$$

Moreover, it is often possible to break down substitution and rearrangement reactions into addition or elimination components.

How electrons redistribute themselves in a reaction

Some form of representation is required which graphically illustrates the dynamic flow of electrons. Reactions are often delineated by the application of 'curved arrows' or 'hooks', which represent the flow of electron pairs or single electrons respectively. This technique has the advantage that it enables all the electrons in the system to be accounted for and ensures that valence rules are not being broken. When used correctly, the 'curved arrow' method enables predictions to be made concerning charge transfer; it can be readily predicted, for example, that the reaction of hydroxide ion with chloramine to yield hydroxylamine involves depletion of negative charge on the oxygen and build-up of charge on the chlorine (eqn [1.5]).

$$Cl-\overset{\overset{H}{|}}{\underset{\underset{H}{|}}{N}}\quad {}^-OH \quad \xrightarrow{-Cl^-} \quad \overset{\overset{H}{|}}{\underset{\underset{H}{|}}{N}}-OH \qquad [1.5]$$

Another representation involves the pictorial use of orbitals where bond formation is the result of overlap of a HOMO (highest occupied molecular orbital) with a LUMO (lowest unoccupied molecular orbital); orbital symmetry must be taken into account in these considerations (Tedder & Nechvatal, 1988). Neither the curved arrow nor the pictorial use of orbitals says anything about how the electrons are transferred.

Mechanism should never be excluded just because it does not fit in with current theory. For example, frontier molecular-orbital theory (Fleming, 1976) appears to indicate that bimolecular substitution at a carbon centre involves inversion of configuration because the LUMO of the bond to the leaving group has a lobe directed to the rear of the carbon (Scheme 1.5) and appears to exclude frontside attack. There are so few authenticated examples of frontside attack, to give retention of configuration (see Chapter 3), that empirical rules have even been advanced for the stereochemical outcome of attack at carbon to be used to deduce the type of mechanism

Scheme 1.5 A frontier molecular-orbital representation of nucleophilic substitution.

involved. Recent thought has indicated that the simple theory, as given above, may not always apply (Harder *et al.*, 1995).

Retention of configuration occurs during nucleophilic displacement at silicon (Sommer, 1965; Holmes, 1990) due to frontside attack of the nucleophile. This is considered to be caused by the electronegativity of the central atom, the high electronegativity of the leaving atom, a high percentage of 's'-character in the central atom and a hard nucleophile (Anh & Minot, 1980). Frontside attack occurs with some reactions of nucleophiles at phosphorus and sulphur atoms. Retention mechanisms, where frontside attack is concerted with leaving group departure, can be explained theoretically (Anh & Minot, 1980) using basis sets with no d-orbitals. Retention of configuration for reactions at saturated carbon is invariably interpreted as a result of double inversion. However, it is *not* impossible that carbon could suffer retention of configuration in an S_Ni-type of process. Frontside attack would be possible when overlap of the attacking HOMO with the frontside LUMO of the central atom outweighs the out-of-phase overlap of the LUMO with the antibonding lobe on the leaving atom (Scheme 1.6).

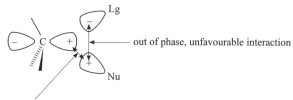

favourable interaction

Scheme 1.6 Non-classical frontside attack is not impossible in the S_N2 reaction.

It is logical that frontside attack might be possible with nucleophilic aliphatic substitution under certain conditions when bond fission is extensive and bond formation is not advanced, so that the leaving atom and the nucleophilic atom do not interfere with each other (Scheme 1.6). One could imagine a spectrum of possibilities where the frontside and rear trajectories are alternately favoured. The trajectories within the reaction complex are governed primarily by Coulombic interactions, orbital interactions and steric requirements.

Three-dimensional reaction diagrams—coupling between bond formation and fission

The classical description of bond making and breaking usually concentrates only on the atoms directly involved, although many other changes occur during this fundamental process. For example, the ionisation step in an S_N1-type mechanism which nominally involves formation of a three-coordinate intermediate from a four-coordinate reactant is often represented as in Eqn [1.6]; the direction of the 'stretching' motion of the C–X bond is commonly called the *reaction coordinate*. In addition to the C–X motion several other changes are involved, for example in the R–C bond length and the R–C–R bond angle. These processes are relatively minor and are often ignored in a description of the reaction coordinate but the degree of 'coupling' of these changes does contribute to a detailed description of the mechanism.

$$
\begin{array}{ccccc}
R & & R & & R \\
\backslash & & \backslash + & & | \\
R-C-X & \longrightarrow & \left[R-C\text{---}\bar{X} \right]^{\ddagger} & \xrightarrow{-X^-} & {}^+C \\
/ & & / & & / \ \backslash \\
R & & R & & R \ \ R
\end{array}
\qquad [1.6]
$$

The majority of reactions involve at least two major bond-making and bond-fission steps. The few reactions where there is only a single major bonding change are generally confined to the gas phase and even these involve hybridization changes. A pathway with a single step and no intermediate is defined as a *concerted mechanism*; it follows that a concerted mechanism has a single transition state. The major bonding changes in a concerted mechanism are not necessarily coupled with each other. Concerted mechanisms do not require transition state structures where the extent of bond formation and bond fission has advanced to the same extent; for example in **1** Nu $\cdots$ C and C $\cdots$ Cl are not necessarily 'half'-bonds. The term *synchronous* is used to denote a concerted mechanism where the bond changes have occurred to the same extent in the transition state. Coupling between bond fission and formation is useful to transmit energy and hence to facilitate a reaction (Bernasconi, 1992a,b). For this reason a concerted mechanism can offer a pathway energetically more favourable than its step-wise counterpart.

It is not possible to measure energies for the whole of a reaction coordinate for solution reactions although this is becoming possible for some *very simple* gas-phase reactions where the experimental determination of Morse curves has also been possible for some time (Mills, 1977). Descriptions of reaction mechanisms are often effected by potential energy diagrams and these can be two-dimensional (Fig. 1.1) or three dimensional (Fig. 1.2). These diagrams are useful as aids to classification and understanding of mechanisms, and refer to the reaction within the common solvation shell of the reaction complex. It should be noted that the diagrams should be *multi*dimensional; it is assumed that a point on the three-dimensional sur-

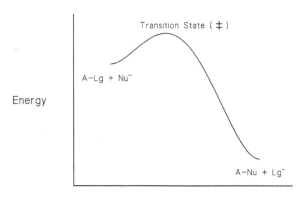

Fig. 1.1 Two-dimensional potential energy diagram for the reaction
$A — Lg + Nu^- \longrightarrow A — Nu + Lg^-$ involving two main bond changes.

face represents the energy minimum for all the degrees of freedom not represented in the coordinates for the two major bonding changes (Jencks, 1972).

The equations of the curves quoted in the literature for these diagrams are defined to give the correct slopes at various boundary conditions. For example, at the left and right edges of the two-dimensional diagram the energies are fixed from known values and the slopes of the lines are zero; in the three-dimensional diagram the surface at the corners should be horizontal. The local curvatures at transition states or energy minima are often given some form of empirical equation (Lewis, 1986; Guthrie, 1990, 1991).

1.6   Energy transfer, redistribution and relaxation

The Rice, Ramsperger, Kassel and Marcus (RRKM) theory (see, for example, Pilling & Seakins, 1995) of unimolecular decomposition assumes that internal energy relaxation occurs on a shorter time-scale than molecular breakdown. Unimolecular decomposition is usually preceded by a process of activation such as reaction between two molecules to produce a high-energy structure. Large amounts of energy may enter a localized region of the substrate molecule in a very short period of time by a particular mechanism and in a non-random fashion. In thermal activation it is usually assumed that collisional events transfer energy randomly into a molecule so that the energy is distributed among the available degrees of freedom— vibration, rotation and translation. The problem of energy flow, even in isolated molecules, has an element of randomness, even when many states of a molecule may be nearly isoenergetic. Sometimes transitions between states occur freely, while under other circumstances the molecule may remain largely in whatever state it starts.

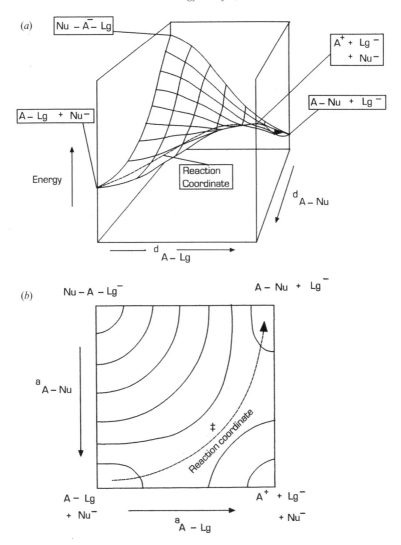

Fig. 1.2 Three-dimensional potential energy diagram for the reaction of Fig. 1.1: (*a*) perspective view; (*b*) energy contour map. These maps and diagrams register energy changes within the reactive encounter complex (Guthrie, 1990).

The validity of transition-state and RRKM theory depends on having sufficiently rapid energy flow between states for statistical estimates of the probability of escape to be valid. If a reaction coordinate involves predominantly one degree of freedom, then its uptake of energy occurs at the expense of the other $3n - 7$ modes of freedom. If this were not the case, then the total activation energy required to achieve reaction would depend on the number of non-reactive degrees of freedom (i.e. on the number of atoms in the molecule) but the activation energies for the same type of reaction in a group of similar molecules do not increase with the size of the molecule.

The rates of intramolecular redistribution of vibrational energy are much lower at low levels of energy uptake. Low-energy vibrations are generally approximately harmonic, whereas anharmonicity of vibrations promotes energy transfer between modes. The classic experiment which supports traditional transition-state and RRKM theory involves the generation of hexafluorobicyclopropyl with a large amount of internal energy (111 kcal mol^{-1}; 464 kJ mol^{-1}) which is initially non-statistically distributed (Rynbrandt & Rabinovitch, 1971). Subsequent decomposition to branched products indicated that 3.5% was formed by non-random decomposition of the newly formed ring (Scheme 1.7).

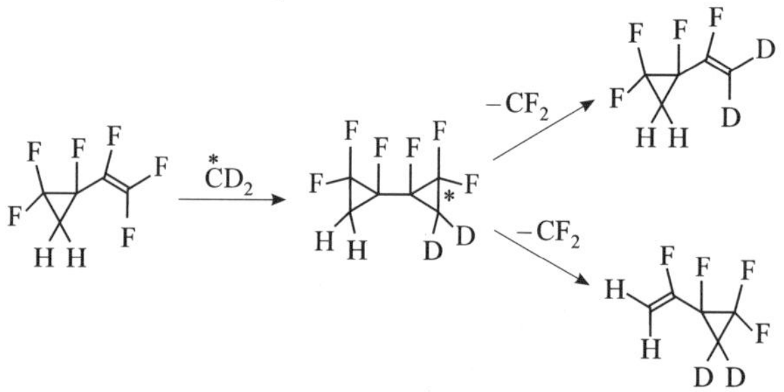

Scheme 1.7 Yields of branched products differ because energy redistribution in the energized molecule lags behind product formation.

Two opposing factors result in the free energies of activation of reactions being independent of the number of degrees of freedom. The probability of a given molecule in an assembly having a high energy increases with increasing number of degrees of freedom (Fig. 1.3). Conversely, the probability of that energy being localized in a degree of freedom such as bond stretching decreases with increasing number of degrees of freedom. These two factors cancel so that, in effect, the probability of a given molecule having the necessary activation energy *and* having it localized in the necessary degrees of freedom is independent of the total number of degrees of freedom available.

Many reactions in solution require a high degree of coupling between several degrees of freedom, which may be intermolecular when solvent molecules interact with the substrate; however, the probability of this localization of energy would appear at first sight to decrease with increasing complexity of the interaction.

Maximum stabilization by a polar solvent for a unimolecular ionic dissociation (Scheme 1.2) is effected by equilibrium solvation of the reactant during the entire dissociation process. The ionization process is, in effect, a competition between electronic coupling—which favours a delocalized transition state of 50% ionic character—and the solvation free energy—which

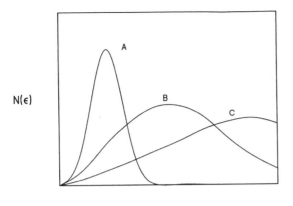

N(ε)

Energy

Fig. 1.3 Probability $N(\epsilon)$ of a given molecule in an assembly having a high energy increases with increasing degrees of freedom (degrees of freedom increase in the order A < B < C).

favours a localized electronic character of the transition state that is purely ionic. The Hammond postulate (see Section 6.2) predicts that the transition state for the ionization step should become more covalent (less ionic) and 'tight' as the solvent polarity increases (Kim & Hynes, 1992; Fig. 1.4).

It is usually assumed that the solvent remains completely equilibrated with the reactant as a bond stretches and breaks and changes its electronic structure on the passage through the transition state. However, the response of reorganization of solvation, for example in orientational polarization, is not sufficiently fast for the solvent to remain equilibrated. The reaction may take place in a 'frozen' solvation environment where the solvent acts as a kind of immobile spectator rather than as the coupled rearranging partner

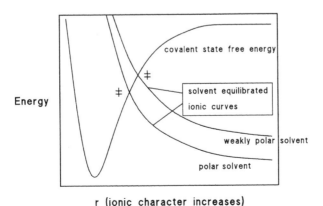

Energy

covalent state free energy

‡

‡

solvent equilibrated
ionic curves

weakly polar solvent

polar solvent

r (ionic character increases)

Fig. 1.4 Effect of solvent polarity on the transition-state structure for a unimolecular ionization, $R - X \longrightarrow R^+ + X^-$.

usually envisioned; this may be true for reactions where it only takes a very short time to cross the barrier (Bernasconi, 1992a,b).

The solvent may be treated as if it were a dielectric continuum, but this neglects an entropic contribution to the free energy of activation of microscopic ordering of solvent molecules by reactant solute. Such interactions could become important for polar solvents and polar reactants.

It is difficult to correlate linear free energy parameters, such as ρ or β (see Chapter 3), with the transition-state ionic character or geometry *on an absolute scale*. The coefficients should be interpreted on a *relative* scale, for example a *trend* in transition-state structures for them to become more reactant-like and less product-like for ionization reactions in more polar solvents.

1.7   Stereoelectronic effects

'*Stereoelectronic control*' is a general term describing the effect of stereochemistry on reactivity (Kirby, 1966). There are equilibrium and kinetic stereoelectronic effects which indicate that some geometrical relationships between atoms are more favourable than others. The unexpected preference of electronegative substituents in the anomeric position of a pyranose, and generally in the 2-position of a tetrahydropyran, for the axial rather than the equatorial position is called the *anomeric* effect. The inversion of configuration associated with the S_N2 mechanism is interpreted in terms of the nucleophile approaching preferably at 180° to the leaving group.

Leaving groups, Lg, depart from a tetrahedral carbon atom substituted with one or more heteroatoms bearing lone pairs of electrons more favourably when these lone pairs are antiperiplanar to the leaving group (Scheme 1.8) compared with when they are orthogonal or synperiplanar. This phenomenon is also known by the acronym ALPH (antiperiplanar lone pair hypothesis) and is related to other stereoelectronic effects such as the preferred *anti*-stereochemistry in E2 elimination reactions or the anomeric effect. The anomeric effect has been generalized to describe any X–C–Y system where X and Y are heteroatoms, and it may be rationalized either by electrostatics resulting from the smallest repulsion between adjacent dipoles or by a frontier-orbital approach which envisages electron donation from the lone pair of X into the antibonding orbital of the C–Y bond which is undergoing fission.

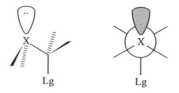

Scheme 1.8 Schematic diagrams to show the conformation of the lone pair on X adjacent to the leaving group (Lg).

Most of the evidence which is compatible with the ALPH is of a preparative nature involving product analysis or, often, the 'failure' of a particular reaction. This approach is largely non-quantitative and often requires assumptions about the mechanism and conformations of intermediates.

The principle of microscopic reversibility (see Section 6.3) applied to the ALPH suggests that the attack of nucleophiles on three-coordinate carbon attached to a heteroatom bearing a lone pair of electrons should occur from the direction which generates an antiperiplanar relationship between the newly formed bond and the lone pair. The six-membered ring product formed in an intramolecular nucleophilic attack on a trigonal carbon should have an equatorial lone pair (Scheme 1.9).

Scheme 1.9 Conformations required by ALPH for ring opening, and formed immediately upon ring closure of six- and five-membered rings. Nu is the attacking nucleophile in ring closure and the leaving group in the ring opening. X is the heteroatom bearing a lone pair.

The lone pair cannot be exactly antiperiplanar to the leaving group without introducing strain into the analogous reaction to produce a five-membered ring. However, five-membered ring opening during, for example, the hydrolysis of oxazolidines, thiazolidines and 1,3-dioxolanes occurs readily and its reverse reaction, ring closure, is often a rapid process instead of the slow one predicted by ALPH.

Further contradictions of the ALPH occur in penicillin chemistry where nucleophilic attack on penicillins (3) occurs from the α-side even though this generates a synperiplanar lone pair on the β-lactam nitrogen (4).

RCONH

RCONH

Nu

CO_2H

CO_2H

3

4

Lone pairs are commonly drawn as occupying orbitals with a specific geometry, for example sp^2 or sp^3 (two lone pairs on an atom are often described as 'rabbit ears'). Electron distributions obtained by X-ray and neutron diffraction difference maps rarely indicate lone pairs as localized entities; whilst it remains graphically attractive, the ALPH can only be regarded as having marginal energy significance.

A major success of ALPH is in the explanation of the reduced lengths of C–X bonds and extended C–Lg bonds often found in X–C–Lg systems when the lone pair on X is assumed to be *anti*-periplanar to the C–Lg bond (Scheme 1.8). However, even here the bond length changes do not vary in a systematic and predictable way (Sinnott, 1988). The importance of orbital alignment in ALPH is expected to vary depending on the degree of bond making and breaking in the transition state. The significance of heteroatom electron-pair 'push' in expelling a leaving group on an adjacent carbon atom (Scheme 1.8) will depend on the stabilities of the reaction/intermediate compared with the leaving group and the incipient electrophilic centre, which control the degree of bond fission between carbon and the leaving group. An early transition state is expected to be less dependent on ALPH than a late one. However, there is little evidence to support the idea that the importance of ALPH does vary with transition-state structure.

1.8   Principle of non-perfect synchronization

The imbalance deduced to occur in the transition state between the degree of bond formation and bond breaking or the amount of charge transfer between atoms has been studied particularly by Claude Bernasconi (1992a,b). He formulated the principle that a product stabilizing factor in a reaction whose development at the transition state is late will increase the activation energy and lower the intrinsic rate constant (k_0) (see Chapter 6). The base-catalysed ionization of nitroalkanes (Eqn [1.7]) requires proton transfer to the base and delocalization of the negative charge formally developed on the carbanion. Linear free-energy relationships (see Chapter 3) suggest that the degree of proton transfer and the accompanying charge development in the transition state are not counterbalanced by an equivalent amount of charge delocalization on the nitro group, which requires reorganization of solvent molecules and changes in geometry/hybridization around carbon.

$$[1.7]$$

Charge delocalization in the transition state lags behind bond fission in the ionization of nitroalkanes, implying a similar lag in development of resonance stabilization of the transition state; this increases the intrinsic barrier (Chapter 6) compared with that of the hypothetical case where synchronous development occurs. Reference to Fig. 6.2 indicates that the rate constants for the deprotonation of acetylacetone by bases (B) are well below those expected from the 'Eigen' plot where diffusion is rate-limiting and the low intrinsic rate constant (rate constant when $\Delta G_0 = 0$) for this reaction ($\log k_0 = 2.75$) could be due to the lag in resonance stabilization behind proton transfer. The deprotonation reactions of nitroalkanes with bases exhibit even lower intrinsic rate constants ($ArCH_2NO_2$ with morpholine has $\log k_0 = -1.22$). The principle of non-perfect synchronization provides an explanation of a relatively long-standing problem where carbon acids exhibit much lower proton-transfer rate constants compared with those for heteroatom acids of similar pK_a (Kresge, 1975a).

1.9   Principle of least nuclear motion

Electrons are much lighter than nuclei and hence their wave properties are more pronounced. Their difference in mass is large enough for their motions to be treated separately (Born & Oppenheimer, 1927). The principle of least nuclear motion states that reactions of a given molecule which involve least nuclear motion are favoured (Hine, 1977). This arises from the fact that the energy required for the displacement of atoms from a stable equilibrium geometry varies approximately as the square of the displacement for small displacements (Hine, 1966, 1977). Reactions are thus favoured in which nuclear motions are minimized. However, least-motion effects are often energetically small and are frequently overridden by other factors. Solvent reorganization is often a major contributor to the activation energy and it is not easy to apply the principle of least nuclear motion to reactions in which solvent motion is significant.

1.10   Is the transition state a molecule?

The concept of 'equilibrium' in the transition state should not be interpreted literally. The implication is *not* that there is a long enough lifetime and a sufficient number of molecules in the transition state to exchange energy and consequently to have a Boltzmann distribution of transition-state molecules. The 'lifetime' of the transition-state 'molecule' is not generally sufficient to allow time for rotations and vibrations (however, see Polanyi & Zewail, 1995). There is *effectively* a Boltzmann distribution in the transition state because a sufficient number of molecules in 'active collision complexes' pass through the various energy levels at the transition state. The active collision complexes are derived from an equilibrium energy distribution of reactants. This is illustrated in the simplified energy surface (Fig. 1.5) drawn in two dimensions which gives a pictorial view of the activation energy and the entropy of activation arising from the ratio of the partition functions (the number of energy levels occupied) for

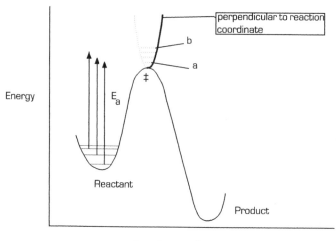

Fig. 1.5 Simplified potential energy surface for a reaction, showing energy contours and reaction coordinate; a and b are vibrational levels in the transition state.

transition state and reactants. Figure 1.6 gives a different view of the reaction surface; it should be emphasized that the reaction pathway is the *minimum* path taken by the systems constituting the reactant state on passage to product. Each trajectory in the surface represents the changing structure of a system with a given starting energy and passes over the 'col' at a variety of levels of energy. Each transit point on the 'col' becomes less probable as its distance from the reaction coordinate increases. The inset shows the number of systems passing at each level as a function of the distance at the 'col' along the line perpendicular to the reaction coordinate. The maximum probability occurs at the junction between the 'col' and the reaction coordinate.

The qualitative picture described above illustrates how a Boltzmann-like distribution of structures can exist at the 'col', and indeed how the transition state can act as if it were a thermodynamic state. The net result of this qualitative description is very important because it illustrates how the transition state can be considered as if it were a molecule with energy freely distributing among the available degrees of freedom. It is likely that the transition state can be considered as a 'molecule' in the reaction conditions for the majority of the cases considered in this text. The situation changes, however, when the numbers of molecules passing over the 'col' become relatively small compared with Avogadro's number; such conditions obtain in gas-phase reactions, in particular those carried out at low pressures (Marcus, 1992; Smith, 1992; Albery, 1993).

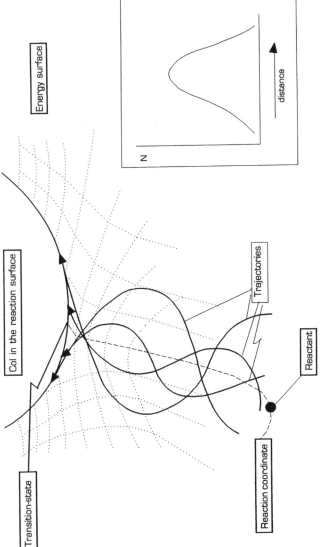

Fig. 1.6 Nominal trajectories of systems with various potential energies are shown crossing the 'col' at different levels of energy. The inset illustrates the numbers of systems (N) with given energy crossing the col as a function of the distance along the line perpendicular to the reaction coordinate.

Further reading

Baggott, J. (1989) Molecules caught in the act, *New Scientist*, 17 June, p. 1958.

Isaacs, N. S. (1995) *Physical Organic Chemistry*, 2nd edn, Longman Scientific, Harlow.

Lowry, T. H. & Richardson, K. S. (1987) *Mechanism and Theory in Organic Chemistry*, 3rd edn, Harper and Row, New York.

Logan, S. R. (1996) *Fundamentals of Chemical Kinetics*, Longman Scientific, Harlow.

Kirby, A. J. (1996) *Stereoelectronic Effects*, Oxford Science Publications, Oxford.

Marcus, R. A. (1992) Skiing the reaction rate slopes, *Science*, **256**, 1523.

Maskill, H. (1985) *The Physical Basis of Organic Chemistry*, Oxford University Press, Oxford.

Pilling, M. J. & Seakins, P. W. (1995) *Reaction Kinetics*, Oxford Science Publications, Oxford.

Truhlar, D. G. & Kreevoy, M. M. (1986) Transition state theory, in *Investigations of Rate and Mechanisms of Reactions*, 4th edn, Part 1, Bernasconi, C. F. (ed.), Wiley-Interscience, New York, p. 13.

Truhlar, D. G., Garrett, B. C. & Klippenstein, S. J. (1996) Current status of transition-state theory, *J. Chem. Phys.*, **100**, 12771.

Zewail, A. (1992) *The Chemical Bond—Structure and Dynamics*, Academic Press, New York.

2 Kinetics and mechanism

2.1 Introduction

The rate constant for a reaction measures the difference in free energy between the transition-state assembly and the reactant state, and gives a parameter for determining energy differences. Knowledge of product structure is very important because the reaction may involve more than one product, attained by a number of competing (parallel) reaction paths. It is necessary to allow for the effects of the competing reactions in order to study the mechanism of one of these, and relative rate constants for formation of products may be obtained from accurate knowledge of product concentrations. Such knowledge of the reactants' identities is also essential because the reactant state sometimes does not simply comprise those species added at the start of the reaction; for this reason the formation of urea from ammonia and isocyanic acid, a classical reaction with apparent simplicity, puzzled chemists for many years (Shorter, 1978).

Intermediates vary in nature from the very reactive, when special methods must be applied for their demonstration, to the relatively unreactive, which can often be isolated from the reaction mixture and studied independently. Since intermediates are discrete molecules they should exist long enough for their detection by some physical or chemical means. Formation and decay of a species which is produced in a large enough quantity to be observed results in an explicit demonstration (Fig. 2.1) of an intermediate. The existence of an intermediate is pertinent to the overall mechanism, whether or not it resides on the reaction pathway to the products or is formed in a 'blind alley' (Eqn [2.1]) which does not lead to products. A 'blind-alley' intermediate can arise from any of the species on the reaction path.

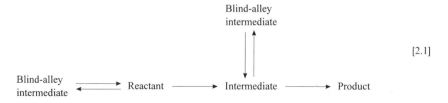

[2.1]

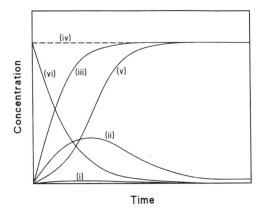

Fig. 2.1 Time dependence of the concentration of intermediates: (i) decomposition is faster than formation; (ii) decomposition and formation rates are commensurate; (iii) decomposition is so slow that the intermediate concentration reaches the concentration (iv) of the initial reactant. In all cases the return rate constant (k_{-1} in Eqn [2.2]) is negligible. The curve (v) represents the progress of product formation when formation and decomposition rates of the intermediate are commensurate. The decomposition of the reactant is represented by curve (vi).

Sensitive instrumental detection techniques must be employed to observe an intermediate formed in small amounts relative to product and reactant. Very reactive intermediates are often formed at concentrations too small to be detected instrumentally because the rapid passage to products prevents a build-up of their concentration.

2.2   Kinetic method

Details of practical kinetics such as measurement of rate constants, determination of empirical equations from pH–rate profiles and the use of 'clock' reactions are important and may be obtained from the Further reading section at the end of this chapter. In deriving rate constants it is often advantageous to manipulate the kinetic data as little as possible, to minimize error magnification; it is also necessary to make sure that the difference between the rates being measured and the error in the rates of control experiments is sufficiently large. In general the protocol of the experimental procedure is best arranged so that either first- or zero-order kinetics will be measured. Mechanisms which are based on complicated kinetic equations should be considered very critically because the complexity substantially increases the possibility of alternative mechanisms.

The UV–vis spectroscopic method for studying kinetics is outstandingly successful and is the instrumental tool of choice for mechanistic studies. Measurement of absorption is often possible at very low concentrations of reactant, in which case the thermodynamics of the system approach ideality and problems associated with solute–solute interactions are therefore excluded. Homogeneity, easily obtained in dilute solution, is essential for 'well-behaved' kinetics which obey simple rate laws. The low concentrations of substrate ensure that changes in pH or ionic strength caused by the progress of the reaction are negligible quantities.

2.3 Rate law and mechanism

Since the rate law is determined by the mechanism it can only be employed to confirm a given mechanistic hypothesis. For example, the mechanism of Eqn [2.2] gives rise to Eqn [2.3] by application of the 'steady-state' assumption, to be discussed in Section 2.4. Equation [2.3] is also consistent with other mechanisms which involve unimolecular decomposition of A or the reaction of A with solvent, the concentration of which remains approximately constant throughout.

$$A \underset{k_{-1}}{\overset{k_1}{\rightleftharpoons}} I \overset{k_2}{\longrightarrow} P \qquad [2.2]$$

$$\text{Rate} = - \,d[A]/dt = d[P]/dt = [A] \cdot k_1 k_2/(k_{-1} + k_2) = k'[A] \qquad [2.3]$$

Apart from confirming mechanistic schemes, rate laws are used in chemical engineering where knowledge of kinetic parameters is important to control the chemical flux in a process.

2.4 Steady state and non-steady state

The concentration of I in Eqn [2.2] can be predicted from k_1, k_{-1} and k_2 and the initial concentration of A. Under conditions where either k_{-1} or k_2 is greater than k_1 the concentration of I is very small compared with that of A and the time dependencies of A and P are first-order. The assumption, usually attributed to Bodenstein, is generally made; this proposes that when the concentration of the intermediate is very much smaller than that of reactants or products, the concentration of intermediate can be considered constant at a 'steady state' over the part of the reaction observable by conventional means, i.e. its rate of formation is equal to its rate of breakdown. Thus Eqn [2.4a] may be derived from Eqn [2.2], and the steady-state condition (Eqn [2.4b]) gives the steady-state concentration of the intermediate (Eqn [2.4c]).

$$d[I]dt = [A]k_1 - [I](k_{-1} + k_2) = 0 \qquad [2.4a]$$

$$\text{Steady-state condition:} \quad k_1[A] = (k_{-1} + k_2)[I] \qquad [2.4b]$$

$$[I] = k_1[A]/(k_{-1} + k_2) \qquad [2.4c]$$

It is simple to utilize Eqn [2.4c] to arrive at the kinetic rate law given earlier (Eqn [2.3]). The rate law (Eqn [2.3]) is extremely useful because the simple mechanism (Eqn [2.2]) is followed by many systems in homogeneous solution. The 'steady-state' assumption breaks down when the rate constants involved in Eqn [2.2] become similar. If [I] approaches the same order of magnitude as the total [A], then the rate laws for [A] and [P] will deviate from first-order kinetics. The reaction occurs in two parts, the first of which involves the build-up of the intermediate I which is mirrored by a decrease in [A] and an increase in [P]; the formation of P includes an induction period because at zero time there is no I available to decompose to form P. The concentration of I starts to fall in the second phase of the reaction, together with a slow decrease in the *rate* of formation of P and the *rate* of consumption of A.

2.5   Rate-limiting step

Discussions involving 'rate-limiting' or 'rate-controlling' steps should only be pursued when intermediates do not build up to concentrations commensurate with those of the reactants during a reaction. Confusing the term 'rate' with 'rate constant' often leads to misunderstandings. For example, it is common to use the phrases 'rate-limiting step' or 'rate-determining step' of a sequential mechanism to define the 'slowest step in the sequence', which would therefore 'control' the overall rate. Use of the term 'reaction flux' instead of 'rate' avoids confusing 'rate' with 'rate constant'.

The conversion of reactant A into product P by a stepwise mechanism may involve the formation of an *unstable* intermediate (I), the rate constant for its formation being k_1 and that for its breakdown k_2 (Eqn [2.5]). The usual steady-state approximation can be applied to this system if the concentration of I is always very low, i.e. $k_2 \gg k_1$. The approximation that the concentration of I remains constant, i.e. $d[I]/dt = 0$, means that the rate of formation of I equals its rate of breakdown (i.e. $k_2[I] = k_1[A]$). In the steady-state phase of the reaction there is *no* slowest step; the *rates* of the two steps are equal. For two consecutive first-order processes (i.e. involving no return step from I to A) it is the relative values of the two *rate constants*, k_1 and k_2, which form the criteria for the rate-controlling step. If the intermediate does not accumulate and is always at low concentration then, by definition, k_1 is always the rate-limiting step because $k_2 \gg k_1$.

$$A \xrightarrow{k_1} I \xrightarrow{k_2} P \qquad\qquad\qquad\qquad [2.5]$$

In a stepwise mechanism with an intermediate which can change back into reactant, the formation of the intermediate is reversible. The rate-limiting step is now the step in the sequence for which the rate constant appears in the numerator of the rate expression but not in the denominator. Furthermore, the rate constant for the corresponding reverse step must also not appear in the denominator. For the mechanistic scheme in Eqn [2.2] the steady-state approximation for the unstable intermediate I (Eqns [2.4]), shows that there is *no* single 'slowest step'; the rates of formation and breakdown of I are the same. If the rate constant k_{-1} is much greater than k_2, the rate is given by Eqn [2.6] and the second step k_2 is rate-limiting. Conversely, if $k_2 \gg k_{-1}$, the rate is given by $k_1[A]$ and the first step is rate-limiting.

$$\text{Rate} = k_1 k_2 [A]/k_{-1} \qquad\qquad [2.6]$$

In the mechanism of Eqn [2.7], reactants A and B react to expel a leaving group (Lg) and form an unstable intermediate I which reacts with X to give product P. The steady-state approximation for the concentration of I gives rise to Eqn [2.8] and indicates that changes in the rate-limiting step can be brought about by changing the ratio of the pseudo-first rate constants $k_{-1}[Lg]$ and $k_2[X]$ in the denominator—which correspond to the relative intrinsic rates of *partitioning* the intermediate I. The partitioning ratio $k_2[X]/k_{-1}[Lg]$ determines whether or not the intermediate is converted back to reactants faster than proceeding on to products. Their relative values can be adjusted by changing the concentrations of Lg or X. At low values of [Lg], $k_{-1}[Lg]$ can be much smaller than $k_2[X]$ and the first step k_1 is rate-limiting with no rate dependence on the concentration of X. At high values of [Lg], $k_{-1}[Lg]$ can become greater than $k_2[X]$ and the second step k_2 becomes rate-limiting with the rate showing a first-order dependence upon [X]. At high values of [X] the first step becomes rate-limiting and the observed rate constant becomes independent of [X].

$$A + B \underset{k_{-1}}{\overset{k_1}{\rightleftharpoons}} Lg + I \xrightarrow{k_2[X]} P \qquad\qquad [2.7]$$

$$\text{Rate} = k_1 k_2 [A][B][X]/(k_{-1}[Lg] + k_2[X]) \qquad\qquad [2.8]$$

When [I] builds up to a significant value compared with the total concentration of A it is then possible to measure *individual* rate constants and there is no need for the concept of the rate-limiting step. When concentrations of the intermediate do not build up, the rate-limiting step is simply that step leading to the transition-state with the highest Gibb's free energy; this is illustrated in Fig. 2.2.

For a reaction with two consecutive first-order processes involving no return of I to A (Eqn [2.5]), the concept of the rate-limiting step is not of much use; more importantly, if $k_2 > k_1$ then the rate constant being measured by analysing for either [A] or [P] is k_1. If $k_2 < k_1$ then analysis for [A] measures k_1 and analysis for [P] measures k_2. It is useful to note that, provided there is no build-up of intermediate species, the experimental rate constant determined by analysis of product formation is the *same* as that determined from reactant decomposition and this result is perfectly valid even for the most complicated of reaction mechanisms.

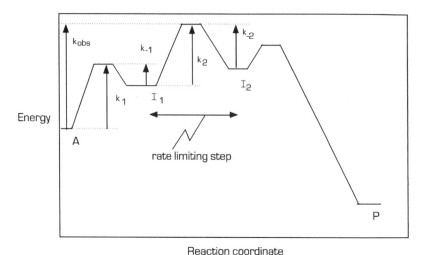

Reaction coordinate

Fig. 2.2 The rate-limiting step for consecutive reactions with reactive intermediates is that leading to the transition state with the highest Gibbs' free energy. The lengths of the vertical arrows are not proportional to the identifying rate constants.

2.6   Curtin–Hammett principle

When a reaction can occur through two forms (A and B) of a molecule, which could be conformers or isomers in rapid equilibrium (Eqn [2.9]), the products (respectively P_A and P_B) do not form in proportion to either *solely* the reactivity of A or B (k_A or k_B respectively) or the relative stability of A or B; the product ratio is given by $[P_A]/[P_B] = K.k_A/k_B$ where $K = (k_{-1})/k_1 = [A]/[B]$. Thus if k_A is much larger than k_B, $[P_A]$ could still be less than $[P_B]$ simply because the equilibrium constant between [A] and [B] could be unfavourable. The situation is best considered in terms of free-energy diagrams (Fig. 2.3).

$$P_A \xleftarrow{\ k_A\ } A \underset{k_{-1}}{\overset{k_1}{\rightleftarrows}} B \xrightarrow{\ k_B\ } P_B \qquad\qquad [2.9]$$

When the rate of interconversion of A and B is greater than the rate of product formation (curve (b) in Fig. 2.3) the ratio of the concentrations [B]/[A] is given by K and the product from the *lowest* energy transition state will be the *most* favoured. In this case P_A will be the favoured product even though k_B is greater than k_A; this result is an expression of the Curtin–Hammett principle (Zefirov, 1977). When the equilibrium is not mobile relative to the rate of product formation (curve (a) in the figure) the ratio [A]/[B] is not given by K; species A will give only P_A and species B will give only P_B simply because the rate of interconversion of A and B is smaller than the decomposition rate of either.

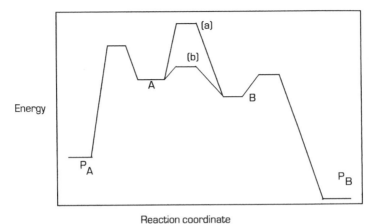

Fig. 2.3 Energy diagram describing the Curtin–Hammett principle (see Eqn [2.9]): (a) rate-limiting interconversion of A and B; (b) mobile equilibrium.

2.7 Which transition state is being observed kinetically?

The identity of the rate-limiting step indicates which transition state is being measured by kinetics; this is an absolute requirement if structures of transition states are to be determined. Favoured techniques employ polar substituent effects and kinetic isotope effects, exemplified as follows. The alkaline hydrolysis of aryl phosphoramidates (Eqn [2.10]) proceeds through the formation of a metaphosphorimidate intermediate (Williams & Douglas, 1972) and the large dependence of the overall rate constant upon leaving-group basicity (see Chapter 3) indicates that the rate-limiting step is the expulsion of the aryl oxide group. If the ionization steps were rate-limiting then subsequent steps would not contribute to the observed rate law and the rate constant would be relatively insensitive to leaving-group basicity.

The base-catalysed bromination of acetone provides an example of the use of the kinetic isotope effect (see Chapter 4, table 4.2) which indicates that the ionization step is rate-limiting. If steps after the ionization (Eqn [2.11]) were rate-limiting the observed rate constant would be governed by an equilibrium isotope effect which would be small (Section 4.1). Moreover, the rate of the reaction is independent of the bromine concentration, which is not consisitent with the bromination step being rate-limiting.

[2.10]

$$\text{(CH}_3)_2\text{C=O} \xrightleftharpoons[\text{BH}^+]{\text{B}} \quad \text{CH}_2=\text{C(CH}_3)\text{O}^- \xrightarrow{\text{Br}_2} \text{BrCH}_2\text{C(CH}_3)\text{=O} \quad [2.11]$$

2.8   Demonstration of intermediates by kinetics

Build-up and decay of intermediates

The most direct way of observing the presence of an intermediate is by use of some physical or chemical analysis technique which effectively monitors concentration as a function of time (Fig. 2.1). The method is excellent provided the lifetime of the intermediate is longer than the relaxation time of the technique in question. The incursion of an intermediate will, in addition, be revealed by a rapid change in an observable parameter which is followed by a slower change. A classic example is the rapid decrease in absorbance at 280 nm on addition of hydroxylamine to furfuraldehyde solution due to the replacement of the carbonyl function by a carbinolamine intermediate; this is followed by the slow appearance of an absorbance at 280 nm due to oxime formation (Jencks, 1959). Acyl-enzyme intermediates may be detected by a rapid initial release of product followed by a slower, steady-state, production (Chapter 10).

Change in rate-limiting step

A change in the rate-limiting step caused by a change in an experimental variable such as concentration or a reactivity parameter (e.g. a different substituent) is evidence for the presence of a reactive intermediate. The demonstration of a mechanism with at least two consecutive steps derives from consideration of Eqns [2.2] and [2.3]; if $k_2 > k_{-1}$ then $k_{obs} = k_1$ but a change in k_2 or k_{-1} so that k_2 becomes less than k_{-1} yields $k_{obs} = k_1 . k_2/k_{-1}$. Because k_2/k_{-1} must be less than unity, the change in rate-limiting step will cause a *decrease* in k_{obs} below the predicted level when the k_1 step is rate-limiting. A change in the rate-limiting step caused by variation of a reactivity parameter is illustrated by the attack of amines on isocyanic acid which generates a non-linear free energy correlation (Eqn [2.12]) (see Chapter 3) where the reaction with amines with low pK_a values involves rate-limiting proton transfer in the second step either to a base B or from an acid HA (Williams & Jencks, 1974). There is a changeover in rate-limiting step as the amine basicity increases (Fig. 2.4). For weakly basic amines which are good leaving groups $k_{-1} > k_2[HA]/[B]$, but for basic amines (high $pK_a^{RNH_3^+}$) proton transfer occurs for the intermediate with a larger rate constant than that for expulsion of amine $(k_2[HA]/[B] > k_{-1})$.

$$\text{RNH}_2 + \text{HNCO} \xrightleftharpoons[k_{-1}]{k_1} \quad \text{RNH}_2^+\text{—C}(\text{O}^-)\text{=NH} \xrightarrow{k_2[HA] \text{ or } k_2[B]} \text{RNHCONHR}$$

$$[2.12]$$

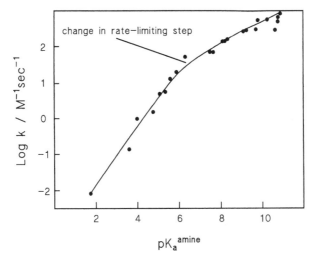

Fig. 2.4 Urea formation from primary amines and isocyanic acid (Eqn [2.12])—evidence for a change in the rate-limiting step (data from Williams & Jencks, 1974).

Amines with high basicity thus have k_1 rate-limiting, which is also evident from the fact that the reaction is no longer associated with general acid–base catalysis.

An intermediate can also be demonstrated when either k_{-1} or k_2 involves a *concentration* term and the overall rate constant, k_{obs}, obeys a non-linear rate law, provided the range of concentration is able to cause a change in the rate-limiting step. An example of this method is provided again for the reactions of Eqn [2.12], where an increase in buffer concentration causes a non-linear increase in the rate constant for reaction of 4-methoxyaniline with isocyanic acid (Fig. 2.5). A change in the rate-limiting step occurs when the rate constant of the second step ($k_2 \times$[buffer]) becomes larger than that of the reversal of the intermediate back to reactants (k_{-1}).

A putative intermediate may be diagnosed by a decrease in rate constant on addition of quantities of the leaving group (Lg), if the k_2 step (Eqn [2.7]) is rate-limiting. An example of this is due to Curran *et al.* (1980), who demonstrated an intermediate in 3-nitrophenyl hippurate hydrolysis in the presence of added 3-nitrophenolate ion (Eqn [2.13] and Fig. 2.6).

[2.13]

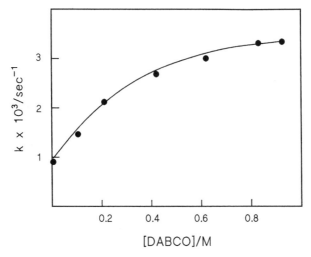

Fig. 2.5 Evidence for a change in the rate-limiting step in the DABCO (diazabicyclo-octane)-catalysed reaction of anilines at constant isocyanic acid concentration (Eqn [2.12]) (data from Williams & Jencks, 1974).

Similar experimental methods may be used for deducing the presence of intermediates in enzyme-catalysed reactions. For example, the hydrolysis of the methyl ester of *N*-acetyl-*L*-phenylalanine is catalysed by α-chymotrypsin but the overall rate of catalysis decreases in increasing amounts of methanol, consistent with the existence of an acyl-enzyme intermediate (Eqn [2.14]). A

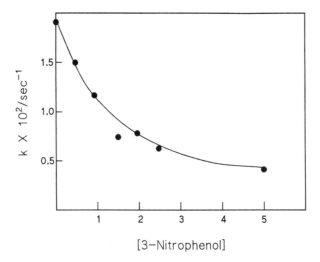

Fig. 2.6 Mass action effect indicating the existence of an intermediate in the hydrolysis of 3-nitrophenyl hippurate in the presence of increasing concentrations of 3-nitrophenolate ion (Eqn [2.13]) (data from Curran *et al.*, 1980).

control experiment indicates that the enzyme does not catalyse significant synthesis of methyl ester from the free carboxylic acid and methanol (Bender & Glasson, 1960).

$$R\text{–}COOCH_3 + H\text{–}O\text{–}CT \underset{-CH_3OH}{\xrightleftharpoons} R\text{–}CO\text{–}CT \xrightarrow{H_2O} R\text{–}COO^- + H\text{–}O\text{–}CT$$

$$R\text{–}CO\text{–} = CH_3CONHCH(CH_2Ph)CO\text{–}; \quad H\text{–}O\text{–}CT = \alpha\text{-chymotrypsin} \qquad [2.14]$$

Change in rate-limiting step caused by change in pH

A change in pH can cause a change in the rate-limiting step if either k_{-1} or k_2 in the standard reaction (Eqn [2.2]) is dependent on hydrogen ion concentration. A break in the pH–rate profile which gives a smaller rate constant than that predicted from extrapolation of the rate from lower pH values is consistent with an intermediate. A good example of this phenomenon is the hydrolysis of 2,4,6-trimethylpyrylium perchlorate (Eqn [2.15] and Fig. 2.7, (Williams, 1971)) which involves a break in the pH–rate profile where $k_{-1}[H^+]$ is equal to k_2. At high pH values the first step, the pH-independent addition of water, becomes rate-limiting because $k_2 > k_{-1}[H^+]$. It is essential in such experiments to prove that the inflection is not due to the ionization of a buffer component which undergoes reaction with the substrate. Moreover, a break can also derive from an ionization step prior to the rate-limiting step and this can be demonstrated by separate pH-titration studies.

$$[2.15]$$

2.9   Change in mechanism

The pH–rate profile for the hydrolysis of isatin (Eqn [2.16] and Fig. 2.8) reveals that sections of the pH-dependence curve (M) suffer an *increase* in slope as the value of the pH increases (Page *et al.*, 1993). The rate at higher values of the parameter is greater than that predicted from extrapolation of the rate at lower values. These regions of pH involve changes in *mechanism* for the reaction; a change in mechanism only occurs when the alternative is more favourable because of the change of parameter. If there is no build-up of species other than substrate and product, the breaks denoted 'R' can be ascribed to changes in the rate-limiting step and the break at low pH in Fig. 2.8 is due to this. A spectral change at pH 12, the position of the break labelled RR, indicates that this is caused by the ionization of the isatin species rather than by a change in the rate-limiting step. At low pH the rate-limiting step is attack of hydroxide ion upon the neutral isatin, but

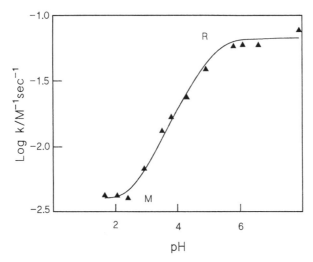

Fig. 2.7 Dependence on pH of the ring opening of a pyrylium ion—evidence for an intermediate (Eqn [2.15]) (data from Salvadori & Williams, 1971). M = change in mechanism; R = change in rate-limiting step.

above pH 7 the pH-independent hydrolysis corresponds to proton-catalysed breakdown of the tetrahedral intermediate (T^-), i.e. a change in the rate-limiting step. Above pH 11 the reaction rate becomes second-order in hydroxide ion and breakdown of T^- now occurs through the dianionic intermediate T^{2-}, i.e. a change in mechanism occurs (at RR in Fig. 2.8).

[2.16]

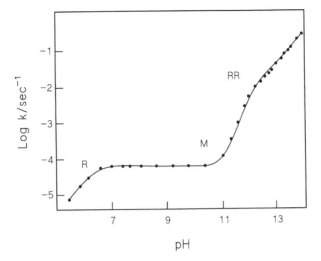

Fig. 2.8 Dependence on pH of the hydrolysis of the γ-lactam, isatin; M = change in mechanism; R = change in rate-linking step (Eqn [2.16]) (data from Page *et al.*, 1993). See the text for details of RR.

A good example of a change in mechanism caused by change in substituent is the hydrolysis of aryl 4-hydroxybenzoates catalysed by hydroxide ions (Cevasco *et al.*, 1985). The rate of hydrolysis is faster for esters with good leaving groups (phenols of low pK_a) than is predicted from the Brønsted plot of esters with weak leaving groups (phenols with high pK_a) (Fig. 2.9). This is indicative of a change in mechanism from a simple bimolecular one for

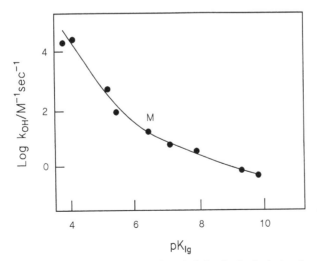

Fig. 2.9 A change in mechanism (M) in the hydrolysis of aryl 4-hydroxy-benzoate esters—evidence for an alternative mechanism to the more usual addition-elimination pathway (Eqn [2.17]) (data from Cevasco *et al.*, 1984).

esters of weakly acidic phenols to an elimination–addition mechanism through an unsaturated intermediate for esters of strongly acidic phenols where the good leaving group is expelled in a unimolecular step (Eqn [2.17]).

$$ HO-\langle \bigcirc \rangle-CO-OAr \ \rightleftharpoons \ O-\langle \bigcirc \rangle-CO-OAr \ \rightarrow \ O=\langle \bigcirc \rangle=C=O + ArO^- $$

$$ \Big| H_2O $$

$$ HO-\langle \bigcirc \rangle-CO-OAr $$

[2.17]

2.10   Rules for change in mechanism or rate-limiting step

Changes in reaction parameters such as the pH, solvent, substituents, concentrations etc. are essentially changes in free energy and if the logarithm of a rate constant is plotted against these parameters a free-energy correlation ensues (see Chapter 3). If the change in reaction parameter causes a break in a free-energy relationship then either a change in the rate-limiting step or a change in mechanism is indicated. A pair of simple rules has been formulated to identify changes in mechanism or changes in the rate-limiting step from the direction of change at breaks in free-energy relationships. Provided there is no build-up of intermediate relative to reactants, the rules do not have serious exceptions. They are:

(1) There is a *change in rate-limiting step* if the observed rate constant on one side of the breakpoint in a free-energy relationship is *less* than that calculated from the correlation on the other side.

(2) There is a *change in mechanism* if the observed rate on one side of the breakpoint in a free-energy correlation is *greater* than that calculated from the correlation on the other side.

It makes no difference, of course, whether the break is examined from higher to lower rate constants, or vice versa. There is an exception to these generalizations if there is a build-up of intermediate, such as when ionization occurs with changing pH or when unreactive adduct formation occurs as in the hydrolysis of isatin at pH 12. It is necessary to demonstrate the absence of a build-up of intermediate in the pyrylium hydrolysis (Eqn [2.15]) before the breakpoint can be accepted as due to a change in rate-limiting step.

2.11 Trapping of intermediates

The existence of non-accumulating reactive intermediates may be demonstrated by reaction with a 'trapping' reagent to yield a product different from the normal one, as in Eqn [2.18], which indicates that the intermediate I can be trapped to form product P_2. To provide convincing evidence of an intermediate, the trapped product must be shown to form faster than from *direct* reaction of the trapping agent T with the reactant A; this indicates that the rate-limiting step *precedes* the trapping step. An example of the trapping method is the demonstration of an isocyanate intermediate in transfer reactions of carbamate esters (Hegarty & Frost, 1973; Eqn [2.19]), where the overall rate constant for decomposition of the carbamate ester is *not* increased by addition of the trapping nucleophile. Since the trapping agent diverts a substantial proportion (50–100%) of the reactant (carbamate ester) to trapped product (urea) with no increase in the rate constant, then the mechanism involves an intermediate (RNCO) with a rate-limiting step prior to its decomposition by trapping agent.

$$
\begin{array}{l}
A \longrightarrow I \xrightarrow{\quad k_2 \quad} P_1 \\
\qquad\qquad\qquad \text{(normal product)} \\[1em]
\text{direct reaction} \qquad \Big\downarrow k_3 \\
\text{with trapping} \qquad \text{trapping agent (T)} \\
\text{agent (T)} \\[1em]
\qquad\qquad\qquad P_2 \text{ (trapped product)}
\end{array}
\qquad [2.18]
$$

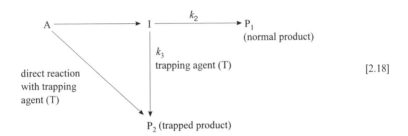

$$ [2.19] $$

A further trapping technique involves measuring the product ratio ($[P_2]/[P_1]$) (see Eqn [2.18]), which is independent of the structure of A if liberation of the structurally variant part of A results in the formation of a common intermediate. For example, the ratio of hippurylhydroxamic acid to hippuric acid is constant when formed from trapping experiments with a *variety* of hippuryl precursors and α-chymotrypsin (CT—OH); this result is diagnostic of a common hippuryl–chymotrypsin intermediate (Eqn [2.20]) (Epand & Wilson, 1963). In this case the formation step of the common hippuryl–chymotrypsin intermediate will have markedly different rate constants but, because the leaving group (Lg) departs before the product-determining step, the ratio of products is constant and independent of the nature of the leaving group, provided the nucleophile concentration remains constant.

$$PhCONHCH_2CO-Lg \xrightarrow{CT-OH} PhCONHCH_2CO-OCT$$

$$\downarrow Lg$$

$$NH_2OH \longrightarrow PhCONHCH_2CONHOH$$

$$H_2O \longrightarrow PhCONHCH_2CO_2H$$

$$[2.20]$$

In many examples cited in the literature, 'trapping' may not be diagnostic of an intermediate simply because of the omission of the important control experiment which is necessary to demonstrate that the overall rate constant for depletion of reactant is independent of the concentration of trapping agent.

Product ratios in trapping experiments can be used to estimate relative rate constants: for example $k_2/k_3 = [P_1]/[P_2]$ in Eqn [2.18], provided the amount of P_2 from the direct reaction with A can be estimated or excluded. One of the rate constants, if known, can be used as a 'clock' to estimate the value of the other rate constant. Such a technique is useful for determining the lifetimes of very unstable intermediates in the presence of trapping reagents; the reactivities of iminium ions, carbenium ions and nitrenium ions have all been studied using as a 'clock' the diffusion-controlled rate constant for reaction of azide ion with these species (Knier & Jencks, 1980; Richard & Jencks, 1982; Fishbein and McClelland, 1987; Amyes & Jencks, 1989).

When there is no build-up of intermediates, the rate constant determined experimentally for reactant decomposition is identical with that obtained from analysis of formation of *any* of the products of trapping.

2.12   Composition of the transition state of the rate-limiting step

The composition of the transition state is given by the rate law connecting the kinetics with the concentrations of the reacting species. The alkaline hydrolysis of esters has a second-order rate law given by Eqn [2.21]. In this case, the transition-state composition includes atoms of the reagent hydroxide ion and of the ester. The solvent composition of the transition state cannot be defined because solvent concentrations cannot be included in the rate law.

$$\text{Rate} = k_2[\text{OH}^-][\text{RCOOAr}] \qquad [2.21]$$

The aminolysis of esters has a much more complicated rate law than simple hydrolysis and each component of an equation such as [2.22] corresponds to a separate mechanism, each with its own transition-state composition; an arbitrary transition-state structure is illustrated (**1–4**) for each of the components in the example which is consistent with that indicated by the rate law. There are, however, several kinetically equivalent structures, all of which are compatible with the rate law.

$$\text{Rate} = k_1[\text{OH}^-][\text{RNH}_2][\text{ester}] + k_2[\text{RNH}_2]^2[\text{ester}] + k_3[\text{B}][\text{RNH}_2][\text{ester}]$$
$$+ k_4[\text{RNH}_2][\text{ester}]$$

[2.22]

The mechanism may involve intermediates, prior to the rate-limiting step, which are formed by the elimination of parts of the reactants. The concentration of these intermediates may depend on the concentration of reactant or catalytic molecules which thus appear in the rate law even though they are not chemically bonded to the central molecule in the transition state of the rate-limiting step. For example, many base-catalysed reactions proceed by the intermediate formation of the conjugate base of the reactant, the decomposition of which is rate-limiting. The observed composition refers to the transition state of the rate-limiting step and therefore includes species which may have been released from the discrete molecule prior to the rate-limiting step (but not any molecules involved after the rate-limiting step); these species *still* appertain to the 'transition state' even though they may not be bonded chemically to the substrate. In the case of the ester aminolysis, we include the base (B, RNH_2, OH^- or H_2O) in its conjugate acid form in the composition of the transition state although it may *not* be bonding in any way with the ester/amide in the transition structure of the rate-limiting step of the currently favoured mechanism (Scheme 2.1). When the leaving group is very active (as in nitrophenyl esters) it is likely that removal of the proton from the ammonium ion is concerted with leaving-group departure, and thus the base will be bonding with the substrate in the transition structure.

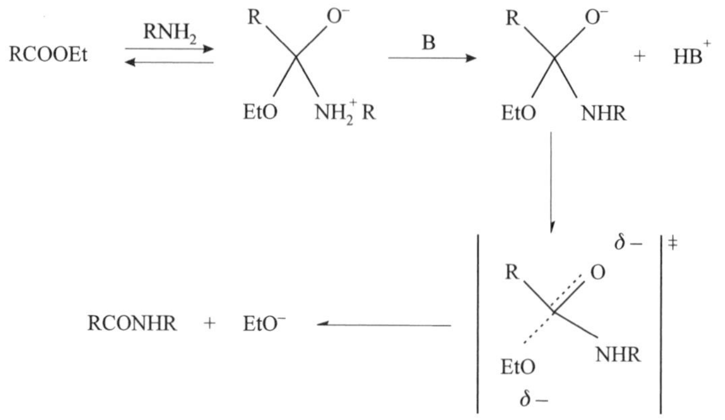

Scheme 2.1 Mechanism of ester aminolysis.

2.13   Kinetic ambiguities

Kinetic ambiguities arise most frequently in reactions where acid–base equilibria are involved. The rate of a bimolecular reaction between the acid (HA) and a nucleophile (Nu) could obey Eqn [2.23].

$$\text{Rate} = k_2[\text{HA}][\text{Nu}] \qquad\qquad [2.23]$$

If HA and Nu are respectively in equilibrium with their respective conjugate base and acid (Eqns [2.24] and [2.25]), Eqn [2.23] may be rewritten as Eqns [2.26a] and [2.26b].

$$K_a^{\text{HA}} = [\text{H}^+][\text{A}^-]/[\text{HA}] \qquad\qquad [2.24]$$

$$K_a^{\text{HNu}} = [\text{H}^+][\text{Nu}]/[\text{HNu}^+] \qquad\qquad [2.25]$$

$$\text{Rate} = k_2[\text{A}^-][\text{HNu}^+]K_a^{\text{HNu}}/K_a^{\text{HA}} \qquad\qquad [2.26a]$$

or

$$\text{Rate} = k_2'[\text{A}^-][\text{HNu}^+] \qquad\qquad [2.26b]$$

The experimental rate law therefore favours neither A^- reacting with HNu^+ nor HA reacting with Nu, and the respective rate constants k_2' and k_2 are related by the equation $k_2' = k_2 K_a^{\text{HNu}}/K_a^{\text{HA}}$. It is tempting to think that these two equivalent rate laws can be distinguished by changing the reaction conditions to favour the concentration of one of the reactive species. For example, increasing the pH will increase the relative concentrations of A^- and HA; it would be incorrect to assume that an observed increase in rate constant with pH is consistent with Eqn [2.26b] because increasing pH has an equal and *opposite* effect on [HNu$^+$] relative to [Nu]. Equations [2.23] and [2.26] are kinetically indistinguishable.

The experimentally determined rate law does not necessarily indicate the nature of the reactive species, although it does describe the composition of the rate-limiting transition state or the concentration of the species on which it is dependent. The reaction of amine with isocyanic acid (HNCO) was thought at one stage to involve reaction of ammonium ion with cyanate ion, even though Chattaway & Chapman (1912) had pointed out the difficulties of forming an N–C bond through such a combination. The reaction of NH$_3$ with EtNCO (the labile proton is essentially 'fixed' as the ethyl group) has an unambiguous bimolecular rate constant and Stark (1965) showed that it is similar in value to k_2 for the reaction of ammonia with isocyanic acid, indicating that the latter mechanism is involved. Although the neutral state is in equilibrium with the ammonium–cyanate ion state, direct reaction from the latter is impossible (Fig. 2.10).

The problems of kinetic ambiguity arise in the above example because the most stable state is the NCO$^-$–NH$_4{}^+$ pair so that these concentrations are the most natural to measure as a reactant state during the reaction. It is often not possible (as in the simple case above) to differentiate between conjugate base acting on the protonated substrate or general acid action

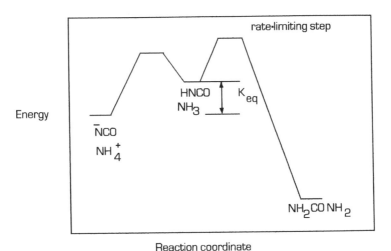

Fig. 2.10 Kinetic ambiguity in the reaction of ammonia with isocyanic acid; cyanate and ammonium ions are required to react through the neutral species NH$_3$ and HNCO. A direct route to the product is not feasible. Relationships: $[H^+][NCO^-]/[HNCO] = K_a^{HNCO} = 10^{-4}$ M; $[NH_3][H^+]/[NH_4^+] = K_a^{NH_4^+} = 10^{-9}$ M; $K_{eq} = K_a^{NH_4^+}/K_a^{HNCO} = [NH_3][HNCO]/[NH_4^+][NCO^-] = 10^{-5}$.

on the unprotonated substrate. The reader is referred to Chapter 5, where pressure, temperature and acidity are employed to distinguish between the kinetic ambiguity resulting from acid-catalysed solvolyses.

Kinetic ambiguity is a consequence of a mobile equilibrium with an alternative state prior to the rate-limiting step; it arises most frequently in reactions involving proton transfer because this is most often a labile process, and it could be regarded as a special case of the ambiguity seen in systems to which the Curtin–Hammett principle is applied. Kinetic ambiguity is an important topic in reactions which require proton transfer and is one which is often forgotten by over-enthusiastic students of mechanism.

2.14 Microscopic reversibility

This concept is simply that, *under the same conditions*, the mechanism for the forward reaction is the same as that for the reverse (Westheimer & Bender, 1962; Burwell & Pearson, 1966; Krupka *et al.*, 1966; Abraham *et al.*, 1971). If the equilibrium constant for a single-step reaction is known, only one of the rate constants is an independent variable. The concept is remarkably useful in a practical way in studies of mechanism because only the most readily accessible rate constant need be studied. Thus the equilibrium for the attack of iodide ion on ammonium ions is not favourable (Eqn [2.27]) and k_1 is consequently very difficult to measure whereas the reverse reaction (k_{-1}), attack of amines on methyl iodide, is easily measured. Details of the mechanism determined for the reaction in the k_{-1} direction apply to the reverse reaction (k_1) (under the same conditions) by the principle of microscopic reversibility. The absolute value of k_1 could also be determined if the equilibrium constant were known ($k_1 = K \cdot k_{-1}$).

$$I^- + CH_3 - NHR_2^+ \underset{k_{-1}}{\overset{k_1}{\rightleftharpoons}} CH_3 - I + R_2NH$$

The hydrolysis of simple monophosphate esters catalysed by alkaline phosphatase involves formation of a phosphoryl enzyme with expulsion of the alcohol (Scheme 2.2) and provides an enzymic example of the application of the principle. Alcoholysis of the phosphoryl enzyme intermediate (to return to ester) must take the path which is the microscopic reverse of the forward step. It is reasonable to assume that when the ROH species is replaced by water as in the scheme the hydrolysis of the phosphoryl enzyme will take a mechanism which is essentially the microscopic reverse of phosphorylation rather than an additional route. Such considerations apply to all the hydrolase enzymes and it is often the case that reaction of water with the enzyme intermediate follows the same mechanisms as acylation or glycosylation in the respective enzyme systems (Chapter 10).

Microscopic reversibility is also important because it affects the value of the observed rate constant. In a reaction forming B from A (Eqn [2.28]), the

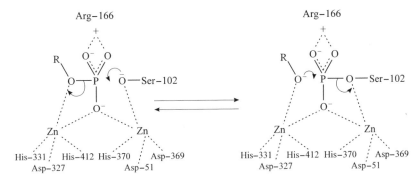

Scheme 2.2 Mechanism of alkaline phosphatase showing hydrolysis of phosphoryl enzyme is essentially the microscopic reverse of phosphorylation (diagram modified from Steitz & Steitz, 1993).

rate constant, k_{obs}, for approach to equilibrium is given by Eqn [2.29] where $K = [B]/[A] = k_1/k_{-1}$.

$$A \underset{k_{-1}}{\overset{k_1}{\rightleftharpoons}} B \qquad\qquad [2.28]$$

$$k_{obs} = k_1(1/K + 1) = k_1 + k_{-1} \qquad\qquad [2.29]$$

Thus the larger the value of K, the closer k_{obs} approximates to k_1. The value of k_{obs} is *always greater* than the forward rate constant. This is very important in practical studies of kinetics but it is often not applicable as many reactions under investigation are chosen because they go to completion ($K > 100$, say) and thus k_{obs} approximates to k_1. Neglect of the law in the case of an unfavourable equilibrium ($K < 1$) could result in measurement of the reverse rate constant (k_{-1}); in this case the analytical methods would have to be very sensitive to demonstrate the relatively small change in concentration in either A or B.

Microscopic reversibility is a topic which is easy to neglect and, no doubt to their embarrassment, even the wisest of mechanistic investigators have been caught out. It is good policy to aim for simple mechanistic hypotheses because most problems arise with complicated mechanisms when it is not easy to keep track of reasonable alternatives.

2.15   Kinetic and thermodynamic control

Thermodynamic control requires that the product, reactants and intermediates are in mobile equilibrium with each other and product distribution

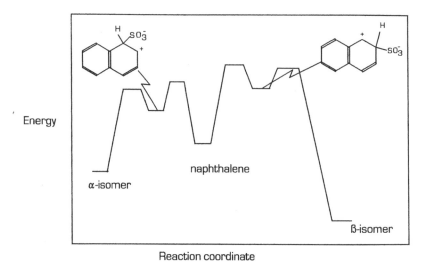

Reaction coordinate

Fig. 2.11 Kinetic versus thermodynamic control: the sulphonation of naphthalene.

is governed by the equilibrium constants between the products. Kinetic control is manifested when product formation is faster than the interconversion between products. A good example is the sulphonation of naphthalene which yields the β-sulphonic acid at high temperature because this is more stable than the α-form. The α-isomer results at lower temperatures although it is the less stable, because the rate constant for its formation exceeds that of the β-form (Fig. 2.11) and there is much slower interconversion between α- and β-isomers.

Reactions which yield different products under kinetically and thermodynamically controlled conditions are of mechanistic interest because whatever properties are responsible for making one product more stable than the other they are not reflected in the relative transition-state energies. The transition-state structure for a kinetically controlled product will not be product-like. In those cases where different products are formed from a common intermediate, the kinetically controlled product is often formed from an exothermic step so that the transition-state structure resembles that of the intermediate rather than the product. For example, kinetically controlled 1,2-addition of HBr to butadiene occurs in preference to the thermodynamically more stable 1,4-reaction because the trapping of the intermediate carbocation by the bromide anion is an exothermic step which has an early transition state with little covalent bond formation (Scheme 2.3). Attack of the bromide ion at the C3 carbon of the allyl cation is electrostatically controlled because this carbon has the highest positive charge density, even though this gives a less stable product than attack at C1.

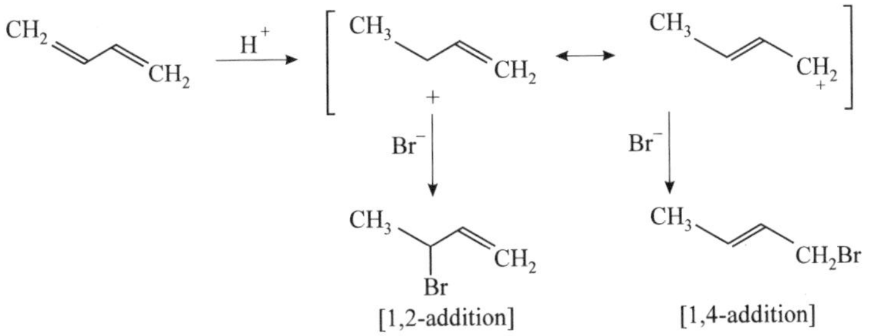

Scheme 2.3 Thermodynamic versus kinetic control in addition of HBr to butadiene.

2.16 Isotope labelling techniques and detection of intermediates

Detection of labelling

Molecular marking is possible with any group but isotopes offer the opportunity, especially useful with enzyme and bio-organic reactions, that their introduction into a molecule will not change the chemistry of the process under investigation. Isotopes, to a first approximation, have identical electronic and steric properties; consequently, the isotope effect on a rate constant is negligible with most atoms other than hydrogen. The substitution of hydrogen isotopes can cause shifts in the rate-limiting step. To our knowledge, no unequivocal case has been observed of isotopic substitution changing a mechanism, although the possibility should not be ruled out for reactions studied by labelling techniques.

The experimental detection of labelling follows the procedure normally employed for structure determination and high-field NMR is often employed; the classic technique uses radioisotopes and appropriate degradative work. Further methods are recorded in Chapter 3 in the sections on isotopologues and in Chapter 4 on the kinetic isotope effect.

Identity of bond fission

The application of isotopic labelling techniques can identify the bond undergoing fission or formation when this is impossible to identify by product analysis. The three fission processes represented by Eqns [2.30]–[2.32] (with labelled oxygen) are typical examples.

$$CH_3CO-OEt + {}^*OH^- \longrightarrow CH_3CO_2^{*-} + HOEt \qquad [2.30]$$

$$CH_3CO-OBu^t + {}^*OH^- \longrightarrow CH_3CO_2^- + HO^*Bu^t \qquad [2.31]$$

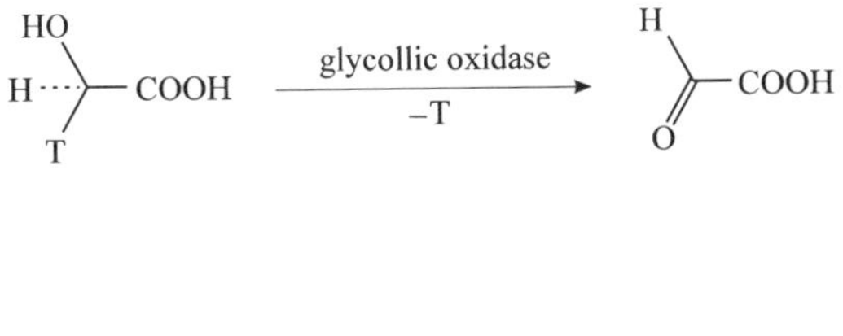

[2.32]

Incorporation of an ^{18}O-isotope into the products indicates which bond to the ester oxygen is being broken. Tritium labelling of one or other of the prochiral hydrogen positions in the glycollic acid and analysis of the isotopic composition of the product (Eqn [2.32]) indicates which of the C–H bonds is undergoing fission (Rose, 1958). In all isotopic labelling experiments it is important to demonstrate that the products do not undergo isotopic exchange under the reaction conditions, otherwise the conclusions could be invalid.

Isotopic labelling often enables decisions to be made concerning intermediates. For example, the hydrolysis of *N*-sulphonylanthranilate in mild acid could involve the mixed sulphuric acid–carboxylic acid anhydride as an intermediate (Hopkins & Williams, 1982a). It is known that fission should occur at the C–O bond in acyl sulphates (as shown in Eqn [2.33]) so that incorporation of labelled oxygen from the water into the sulphate product rather than into the anthranilic acid indicates that the anhydride is not an intermediate.

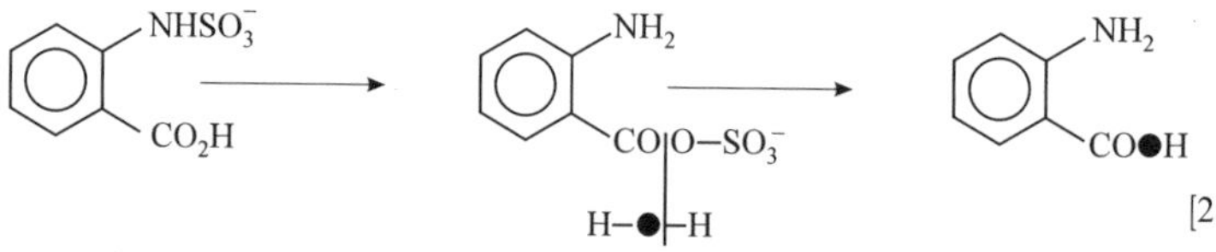

[2.33]

Labelling the substrate with two isotopes (double labelling) is a technique which can be employed to detect a symmetrical intermediate through determination of the position of bond fission. For example, phthalic anhydride (**6**) is a symmetrical intermediate in the hydrolysis of phthalamic acid (**7**) (Bender *et al.*, 1958); the labelled oxygen incorporated from solvent water into the phthalic acid product (**8**) is equally distributed in the carboxylic acid groups arising from the amide and the carboxylic acid in the original phthalamic acid. The anhydride mechanism (Scheme 2.4) would be excluded if the product were labelled only on the carboxylate derived from the amide.

It is worthwhile mentioning that accuracy in the application of isotopic labelling increases with increased isotopic content; although this is self-evident, it contrasts with the low isotopic contents needed for the highest accuracy in heavy-atom *kinetic* isotope effect studies (see Chapter 4).

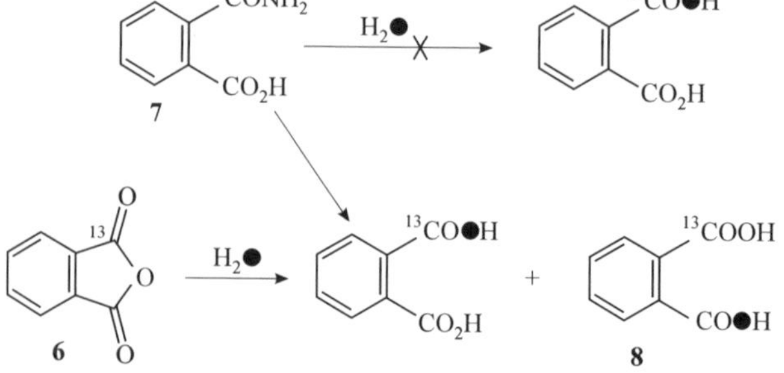

Scheme 2.4 Demonstration of intermediates in phthalamic acid hydrolysis by double labelling.

Isotopic exchange

Proton exchange with solvent

Rates of exchange of tritium or deuterium labels for hydrogen, usually with solvent, indicate the lability of a proton from a molecule and provide a tool to estimate the pK_a for the conjugate acid. The tritium label is excellent for even the slowest processes because of the sensitivity and accuracy obtained with radioisotope counting techniques. The pK_a is obtained by assuming that the recombination of conjugate base and proton is a diffusion-controlled process and is thus independent of reactant structure.

$$CH_3CO\!-\!SEt \underset{k_{-1}[HB^+]}{\overset{k_1[B]}{\rightleftarrows}} \bar{C}H_2CO\!-\!SEt \xrightarrow[DB^+]{\text{diffusion}} CH_2DCO\!-\!SEt \qquad [2.34]$$

The pK_a of ethyl thioacetate ($\sim$21), obtained using the exchange method ($K_a = k_1/k_{-1} = k_{\text{exchange}}/k_{\text{diffusion}}$), shows that thiol ester carbanions (Eqn [2.34]) have concentrations sufficient for them to act as intermediates in many coenzyme A reactions (Amyes & Richard, 1992).

The Cannizzaro reaction of aldehydes involves no exchange of the transferring hydrogen (Eqn [2.35]) with solvent protons, and the proton transfer is not involved in its mechanism. It is thought that the hydrogen is transferred directly, with its bonding electrons in a hydride ion transfer process. For many years the glyoxalase reaction was thought to involve an internal Cannizzaro-type process because of its apparent lack of proton exchange with solvent (Scheme 2.5(A)). The discovery (Hall *et al.*, 1976) that a small amount of exchange of the transferring hydrogen occurs with the solvent indicates that, instead, a proton transfer mechanism must be operating (Scheme 2.5(B)). The ion pair formed between substrate and base catalyst survives long enough in the encounter complex for 'internal' reprotonation to occur at another site within the molecule at a rate which is faster than exchange with solvent. Such a mechanism forms the basis of Cram's 'conducted tour' system (Scheme 2.6), where the transfer process involves the proton migrating across five atoms (a 1,5-proton transfer) faster than exchange can occur with solvent.

$$RCDO \;+\; RCDO \xrightarrow{\;OH^-\;} RCD_2OH \;+\; RCO_2H$$

$$[2.35]$$

$$RCO_2^- \;+\; RCD_2OH$$

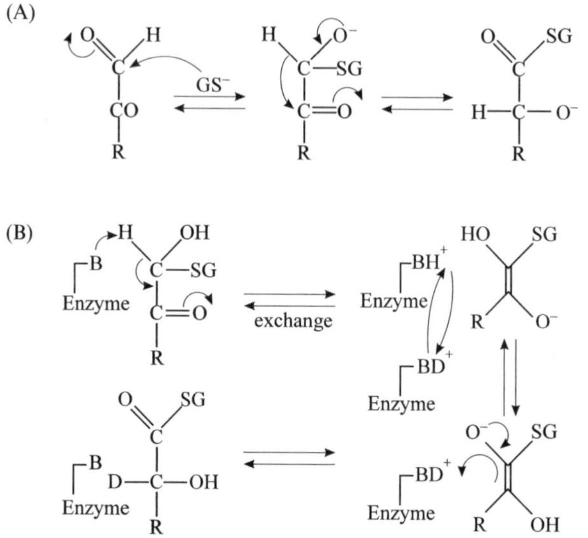

Scheme 2.5 Glyoxalase mechanisms: (A) hydride ion transfer mechanism; (B) proton transfer mechanism, illustrating exchange with solvent.

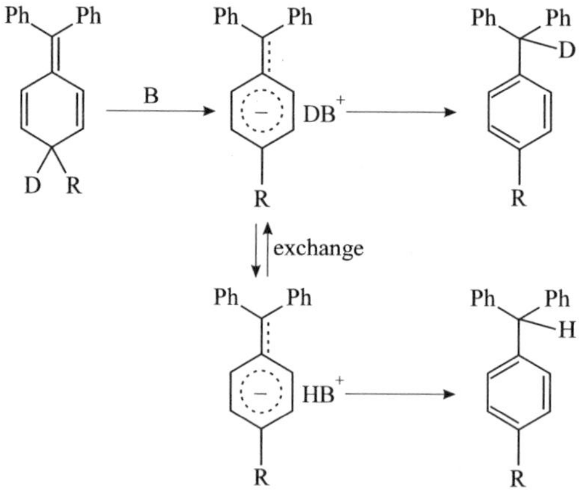

Scheme 2.6 Cram's 'conducted tour' mechanism.

The isomerization of a triose in the presence of base is expected to involve exchange of the transferred proton with solvent as in the Lobry de Bruyn–van Eckenstein reaction (Scheme 2.7). Originally no hydrogen exchange was observed between the carbon and the solvent in the triose phosphate isomerase catalysis and it was therefore assumed that a hydride ion transfer mechanism occurred. However, Rieder & Rose (1959) observed a small proportion of exchange and a mechanism was proposed involving transfer of the proton between C1 and C2 positions prior to exchange; exchange of this proton with that of the solvent is slower than the transfer step (see Chapter 10, Schemes 10.5 and 10.6).

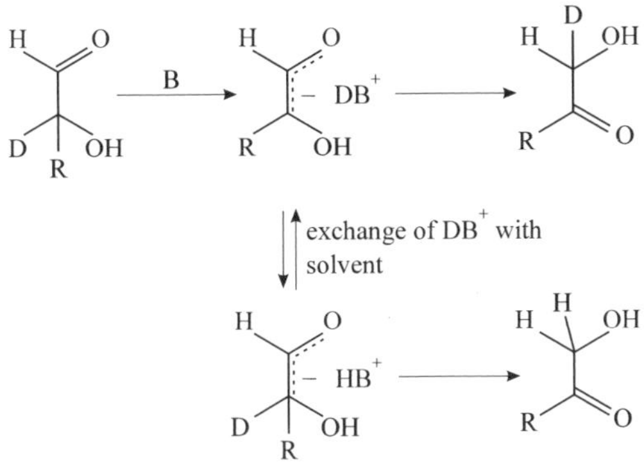

Scheme 2.7 A mechanism of the Lobry de Bruyn–van Eckenstein reaction for proton exchange in a triose (R = —CH$_2$OH).

Detection of intermediates—positional isotopic exchange (PIX)

Isotopic exchange can be used to indicate the formation of an intermediate occurring by addition or elimination: the reactant composition is examined, at increasing times, for evidence of exchange of the isotope. The method is general (Scheme 2.8) and the exchanging isotope enters through either an associative or dissociative intermediate. The exchangeable isotope B or C is invariably of a molecular species rather than the atom itself. A classical example of the technique for an addition intermediate is provided by the alkaline hydrolysis of esters (Scheme 2.9). This was the original PIX experiment (Bender, 1951); in our opinion, this outstanding work marked the beginning of modern mechanistic bio-organic chemistry.

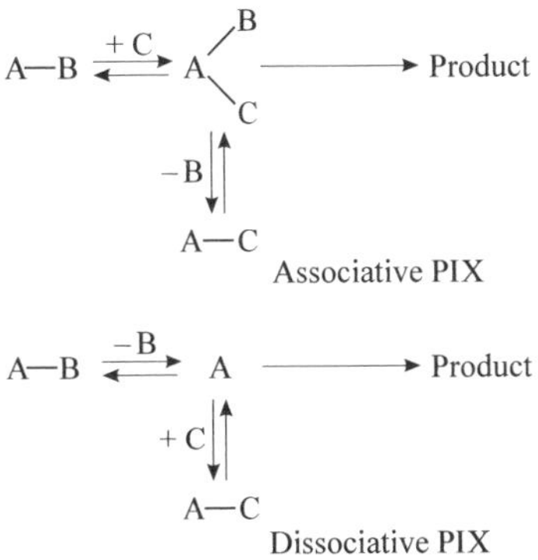

Scheme 2.8 Positional isotopic exchange to detect intermediates—general cases.

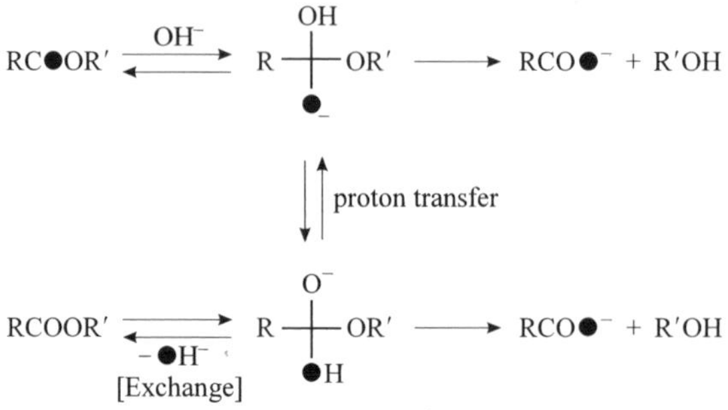

Scheme 2.9 Associative mechanism for isotopic exchange during alkaline hydrolysis of esters.

Incorporation of isotope into the carbonyl oxygen of the ester as the reaction proceeds requires that the proton transfer step (Scheme 2.9) is faster than or comparable in rate with decomposition of the intermediate. Demonstration of the addition species as an intermediate only proves its existence, it does not prove that the species lies on the reaction path between reactant and product. It is unlikely that in this case the adduct is a 'blind-alley' intermediate because of the symmetrical nature of the intermediate and because the less basic alkoxide ion would be expelled from the adduct in preference to the hydroxide ion.

Positional isotope exchange has been employed extensively in studies of ion-pair intermediates resulting from a dissociative pathway in nucleophilic aliphatic substitution and in electron-deficient rearrangements. Goering and his co-workers (Goering *et al.*, 1961, 1963) studied the solvolysis of the 4-nitrobenzoate ester of a benzhydryl alcohol (**9**) and found (Eqn [2.36]) that positional exchange occurs, indicating a discrete ion-pair intermediate (**10**) with a significant lifetime compared with the rotation of the carboxylate ion group within it. It is not possible to exclude a mechanism for exchange involving concerted attack (**11**), although such a mechanism would be contrary to empirical rules governing ring closure such as strain energy in four-membered rings (Page, 1973) (see, however, Chapter 1, Scheme 1.6).

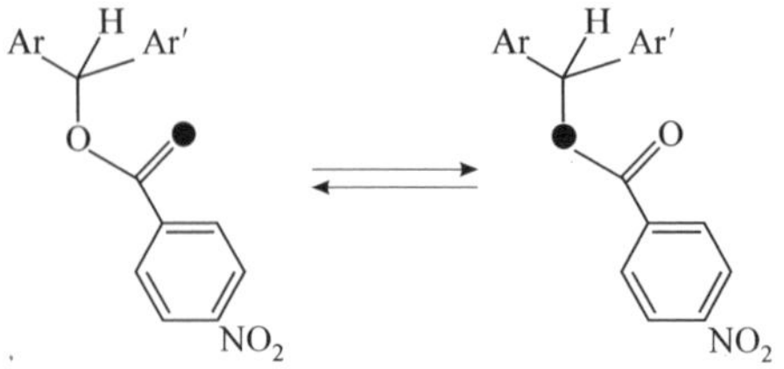

[2.36]

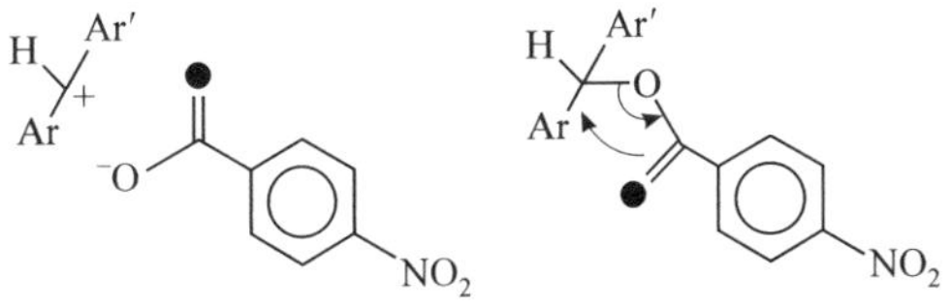

10   **11**

The ethanolysis of the norbornyl brosylate (**12**) can be followed readily by ^{17}O-NMR spectroscopy (Chang & le Noble, 1983) and undergoes positional isotopic exchange (Eqn [2.37]) indicating the existence of ion pairs similar to **10**. Chapter 3 describes a Fourier-transform infrared (FTIR) spectroscopic method which could be employed to carry out this type of experiment.

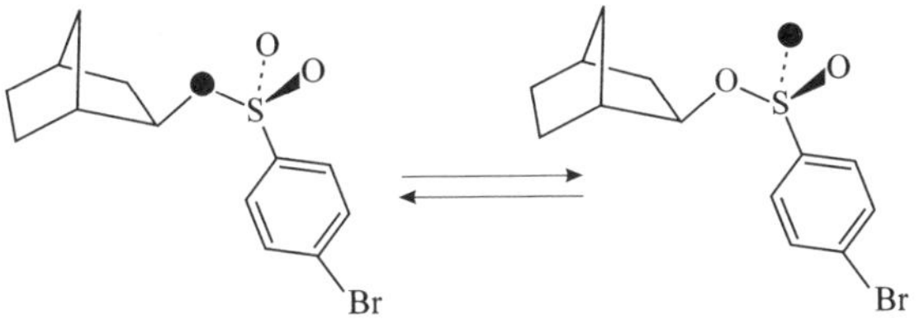

[2.37]

12

Further reading

Bernasconi, C. F. (ed.) (1986) *Investigations of Rate and Mechanisms of Reactions in Solution*, 4th edn, Wiley–Interscience, New York.

Bunnett, J. F. (1986) Kinetics in solution, in *Investigations of Rate and Mechanisms of Reactions*, 4th edn, Bernasconi, C. F. (ed.), Wiley–Interscience, New York, p. 171.

Bunnett, J. F. (1986) From kinetic data to reaction mechanism, in *Investigations of Rate and Mechanisms of Reactions*, 4th edn, Bernasconi, C. F. (ed.), Wiley–Interscience, New York, p. 251.

Cox, B. G. (1994) *Modern Liquid Phase Kinetics*, Oxford Science Publications, Oxford.

Hammett, L. P. (1970) *Physical Organic Chemistry*, 2nd edn, McGraw-Hill, New York.

Jencks, W. P. (1969) *Catalysis in Chemistry and Enzymology*, McGraw-Hill, New York.

Laidler, K. J. (1987) *Chemical Kinetics*, 3rd edn, Harper and Row, New York.

Maskill, H. (1985) *The Physical Basis of Organic Chemistry*, Oxford University Press, Oxford.

Moore, J. W. and Pearson, R. G. (1981) *Kinetics and Mechanism*, 3rd edn, Wiley, New York.

Wentrup, C. (1986) Tracer methods, in *Investigations of Rate and Mechanisms of Reactions*, 4th edn, Bernasconi, C. F. (ed.), Wiley–Interscience, New York, p. 613.

3   The effect of changes in reactant structure

3.1   Introduction

Our knowledge of the structure of transition states largely derives from observations of changes in reactivity as a function of the variation of some structural parameter in the molecule (an internal effect). The magnitude of the effect depends on the coupling between the structural change and the site of reaction. Insulation of the parameter from the site, for example by a long alkane chain, reduces its effect and thus its utility in determining mechanism. An exactly analogous method, namely the correlation of the effect of external parameters on reactivity (Chapter 5), suffers from the problem that the bulk parameter (e.g. pressure or dielectric constant) is a macroscopic property and its effect on the reaction site is less direct than changes in the molecular structure of the reactants. In general, internal effects are considered to be more diagnostic of transition-state structure than are external effects. The application of both structural effects (Chapters 3 and 4) and external effects (Chapter 5) is carried out by correlating the free energy of the reaction under investigation and that of a standard process. When the correlations are linear, as they often are, they are given the term LFER (linear free energy relationships).

3.2   Comparison of known with unknown—the Leffler approach

Structures of transition states may be deduced by comparing the effects on rates and equilibria of variation of reactant parameters (substituent, solvent, etc.—Leffler, 1953; Leffler & Grunwald, 1963). Introduction of a substituent perturbs the energy of a reactant; with reactants and products these perturbations are usually known or measurable changes. Most experimental techniques measure differences; a change in the rate constant of a reaction brought about by a change of substituent represents the *difference* in the effect on reactants compared with the transition state. A larger rate constant could result from a less stable reactant, a more stable transition state or a combination of these. The slope of the free-energy plot of rate versus equilibrium parameters is a 'similarity' coefficient (Pross & Shaik, 1989) because, on a scale of 0 to 1, it relates the structure of the transition state with those of product and reactant.

3.3 Polar and steric substituent effects

A polar substituent changes the energy of the transition state of a reaction by modifying the change in charge brought about by the bond changes (Hine, 1960). Thus substituents which withdraw electrons and 'spread' negative charge will tend to make the reaction go faster if there is an increase in negative charge at the reacting bond on going from reactant to transition state. The polar substituent is required to be suitably placed in the molecule for the effect to be transmitted to the reaction centre through the bond framework, space or solvent.

Polar substituent effects are used extensively in kinetic investigations of mechanism and a number of related techniques are employed, for example the use of Hammett, Brønsted, Taft and Charton correlations. None of these methods is intrinsically different as each is based on 'similarity'. The criterion of use is that the standard reaction should resemble the reaction being studied as closely as possible, and each of the above approaches uses a different standard model reaction (Table 3.1).

Table 3.1 Summary of useful linear free-energy relationships.

Name	Equation	Standard reaction
Hammett	$\log k = \rho.\sigma + C$	$X\text{-}C_6H_4\text{-}COOH \rightleftharpoons X\text{-}C_6H_4\text{-}COO^- + H^+$
Roberts	$\log k = \rho'.\sigma' + C$	$X\text{-}(\text{bicyclic})\text{-}COOH \rightleftharpoons X\text{-}(\text{bicyclic})\text{-}COO^- + H^+$
Grob	$\log k = \rho_I^Q.\sigma_I^Q + C$	$X\text{-}(\text{bicyclic})\text{-}NH^+ \rightleftharpoons X\text{-}(\text{bicyclic})\text{-}N + H^+$
Charton	$\log k = \rho_I.\sigma_I + C$	$RCH_2CO_2H \rightleftharpoons RCH_2CO_2^- + H^+$
Taft	$\log k^R/k^{Me} = \rho^*.\sigma^* + \delta.Es$	$RCO_2Et + OH^- \longrightarrow RCO_2^- + EtOH$
		$RCO_2Et + H_3O^+ \longrightarrow RCO_2H + EtOH$
Brønsted	$\log k = \beta.pK_a + C$	$H\text{-}A \rightleftharpoons H^+ + A^-$
Hansch	$\log k = a.\pi + C$	Partitioning n-octanol–water

The 'similarity' coefficients obtained from the correlations described above can be used to determine transition-state structure, provided there is sufficient background knowledge of those values relating to known mechanisms. Table 3.2 illustrates Hammett ρ-values for a selection of reactions.

Table 3.2 Examples of Hammett correlations.[a]

Reaction	ρ	Substituent constant[b]
$ArH + Cl_2 \rightarrow ArCl + HCl$	-8.06	σ^+
$ArPhCHCl + EtOH \rightarrow ArPhCHOEt + HCl$	-4.1	σ^+
$ArCHO + HCN \rightarrow ArCH(OH)CN$	2.33	
$ArNMe_2 + MeI \rightarrow ArNMe_3^+ + I^-$	-3.3	σ^+
$ArCH_2Cl + H_2O \rightarrow ArCH_2OH + HCl$	-2.18	σ^+
$ArO^- + EtI \rightarrow ArOEt + I^-$	-0.99	
$ArCH_3 + Cl_2 \rightarrow ArCH_2Cl + HCl$	-1.25	
$ArCO_2Me + OH^- \rightarrow ArCO_2^- + MeOH$	2.23	
$ArCO_2H \rightleftharpoons ArCO_2^- + H^+ (H_2O)$	1.00	
$ArCO_2H \rightleftharpoons ArCO_2^- + H^+ (CH_3OH)$	1.54	
$ArCO_2H \rightleftharpoons ArCO_2^- + H^+ (EtOH)$	1.96	
$ArCH_2CO_2H \rightleftharpoons ArCH_2CO_2^- + H^+ (H_2O)$	0.47	
$ArCH_2CH_2CO_2H \rightleftharpoons ArCH_2CH_2CO_2^- + H^+ (H_2O)$	0.22	
$ArNH_2 + HCO_2H \rightarrow ArNHCHO + H_2O$	-1.43	σ^-
$ArSO_2OCH_3 + OH^- \rightarrow ArSO_3^- + CH_3OH$	1.25	
$ArCOCl + H_2O \rightarrow ArCO_2H + HCl$	1.78	
$ArCOCl + C_6H_5NH_2 \rightarrow ArCONHC_6H_5 + HCl$	1.22	
$ArCONH_2 + H^+ + H_2O \rightarrow ArCO_2H + NH_4^+$	-0.48	
$ArCONH_2 + OH^- \rightarrow ArCO_2^- + NH_3$	1.36	
$ArO^- + (CH_2CH_2O) \rightarrow ArOCH_2CH_2O^-$	-0.95	σ^-
2,4-Dinitrophenyl aryl ether $+ CH_3O^-$ $\rightarrow ArOCH_3 +$ 2,4-dinitrophenol	1.45	
$PhCOCl + ArNH_2 \rightarrow PhCONHAr + HCl$	-2.78	
2,4-Dinitrofluorobenzene $+ ArNH_2$ $\rightarrow$ 2,4-dinitrophenylarylamine $+$ HF	-4.24	σ^-
$ArNMe_2 + CH_3I \rightarrow ArN^+Me_3I^-$	-3.3	σ^-
$ArNCO + CH_3OH \rightarrow ArNHCOOCH_3$	2.46	
$ArCOCH_3 + Br_2 \rightarrow ArCOCH_2Br + HBr$	-0.46	
$2ArCHO + OH^- \rightarrow ArCH_2OH + ArCO_2^-$	3.63	

[a] Fuller tables of Hammett correlations are given in Jaffé (1953a,b).
[b] Regular Hammett relationship except where stated (Eqn [3.1b]).

The importance of a general method for predicting rate and equilibrium constants has led physical organic chemists to search for parameters which are apparently independent of standard models. Pharmaceutical chemists have invested immense efforts to obtain parameters to predict binding constants of drugs (see Appendix A.2) from standard parameters obtained from partition coefficients. The electronic and steric parameters such as σ or E_s, while seemingly reaction-independent, are nevertheless derived from reaction models and the problem in using such correlations is that the limitations imposed by the standard models are neglected.

3.4   Hammett's equation

Transmission of the polar effect

Linear correlations between the logarithms of rate constants and pK_a values of substituted benzoic acids were discovered independently by Hammett and Burkhardt in the 1930s (Hammett & Pfluger, 1933; Hammett, 1937; Burkhardt *et al.*, 1936) and have since been used extensively as a means of estimating the relative polarity of substituents. The measure of polarity, the Hammett σ-value for the substituent, is simply the effect of that substituent on the dissociation constant of the substituted benzoic acid (Eqn [3.1a]) and a typical Hammett equation is given by [3.1b].

$$\underset{X}{\text{⬡}}-CO_2H \quad \xrightleftharpoons{K_a^X} \quad \underset{X}{\text{⬡}}-CO_2^- \;+\; H^+ \qquad\qquad [3.1]$$

$$\sigma = \log K_a^X - \log K_a^H = pK_a^H - pK_a^X \qquad\qquad [3.1a]$$

$$\log k^X = \rho.\sigma + C \qquad\qquad [3.1b]$$

The ionization of substituted benzoic acids (Eqn [3.1]) is a reasonable model for the alkaline hydrolysis of substituted benzoate esters (Eqn [3.2]) and a plot of $\log k_{OH}$ versus σ is linear (Fig. 3.1). Hammett relationships also include equilibria and Fig. 3.2 illustrates the linear dependences of pK_a for various acids upon the σ-value. The ρ-value indicates the susceptibility of the reaction to change in substituent and a positive ρ-value registers that electron-withdrawing substituents increase the rate or equilibrium constant, consistent with an increase in negative charge at the reaction centre. The ρ-value measures the change in charge *relative* to that in the ionization of benzoic acids ($\rho = 1.0$).

$$\underset{X}{\text{⬡}}-CO_2Et \quad \xrightarrow{k_{OH}^X \; [OH^-]} \quad \underset{X}{\text{⬡}}-CO_2^- \;+\; HOEt \qquad\qquad [3.2]$$

The Hammett ρ for reaction rates usually bears no exact and direct relationship to the transition-state structure because the model and the reaction or equilibrium being studied are often dissimilar. A positive ρ-value indicates the generation of negative charge or the removal of positive charge at the reaction centre on going from the reactant to the transition state. The magnitude of the ρ-value is often used to indicate whether there is a large or a small change in charge but this implicitly uses knowledge of ρ-values for known reactions. For example, the alkaline hydrolysis of substituted phenyl phosphoramidates (Scheme 3.1) has a ρ-value of 2.84 (Williams & Douglas, 1972), which is much more positive than the value of approximately unity

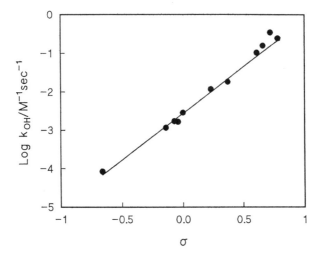

Fig. 3.1 Hammett dependence of the second-order rate constant for the attack of hydroxide ion on ethyl esters of substituted benzoic acids.

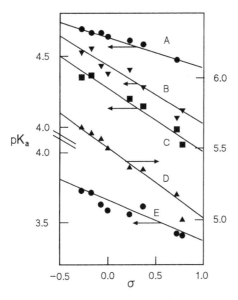

Fig. 3.2 Transmission of the polar effect through various bonds: Hammett dependencies of the ionization of substituted carboxylic acids. A, arylpropionic acids/H_2O; B, substituted cinnamic acids/H_2O; C, arylacetic acids/H_2O; D, arylthioacetic acids ($ArSCH_2CO_2H$)/50% dioxane/H_2O; E, arylthioacetic acids/H_2O.

observed for the attack of hydroxide ion on substituted phenyl phosphate triesters, where an associative mechanism is thought to hold. The mechanism of hydrolysis for the amidates is deduced to be different from the associative processes and a dissociative pathway involving a metaphosphoramidate (intermediate **1**) is postulated (Scheme 3.1) which is compatible with generation of a much larger negative charge on the aryl oxygen leaving group in the transition state.

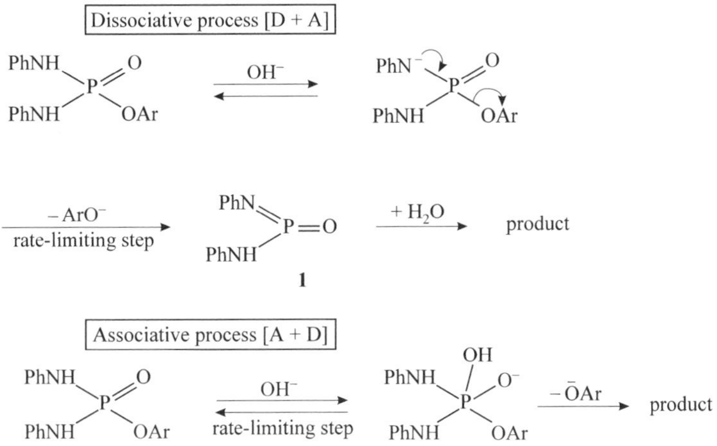

Scheme 3.1 Dissociative $(D+A)$ and associative $(A+D)$ processes for phosphoramidate hydrolysis.

Values of σ for individual substituents are collected in Appendix A.2, Table 3.a1. The sum of all the σ-values may be employed when there are multiple substituents because the effects are usually additive. The Hammett relationship is best limited to systems where substituents are attached to the reaction centre via aromatic rings; systems with *ortho*-substituents should be treated with care because of the extra steric effect on the reaction which is not felt by the standard ionization reaction. However, if a series of reactants has a constant *ortho*-group, such as all chloro or all nitro groups, then a Hammett relationship usually holds for substituents in the other positions; even benzene rings 'unsubstituted' at the 2- and 6-positions possess an *ortho*-substituent—namely hydrogen.

The Hammett ρ-values for ionization of phenylacetic and phenylpropionic acids indicate that transmission of the polar effect is attenuated by the intervening methylene groups. The attenuation factor falls roughly as the power of the number of intervening atoms; thus $\rho_n = \rho_0.f^n$ where f is the attenuation factor (Appendix A.2, Table 3.a2) for one group, n is the number of intervening groups and ρ_0 is the ρ-value when no group intervenes.

Resonance effects

Transmitting a polar effect through the aromatic nucleus can involve both inductive (I) and resonance (R) paths (Eqn [3.3]). The *meta*-substituent constant σ_m has no significant resonance interaction $(\sigma_R = 0)$; thus

$\sigma_I \simeq \sigma_m$. The resonance interaction is also small in the ionization of carboxylic acids, probably due to cancellation of the similar energies of the resonance contributions in acid and base (Scheme 3.2).

$$\sigma = \sigma_I + \sigma_R \qquad\qquad [3.3]$$

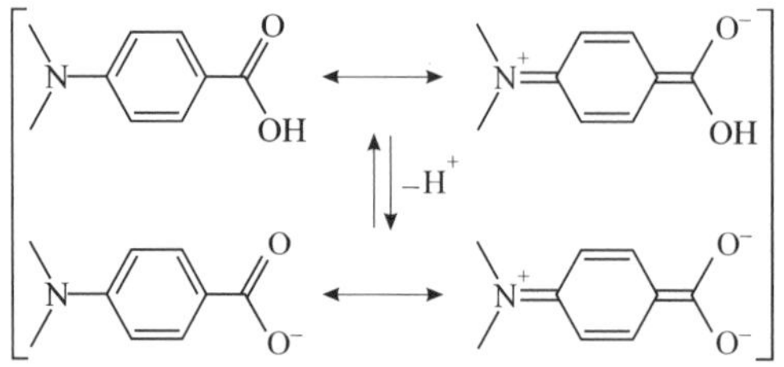

Scheme 3.2 Cancellation of resonance contributions in the ionization of a benzoic acid.

Ionization of phenols or anilinium ions involves substantial resonance stabilization of product base but not of the reactant acid (Scheme 3.3) and the relevant pK_a values do not correlate well with σ-values as defined in Eqn [3.1]. Rate constants for reactions with a developing negative charge on a phenolic oxygen often correlate well with the pK_a of the corresponding phenol and a parameter (σ^-) may be defined $(\sigma^- = (pK_a^{PhOH} - pK_a^{XC_6H_4OH})/$ 2.23); the value 2.23 is a normalizing factor and is the ρ-value for the ionization of *meta*-substituted phenols against regular σ-parameters (Table 3.a3 in Appendix A.2 collects σ^- constants for a selection of substituents).

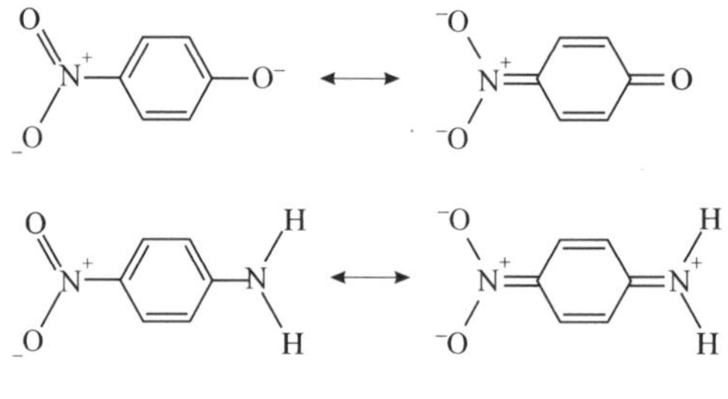

Scheme 3.3 Effect of electron-withdrawing resonance interactions on ionization of phenols and anilines.

Poor correlations are often obtained with σ for reactions where a positively charged centre is formed which can resonate with an electron-*donating* substituent because of the additional resonance transmission for these substituents. For example, Scheme 3.4 shows how a transition state may obtain extra stabilization by resonance interaction between a carbenium ion and 4-methoxy and 4-dimethylamino groups. A parameter (σ^+) may be defined which yields good correlations; this is derived from the solvolysis of $ArCMe_2Cl$ in 90% acetone/water at 25 °C $(\sigma^+ = (\log k_X - \log k_H)/-4.54;$

see Appendix A.2, Table 3.a4 for σ^+-values for a selection of substituents); the quantity -4.54 is the ρ value for the standard solvolysis reaction for *meta*-substituents.

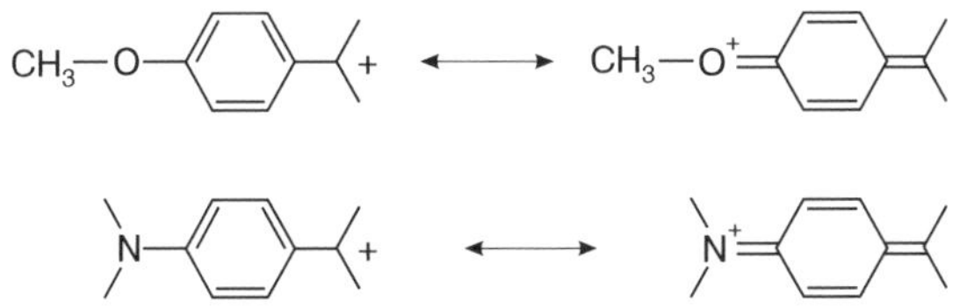

Scheme 3.4 Electron-donating resonance interactions.

Various methods have been employed to measure resonance participation. The r parameter of Yukawa & Tsuno (1959) (Eqn [3.4]) measures the extent of the interaction:

$$\log k/k_0 = \rho[\sigma + r(\sigma^+ - \sigma)] \qquad [3.4]$$

Table 3.a5 in Appendix A.2 gives some applications. This concept has also been extended (Humffray & Ryan, 1967) to reactions where negative charge builds up and an equation analogous to Eqn [3.4] is used to correlate the data.

The Jaffé dual pathway treatment

Transmission of the polar effect from substituent to reaction centre in benzenoid systems is effectively by a single pathway even though it may involve more than one route. For example, the *meta*-substituent can in principle transmit its effect to the *ipso*-position via the 2-position or via the 4-, 5- and 6-positions, to the same extent whichever *meta*-position is taken in a monosubstituted phenyl group (Scheme 3.5). In the case of the benzothiazole example the routes are not the same when the two *meta*-positions are interchanged (Scheme 3.5), and the two routes contribute differently to transmission. A special equation (Eqn [3.5]; Jaffé, 1953, 1954) is employed

$$\log k_X/k_H = \rho_A \sigma_A^{\,X} + \rho_B \sigma_B^{\,X} \qquad [3.5]$$

to enable a correlation to be made with σ. Equation [3.5] predicts the substituent effect if it is transmitted to the reaction centre by two pathways (A and B), for example by the substituent having an effect on both the nucleophile and electrophile in an intramolecular reaction. The classical treatment is to plot $(\log k_X/k_H)/\sigma_A^{\,X}$ versus $\sigma_B^{\,X}/\sigma_A^{\,X}$ to give a straight line with slope ρ_B and an intercept of ρ_A. However, the form of this equation ensures that the correlation is often quite poor owing to error magnification in the arithmetical manipulation. Good curve-fitting computer programs are available which are preferable to the graphical method to obtain the ρ_A- and ρ_B-parameters. If σ_A and σ_B correlate with each other a false Jaffé correlation will be obtained; the criterion for significance is that the correlation coefficient between the σ_m- and σ_p-values employed in the relationship should be less than 0.90 (Fersht & Kirby, 1967).

Transmission via effectively a single path

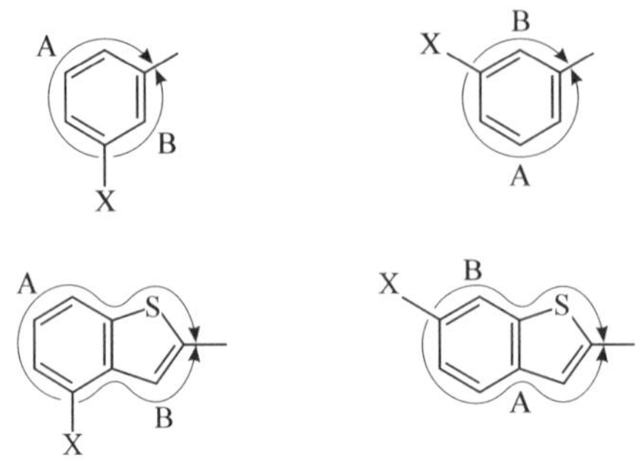

Transmission via two discrete paths A and B

Scheme 3.5 The effect of multiple transmission routes.

Inductive pathways

Equation [3.3] assumes that resonance and inductive transmission are mutually exclusive and they may be separated by employing standard reactions possessing no resonance transmission. The following methods (Eqns [3.6]–[3.8]—see Table 3.1 for definitions of the pK_a values) have been employed to define inductive polar effects:

Roberts & Moreland (1953):

$$\sigma' = (pK_a^H - pK_a^X)/1.46 \qquad [3.6]$$

Grob (1976):

$$\sigma_I^Q = pK_a^H - pK_a^X \qquad [3.7]$$

Charton (1964) defined a σ_I-value (Eqn [3.8]) obtained from the ionization constants of substituted acetic acids which are very well documented for a large variety of substituents.

$$\sigma_I = (pK_a^H - pK_a^X)/3.95 \qquad [3.8]$$

Since the σ_I values as defined above are used extensively, a relatively comprehensive table is included in Appendix A.2 (Table 3.a6). Table 3.3 provides a comparison of the various σ-values for standard substituents.

The apparently arbitrary denominators in Eqns [3.6]–[3.8] are normalizing factors to generate values of the various σ-parameters reasonably close to those of the original Hammett σ-values (see Table 3.3). Resonance contributions probably constitute a small proportion of most σ-values (van Bekkum *et al.*, 1959). An average σ for substituents is obtained by fitting Hammett equations to a standard series of reactions which are thought to possess no resonance interaction; the average values of σ from these correlations provides a set of purely inductive coefficients (σ^n).

Table 3.3 Comparison of σ-values.

Substituent	σ^I	σ_{meta}	σ_{para}	σ_I	$\sigma^n(para)$	$\sigma^n(meta)$
H	0	0	0	0	0	0
OH	0.283	0.121	−0.370	0.250	−0.178	0.095
COOEt	0.297	0.370	0.450	0.340	0.385	0.321
Br	0.454	0.391	0 232	0.450	0.265	0.391
CN	0.579	0.560	0.660	0.580	0.674	0.613

Taft polar (σ^*) and steric (E_s) parameters

The hydrolysis of substituted ethyl formate esters X—COOEt may be employed as a standard reaction for defining polar and steric effects (Taft, 1956; Shorter, 1972). Equation [3.9] describes the results when the two effects are mutually exclusive.

$$\log k^R/k^{Me} = \rho^*\sigma^* + \delta E_s \qquad [3.9]$$

The constants σ^* and E_s are the polar and steric substituent constants respectively and ρ^* and δ are the corresponding reaction constants, which indicate the susceptibility of the reaction to these effects.

It is assumed that the transition states for the specific acid- and base-catalysed hydrolysis of the formate esters have the same steric requirements, and thus the susceptibility parameters δ_H and δ_{OH} in Eqns [3.10] and [3.11] are identical.

$$\log k_H^R/k_H^{Me} = \rho_H^*\sigma^* + \delta_H E_s \qquad [3.10]$$

$$\log k_{OH}^R/k_{OH}^{Me} = \rho_{OH}^*\sigma^* + \delta_{OH} E_s \qquad [3.11]$$

These equations may be combined to give Eqn [3.12].

$$\log k_{OH}^R/k_{OH}^{Me} - \log k_H^R/k_H^{Me} = (\rho_{OH}^* - \rho_H^*)\sigma^* \qquad [3.12]$$

The difference $\rho_{OH}^* - \rho_H^*$ is defined as 2.48 because this is the difference between the ρ-values for hydroxide ion- and proton-catalysed hydrolysis of ethyl benzoates. Since ρ_H for acid-catalysed hydrolysis of ethyl benzoates is near to zero, the value of E_s may be determined if an arbitrary standard value of $\delta = 1$ is defined for this reaction. Thus δ is *defined* by Eqn [3.13], and values of σ^* and E_s are given in Appendix A.2 (Tables 3.a7 and 3.a8 respectively). Values of $\delta > 1$ for a reaction indicate steric requirements greater than that for hydrolysis of ethyl formates, whereas $\delta < 1$ indicates a smaller steric requirement.

$$\log k_H^R/k_H^{Me} = \delta E_s \qquad [3.13]$$

Which σ-values to use?

Since all the types of σ-value recorded here are meant to register an absolute polar effect, it would appear superfluous to have such a wide choice of parameters. In practice, most types of σ given here are employed, and the choice depends on the resemblance between the reaction under investigation and the standard reaction by which the σ-value is defined. The regular Hammett σ is generally preferred for reactions where the substituent change is in a benzenoid system, whereas Taft σ^* and E_s are useful in studies where the steric effect is important as in aliphatic systems. The very large number of Charton σ_I-values available from the pK_a values of acetic acids (see Appendix A.2, Table 3.a6) makes this a useful parameter for aliphatic systems.

3.5 Brønsted's equation

Introduction

Brønsted (1928) noted the linearity of a plot of the logarithm of the rate constant for a proton transfer reaction against the pK_a of the acid acting as a variable proton donor (k_{HA}) or against the pK_a of the conjugate acid of the base acting as a variable proton acceptor (k_B). The slopes of these plots are classically given the symbols α and β, for general acid and base catalysis, respectively (Scheme 3.6) and indicate how the energy of the transition state varies with substituents compared with the acid dissociation reaction. The ionization of acetone (Scheme 3.7 and Fig. 3.3) is catalysed by base and the slope of a plot of $\log k_f$ versus pK_a^{HB} will be the same as that against $\log K$, because pK_a^{HS} is constant and independent of the base (B). This relationship is very important as it is often very difficult to measure the equilibrium constant K explicitly; moreover changes would be even more difficult to determine accurately. If the plot of $\log k_f$ (proportional to the free-energy change from reactant state to transition state) against $\log K$ (a measure of the free energy of the overall reaction) has a slope of unity, the transition-state structure is product-like because the transition-state energy will change (as a function of substituents) in the same way as does that of the product. This is especially true if one of the rate constants, for example k_r, is at the diffusion-controlled limit. If the plot has zero slope, then the transition-state structure is reactant-like because $\log k_f$ represents little change in energy.

General acid catalysis: $HA + S \xrightarrow{k_{HA}}$ products

$$\log k_{HA} = \alpha.pK_a^{HA} + C$$

General base catalysis: $B + HS \xrightarrow{k_B}$ products

$$\log k_B = \beta.pK_a^{HB} + C$$

Scheme 3.6 The Brønsted rate laws.

$$CH_3COCH_3 \xrightleftharpoons{K_a^{SH}} CH_3COCH_2^- + H_3O^+$$

$$B + CH_3COCH_3 \xrightleftharpoons[k_r]{k_f} CH_3COCH_2^- + HB^+$$

$$B + H_3O^+ \xrightleftharpoons[K_a^{HB}]{} H_2O + HB^+$$

$$K = k_f/k_r = \frac{[CH_3COCH_2^-][HB^+]}{[CH_3COCH_3][B]}$$

$$= \frac{[H^+][CH_3COCH_2^-]}{[CH_3COCH_3]} \cdot \frac{[HB^+]}{[H^+][B]}$$

$$= K_a^{HS}/K_a^{HB}; \quad \log K = pK_a^{HB} - pK_a^{HS}$$

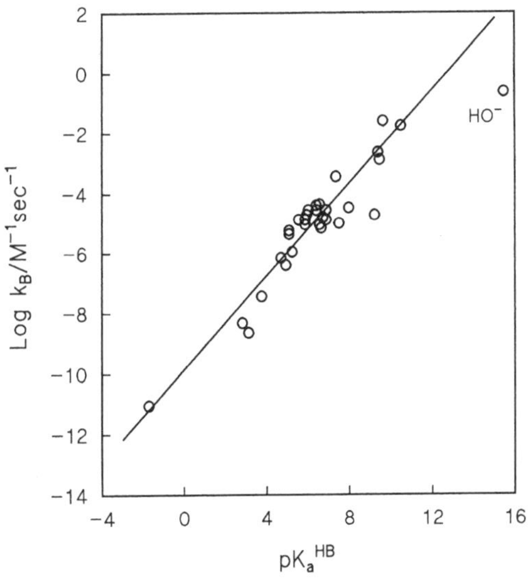

Fig. 3.3 Rates of ionization of acetone by different bases; data are not corrected statistically and are obtained from Bender & Williams (1966). The reaction is measured by the iodination of acetone and the deviation of the hydroxide term is discussed in Chapter 6.

Scheme 3.7 Brønsted equilibria in the ionization of acetone.

The values of the Brønsted coefficients α and β for the reverse and forward reactions respectively are related by Eqn [3.14] in Scheme 3.8. The value of β refers to the simple degree of protonation in the catalyst. Bell (1959) illustrated this relationship by a simple crossing of two Morse curves (SH could be equivalent to CH_3COCH_2—H) for proton dissociation (SH $\rightleftharpoons$ S$^-$ + H$^+$) and another for association (B$^-$ + H$^+$ $\rightleftharpoons$ BH) (Fig. 3.4).

$$\log k_f = \beta.pK_a^{HB} + C$$
$$\log k_r = \alpha.pK_a^{HB} + D$$
$$\log K = (\beta - \alpha)pK_a^{HB} + C - D \qquad [3.14]$$
$$\beta - \alpha = 1$$

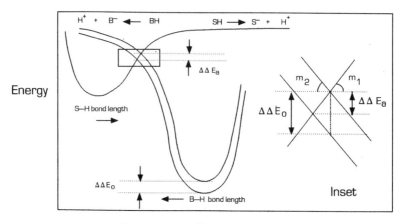

Reaction coordinate

Fig. 3.4 Origin of a free-energy relationship for the reaction where the base is varied $B^- + SH \longrightarrow BH + S^-$.

Scheme 3.8 Relationship between β and α for a proton transfer reaction.

When the base, B^-, is varied (and HS is kept constant) the right-hand Morse curve will alter vertically and if the curves in the cross-over region are linear the change in energy of the transition state (the crossing point) will be related to the change in overall energy by Eqn [3.15]. In the example given, where there is base catalysis, the ratio $m_1/(m_1 + m_2)$ is equal to β and a reactant-like transition state has $m_1 \approx 0$ and $\beta \approx 0$ whereas a product-like transition state has $m_2 \approx 0$ and $\beta \approx 1$.

$$\Delta\Delta E_a = \{m_1/(m_1 + m_2)\}\Delta\Delta E_0 \qquad [3.15]$$

The extended Brønsted relationship

The Brønsted relationship for proton transfer reactions is directly related to the ionization process, which formally involves the removal or addition of a unit charge. Bases, by definition, are proton acceptors but they may also be nucleophiles by acting as electron-pair donors to carbon or other electrophilic centres. Formally, this process could involve the addition or removal of unit charge and it is therefore not surprising that there are often good correlations between the logarithm of the rate constants and the corresponding pK_a. Similarly, the reverse process, the expulsion of a leaving group from carbon or another centre, generates a base. For example, nucleophilic substitution of 2,4-dinitrochlorobenzene by phenolate ions may be correlated with the ionization of the phenol (Knowles *et al.*, 1961; Eqn [3.16]). To distinguish the exponents of the extended Brønsted correlation from the acid–base parameters they are given the symbol β_{nuc} or β_{lg}, depending on the group (nucleophile or leaving group) wherein lies the structural variation.

$$[3.16]$$

$$\left[\overset{+}{H} + ArO^- \rightleftharpoons ArOH \right]$$

Effective charge

The effect of substituents on the ionization equilibrium (Eqn [3.17]) will depend on the overall change in charge on the oxygen atom. The absolute charge on the oxyanion will differ from the formal value of unity because charge is spread to a greater or lesser extent over the whole molecule and by solvation. A plot of the logarithm of the equilibrium constant for Eqn [3.17] against pK_a^{ArOH} manifestly has unit slope, whereas that of the equilibrium constant for Eqn [3.18a] has a slope β_{eq} against the pK_a^{ArOH}. The value β_{eq} indicates the sensitivity to change in polar substituents for the equilibrium *relative* to that of the ionization equilibrium, and measures the change in relative charge on the oxygen atom. Acyl group transfer (Eqn [3.18]) has a β_{eq} of 1.7, indicating that there is a greater sensitivity to polar substituents than in the ionization reaction and consistent with a change in relative charge of 1.7 compared with the unit change in the standard ionization equilibrium. Since the $Ar-O^-$ species may be defined as possessing 1 unit of negative charge on its oxygen, in comparison with zero in the neutral phenol (ArOH), the charge on the phenolic oxygen of the reactant ester is $+0.7$. In other words the acyl group (acylium ion, RCO^+) is effectively more electron-withdrawing than the hydrogen (proton) when covalently linked to an aryl oxygen. Since the charges measured by the β-values are not electronic charges they are distinguished by the term *effective charge* (Jencks, 1972; Williams, 1984, 1991; Thea & Williams, 1986).

$$H_2O + ArO-H \quad \underset{}{\overset{K_a^{ArOH}}{\rightleftharpoons}} \quad ArO^- + H_3\overset{+}{O} \qquad [3.17]$$

$$NR_3 + ArO-COCH_3 \quad \underset{k_{lg}}{\overset{k_{nuc}}{\rightleftharpoons}} \quad ArO^- + CH_3CO-\overset{+}{NR_3} \qquad [3.18a]$$

The value of β_{eq} and hence of the effective charge on atoms/groups may often be determined indirectly by a comparison of related equilibrium constants. For example, the β_{eq} for the equilibrium constants of Eqn [3.17] is independent of the nucleophile (Eqn [3.18b]). The equilibrium constants for Eqn [3.18b] may be calculated from those of the related acyl transfer reactions K_1 and K_2 in Scheme 3.9. The calculated β_{eq} is independent of the nature of the nucleophile because K_2 does not involve an aryloxide ion and therefore β_{eq} depends only on the group and the nature of the molecule from which it is transferred.

$$CH_3CO\!-\!OAr + Nu^- \xrightleftharpoons{\quad K_{eq} \quad} ArO^- + CH_3CO\!-\!Nu$$

$$CH_3CO\!-\!OCOCH_3 + ArO^- \xrightleftharpoons{\quad K_1 \quad} CH_3CO_2^- + CH_3CO\!-\!OAr$$

$$CH_3CO\!-\!OCOCH_3 + Nu^- \xrightleftharpoons{\quad K_2 \quad} CH_3CO_2^- + CH_3CO\!-\!Nu \qquad \text{[3.18b]}$$

$$K_{eq} = K_2/K_1$$

$$\log K_{eq} = \log K_2 - \log K_1$$

$$\beta_{eq} = d \log K_{eq}/d \log pK_a^{ArOH} = -d \log K_1/d \log pK_a^{ArOH}$$

Scheme 3.9 The value of the Brønsted β_{eq} is independent of nucleophile.

The value of β_{eq} and the associated effective charges also measure polarity and the values known for many group transfer equilibria (Table 3.4) correlate with the electropositive character of the group.

Effective charge in the transition state

Since the transition state can be considered as if it were an 'equilibrium' state, it is possible to define its effective charges in the same way as those just considered for equilibrium reactions. Equation [3.19] has 'rate constants' for forward (k_+) and return (k_-) breakdown of the transition state 'species' ($\ddagger$) which are essentially invariant because they register the collapse of the transition structure. These 'rate constants' are independent of substituent changes and therefore have zero β-values associated with them. The 'equilibrium constants' for formation of the transition state (k_1/k_-) and for its breakdown to products (k_+/k_{-1}) will therefore vary only according to changes in k_1 and k_{-1}. The polar effect on the rate constants k_1 and k_{-1} (corresponding to β_1 and β_{-1} respectively) will therefore measure changes in effective charge from ground or product states to the transition state.

$$\text{Reactant} \xrightleftharpoons[\; k_- \;]{\; k_1 \;} \ddagger \xrightleftharpoons[\; k_{-1} \;]{\; k_+ \;} \text{product} \qquad \text{[3.19]}$$

Effective charge and transition-state structure

The elucidation of transition-state structures from linear free-energy relationships is based on similarity, as discussed in Section 3.2: the effect of substituents on the equilibrium free-energy difference between reactants

Table 3.4 Values of effective charges (β_{eq}) for some group transfer reactions.[a]

+0.36 Ar—O—PO₃²⁻	+0.74 Ar—O—PO₃H⁻	+0.83 Ar—O—PO₃H₂
+0.25 Ar—O—POPh₂	+0.33 Ar—O—PO(OPh)₂	+0.87 Ar—O—PO(OEt)₂
+0.3 CH₃—S—Ar	+1.47 CH₃N⊕...X	+0.3 CH₃—Se—Ar
−0.16 R—S—Ar	+0.04 to +0.2 R—O—Ar	+0.86 to +1.0 X—R—NHR₂⁺
+0.59 Ar₃C—NH₂R⁺—X	−1.0 RS—S—Ar	+0.4 CH₃CO—S—Ar
+0.3 Ar—O—CONH⁻	+0.8 Ar—O—CONH₂	+0.4 X—R—O—CO₂⁻
+1.07 ²⁻O₃P—N⊕...X	+1.25 ⁻O₃S—N⊕...X	+1.6 CH₃CO—N⊕...X
+0.7 CH₃CO—NHAr	+0.5 X—RNH—CHO	+0.48 R₂C̄CO—O—Ar
+0.7 CH₃CO—O—Ar	+0.3 RNH—CO₂⁻	+1.4 Ar—O—CSNHAr
+0.7 Ar—O—CSNAr⁻	+0.8 Ar—O—SO₂R	+0.7 Ar—O—SO₃⁻

[a] Data mainly from Williams (1991). Figures are the values of β_{eq} for transfer of the **functional** group to an acceptor nucleophile. The residue bearing the substituent variation is either the aryl group, or is marked with an X.

and products of known structure should be related to their effect on the free-energy difference between reactants (of known structure) and the unknown transition-state structure.

The similarity principle (Pross & Shaik, 1989) was first exemplified by use of polar substituent effects such as the Brønsted β or Hammett ρ (Leffler, 1953; Leffler & Grunwald, 1963). These reactivity parameters have often been misinterpreted because they were originally demonstrated for an idealized reaction (Eqn [3.20]). The Leffler parameter is defined as $\alpha = \mathrm{d}\log k_1 / \mathrm{d}\log K_{eq}$. When the transition-state structure resembles that of the reactant, the polar effect on the change from reactant to transition state should be small compared with the effect on the overall reaction and α should then be approximately zero. Conversely, if the transition-state structure resembles that of the products, the value of α should approximate to unity.

$$A \underset{k_{-1}}{\overset{k_1}{\rightleftarrows}} B \qquad \{K_{eq} = k_1/k_{-1} = [\text{B}]/[\text{A}]\} \qquad [3.20]$$

The application of linear free-energy relationships, especially the interpretation of ρ and β as measures of transition-state structure, is not without criticism. The discovery that Leffler exponents, α, *exceed* unity for some reactions (Bordwell *et al.*, 1969; Bordwell & Boyle, 1971, 1972) reinforced the arguments of sceptics. This remarkable anomaly is discussed in Chapter 6 and gives more information about the transition state, rather than invalidating Leffler's principle. It was even suggested that linear free-energy relationships have little role to play in probing mechanism. However, the various polar effects manifestly derive from changes in electronic charge and steric requirements and must be directly related to mechanism, although not in a simple way. It should be emphasized that observed polar effects are not just the result of change in electronic structure of a discrete bond in an isolated molecule, but result from changes in the overall structure of the *states*.

Complete knowledge of a reaction mechanism requires study of *all* major bonding changes in the system. There is little advantage in using the term 'advanced' or 'late' for a transition state without reference to the state of a particular bond, because fission or formation of one bond could be 'advanced' while that of another could be 'late' in the *same* transition state. The problems of interpretation arise because the various methods of probing transition-state structure usually refer to a particular bonding arrangement and not to *overall* structure (see Chapter 5). Leffler's parameter, α, has been shown to be the same for a single bond fission in reactions where the technique observes from *both* ends of the same bond (Hill *et al.*, 1982).

Brønsted or Hammett kinetic parameters for bond changes not involving proton transfer are not particularly useful for deducing charge unless they have been 'calibrated' by the corresponding equilibrium parameter for the overall reaction. Qualitative knowledge of change in charge can most often be derived from the rate law of the reaction and does not always require recourse to linear free-energy relationships. For example, the reaction of dimethylanilines with methyl iodide (Eqn [3.21]) has a ρ-value (-3.3) which simply *confirms* that positive charge builds up on the nitrogen; this conclusion is self-evident from our basic knowledge of the electronic structures of amines and alkyl halides and the product quaternary ammonium salts. When it is compared with the ρ-value of -4.2 for the equilibrium protonation of N,N-dimethylanilines it may then be deduced that significant C–N bond formation and charge development have occurred in the transition state (Eqn [3.21]).

$$[3.21]$$

The magnitude of ρ or β can sometimes be used to distinguish between mechanisms. For example, fission of an aryloxy bond in the transition state of the reaction of an aryl ester gives rise to a much larger Hammett or Brønsted parameter than when the aryloxy bond is *intact* in the transition state. Experience of the magnitude of β- or ρ-parameters for a range of known mechanisms enables us to judge approximately whether there is large or small charge development for a given reaction; for example, the ρ-value of 2.5 in the alkaline hydrolysis of methyl esters of substituted benzoic acids results from a large build-up of negative charge on the ester's carbonyl oxygen, and a reduction in the positive charge density on the carbonyl carbon. The parameters ρ, ρ^*, β and the like tend to have particular ranges of values for individual reactions. For example, nucleophilic attack of hydroxide ion on aryl esters has a ρ-value in the region of $+1$ whereas a reaction involving substantial fission of an aryloxy bond, such as in the alkaline hydrolysis of aryl acetoacetates or aryl carbamates, has a much higher ρ-value (in the region of 2 to 3). The direct application of the Hammett equation suffers from the difficulty that many reactions do not resemble the ionization of carboxylic acids used as the defining standard for σ. The Brønsted equation is a direct comparison for general acid–base catalysis but is less relevant in its extended form. The Hammett ρ-value is a poor indicator of detailed transition-state structure unless there is a fairly large body of comparable values for similar reactions; used correctly, the Hammett correlation is capable of giving definitive answers in selected cases.

Charge development in a reaction is a good indication of transition-state structure but it does not directly give an estimate of change in bonding because it is affected by other factors such as solvation (Chapter 5). The change in charge on an atom may be followed from reactant, through transition state, to product. For example, rate constants for phenolate ion attack on aryl diphenylphosphate esters (Scheme 3.10) generate a β_{lg} of -0.52 and the overall β_{eq} for the equilibrium reaction is -1.30 indicating that, for that transition state, the effective charge on the aryl oxygen is some 40% of the difference between that in ground and product states. This does not necessarily mean that the bond order (as for an isolated molecule in the gas phase) is 0.4, because there could be a contributing solvation change (see Chapter 5).

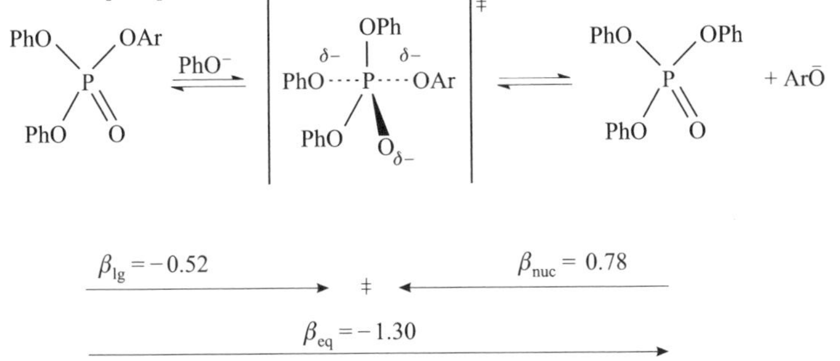

Scheme 3.10 Relationship between β and effective charge for an ester reaction.

Maps of effective charge and conservation of effective charge

Reaction maps, where effective charge is assigned for each atom as it passes through the various stages of the mechanism, may be constructed and are particularly effective in demonstrating the charge structure of the various transition states. Schemes 3.11 and 3.12 illustrate effective charge maps and show how β-values may be estimated making use of the principle of conservation of effective charge.

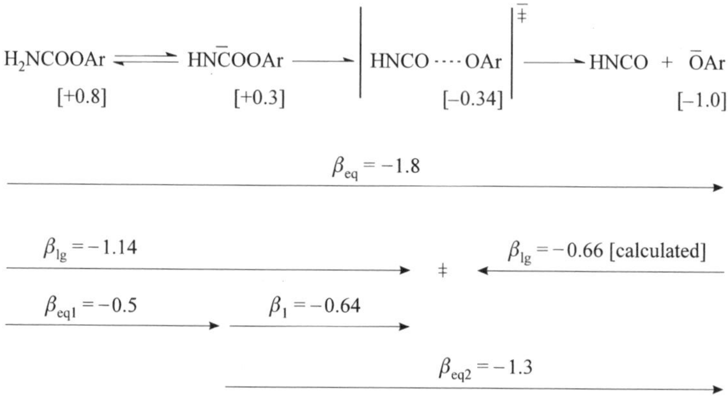

Scheme 3.11 Effective charge map for aryl oxygen in the reactions of carbamate esters ([e] = effective charge) (Al-Rawi & Williams, 1977).

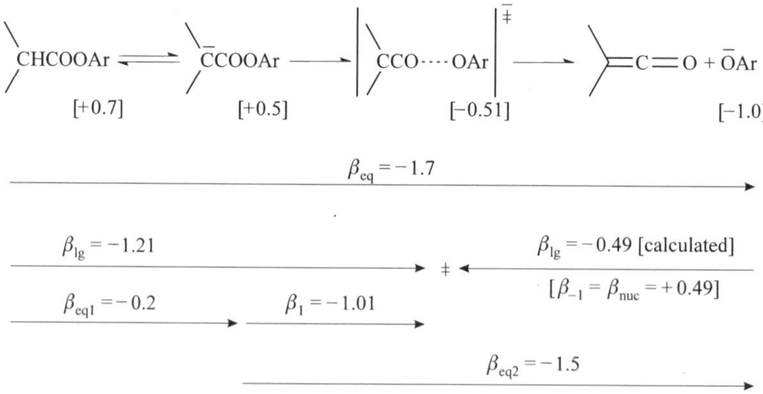

Scheme 3.12 Effective charge map for aryl oxygen in the reactions of fluor-enoyl esters (Thea & Williams, 1986).

Assuming that the effective charge changes shown in Schemes 3.11 and 3.12 are *conserved*, the β-values relating rate and equilibria are given in Eqn [3.22]. The principle of conservation of effective charge is very useful as it

enables the calculation of effective charges for reactions not easily accessible
to direct measurement.

$$\beta_{eq} = \beta_{eq1} + \beta_{eq2} = \beta_{eq1} + \beta_1 - \beta_{-1} \tag{3.22}$$

An example of the application of the principle of effective charge conservation is in the attack of phenolate ion on isocyanic acid (Scheme 3.11)
and on a ketene (Scheme 3.12), which are not easy to measure kinetically;
the respective values of $\beta_{nuc}(= -\beta_{lg})$ may be obtained from β_{eq2} and β_1.
Care must be taken in interpreting the results when comparing changes in
effective charge in a reaction where the bonding changes differ structurally.
The change in charge for bonds of different character is *not* conserved (for
example in a reaction where fission of N—P and formation of ArO—P
occur). When the bonds have the same character and the overall charge
change is therefore the same, the effective charge *is* conserved. Thus attack
of phenolate ions on phenyl esters has the reaction map (shown in Scheme
3.13). In this case the decrease in charge on attacking phenolate ion in the
transition state (+1.37) is not balanced by the observed increase in charge
on the leaving phenyl oxygen (−0.33); a change in charge on the components of the CH_3CO group is required if the overall effective charge is
to be conserved. The magnitude of the charge is based on the standard
charge change in the calibrating equilibrium, even though the atom
structure in the carbonyl function does not resemble the proton in the
ionization reaction used for standardization. The positive effective charge
on the atom adjacent to acyl functions (see Table 3.4) is balanced by an
equal negative charge on the acyl function in order to conserve charge.

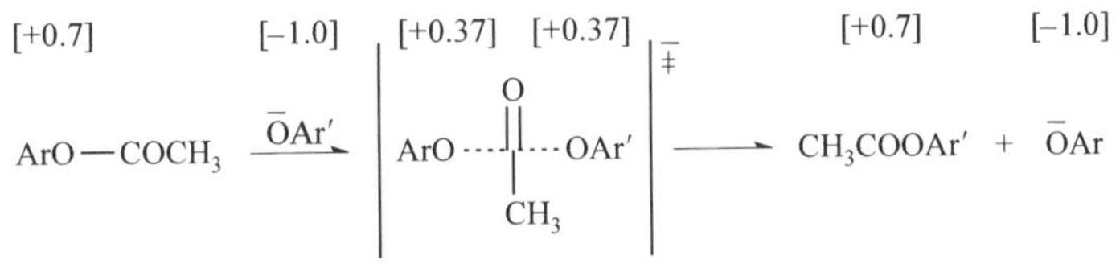

$$[+0.7] \qquad [-1.0] \quad [+0.37] \quad [+0.37] \qquad\qquad [+0.7] \qquad [-1.0]$$

$$ArO—COCH_3 \xrightarrow{\ \overline{O}Ar'\ } \left| ArO\cdots\overset{\overset{\textstyle O}{\|}}{\underset{\underset{\textstyle CH_3}{|}}{C}}\cdots OAr' \right|^{\ddagger} \longrightarrow CH_3COOAr' + \overline{O}Ar$$

Scheme 3.13 Reaction map for attack of phenolate ions on aryl acetates
(Williams, 1991).

Charge balance

The relative extent of individual bond changes in a transition state is called
the 'balance'. For example, in a reaction such as nucleophilic substitution
involving two major bond changes, the transition state could involve a *range*
of structures from that where bond formation is largely complete and fission
incomplete (B in Fig. 3.5) to that where formation is incomplete and fission
is complete (C in Fig. 3.5).

The term 'synchronous' (Jencks, 1988) is reserved to denote a mechanism where formation and fission occur to the same extent in the transition

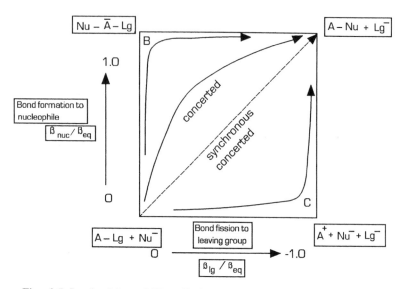

Fig. 3.5 Jencks–More O'Ferrall diagram for the reaction A–Lg + Nu ⟶ A–Nu + Lg (see text for details).

state of a concerted mechanism, i.e. the transition state is anywhere on the diagonal broken line in Fig. 3.5. The ratio $\alpha_{(formation)}/\alpha_{(fission)}$ may be taken as a formal indicator of the degree of balance, where deviation from unity indicates the amount of *imbalance* in the transition state; for example, for reactions such as in Scheme 3.14 which do not involve identical formation and fission processes the ratio equals $0.2/0.87 = 0.23$. Scheme 7.18 (p. 156) illustrates the general base-catalysed cyclization of aryl esters of uridine-3-phosphoric acid. The balance of charge is obtained by comparing the Leffler α-value ($\alpha_{formation}$) for formation of the H—B bond and α_{lg} for the fission of the ArO—P bond. It is not acceptable to use effective charge directly to compare the two bonding changes because of the gross dissimilarity in the nature of the bonds undergoing change.

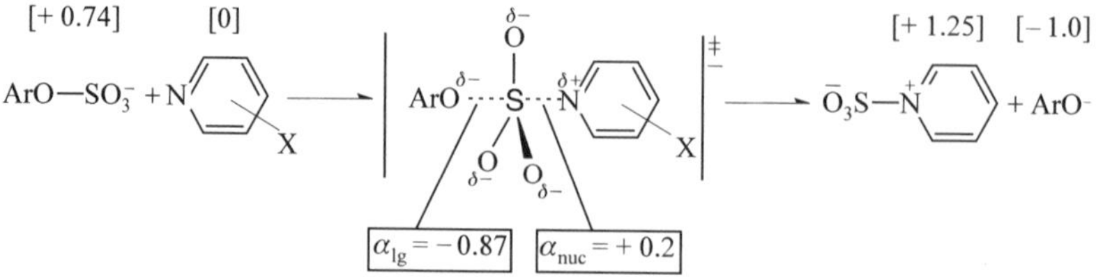

Scheme 3.14 Balance of effective charge in a sulphyl group (sulphur acyl) transfer reaction.

Identity reactions

Identity reactions occur when a group expels an identical leaving group in a displacement reaction, and are very useful for discussions of *concerted* nucleophilic displacements because the transition state is *required* to lie on the 'tightness' diagonal shown in the reaction map (Fig. 3.6). The 'tightness' diagonal (Kreevoy & Lee, 1984) is sometimes called the 'disparity' mode (Grunwald, 1985); essentially it maps movement on the reaction surface perpendicular to the reaction coordinate.

The τ-value (on a scale of 0 to 2) registers the position of the transition state on the 'tightness' diagonal; τ may be determined from a single Brønsted experiment for an identity reaction, either from β_{lg} or from β_{nuc}. Effective charges on the entering and leaving groups are identical for concerted reactions which have transition states on the tightness diagonal.

The position of the transition state (and hence its electronic structure) can be readily discussed in terms of the various energies of the two hypothetical intermediate states at the top left and bottom right corners of Fig. 3.6. Identity rate constants (k_{ii}) may be studied largely by use of isotope replacement techniques (Kreevoy & Lee, 1984; Lewis & Hu, 1984) but are difficult to measure accurately; effective charge studies with quasi-symmetrical reactions offer a simple approach to measuring identity rate constants. The diagram in Fig. 3.6 has the advantage that it can be constructed without recourse to the explicit measurement of identity rate constants; the equations which enable this to be done are given in Scheme 3.15.

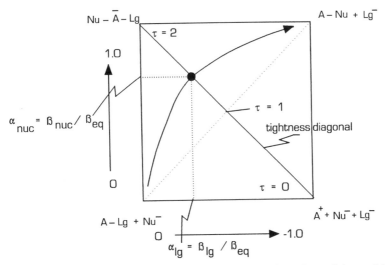

Fig. 3.6 Lewis–Kreevoy plot for the reaction shown in Fig. 3.5; see Scheme 3.15 for equations relating the parameters in this Figure.

$$\mathrm{d}\log k_{\mathrm{ii}}/\mathrm{d}pK_{\mathrm{i}} = \delta = \tau - 1$$

$$\mathrm{d}\log K_{\mathrm{i}}/\mathrm{d}pK_{\mathrm{a}} = \beta_{\mathrm{eq}} \qquad\qquad \mathrm{d}\log k_{\mathrm{ii}}/\mathrm{d}pK_{\mathrm{a}} = \beta_{\mathrm{ii}}$$

$$\beta_{\mathrm{ii}} = \beta_{\mathrm{nuc}} + \beta_{\mathrm{lg}} \qquad\qquad \beta_{\mathrm{eq}} = \beta_{\mathrm{nuc}} - \beta_{\mathrm{lg}}$$

$$\tau = 2\alpha_{\mathrm{nuc}} = 2(1 - \alpha_{\mathrm{lg}})$$

$$\alpha_{\mathrm{nuc}} = \beta_{\mathrm{nuc}}/\beta_{\mathrm{eq}}; \ \ \alpha_{\mathrm{lg}} = \beta_{\mathrm{lg}}/\beta_{\mathrm{eq}}$$

Scheme 3.15 Equations relating α, β and τ for identity reactions.

3.6   Transition-state acidity and transition-state complexation

The ratio of the rate constants for a reaction involving transition-state compositions with and without a proton enables a hypothetical acidity ($K_{\mathrm{a}}^{\ddagger}$, Scheme 3.16) to be determined for the proton in the transition state of an acid-catalysed reaction (Kurz, 1963, 1992). An example of this model involves a reaction which is acid- and water-catalysed, or one occurring by water and hydroxide ion pathways. Comparison of this ratio with acidities calculated or measured for standard models offers another way of investigating the transition state.

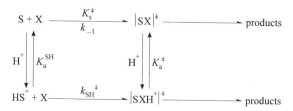

Scheme 3.16 Protonation of a transition state $|\mathrm{SX}|^{\ddagger}$.

Scheme 3.16 illustrates the method for a reaction of substrate S with reagent X which also involves an acid-catalysed process. Equations [3.23]–[3.25] govern the pseudo-thermodynamic cycle and enable the parameter $K_{\mathrm{a}}^{\ddagger}$ to be determined.

$$K_{\mathrm{a}}^{\ddagger} = a_{\mathrm{H}}\, a_{\mathrm{S}}^{\ddagger}/a_{\mathrm{SH}}^{\ddagger} = K_{\mathrm{S}}^{\ddagger}\cdot K_{\mathrm{a}}^{\mathrm{SH}}/K_{\mathrm{SH}}^{\ddagger} \tag{3.23}$$

The ratio of rate constants for the uncatalysed (k_{S}) and acid-catalysed (k_{SH}) reactions gives:

$$k_{\mathrm{S}}/k_{\mathrm{SH}} = K_{\mathrm{S}}^{\ddagger}/K_{\mathrm{SH}}^{\ddagger} \tag{3.24}$$

$$K_{\mathrm{a}}^{\ddagger} = K_{\mathrm{a}}^{\mathrm{SH}}\cdot(k_{\mathrm{S}}/k_{\mathrm{SH}}) \tag{3.25}$$

The technique is exemplified by the comparison of the solvolysis of methyl chloride (S) in water (k_{SH}) and in alkaline solution (k_{S}) (**2** and **3**) (Kurz, 1972) where X = hydroxide ion. The $pK_{\mathrm{a}}^{\ddagger}$ value of 11.9 (obtained from the ratio of k_{S} and k_{SH} and the pK_{a} of water) is compared with the $pK_{\mathrm{a}}^{\ddagger}$ values *expected* for a structure with full bond association ($\mathrm{H_2\overset{+}{O}}\!-\!\mathrm{CH_3}\!-\!\mathrm{Cl}$, -0.9) and for zero bond association ($\mathrm{H_2O.CH_3}\!-\!\mathrm{Cl}$, 14.8) in the transition state. The standard values are estimated from the bulk values for $\mathrm{H_2\overset{+}{O}}\!-\!\mathrm{R}$ and $\mathrm{H_2O}$, assuming the effect of the negative charge on the chlorine in the former case and the dipole of CH_3Cl in the latter (H_2O^+—$CH_3.Cl^-$ and $H_2O.CH_3Cl$). A $pK_a^{\#}$ value of 12.1 is observed for solvolysis of methyl toluenesulphonate by water and hydroxide ion. These data indicate that the bond formation is not far advanced in both cases (**4** and **5**) (Kurz & Lee, 1975).

The method requires that pairs of reactions are available which differ in their transition-state compositions by a single proton. Interpretation of the results is complicated because the structure of the 'protonated' transition state probably differs from that of the unprotonated one.

Kurz's method can be employed generally to determine the hypothetical equilibrium constant between a transition state and a host molecule such as cyclodextrin (Tee, 1994; see Chapter 8) and can provide a comparison between the binding of the host with the transition state and with the reactant.

3.7   Topology of transition states

Introduction

Stereochemical analysis of the product of a reaction can lead to a distinction between particular topologies in the transition state, but this can only give qualitative information about the extent of bond fission or formation in the transition state (Cornforth, 1969; Verbit, 1970; Billups *et al.*, 1986). Stereochemical techniques have classically relied on gross variation in group structure and the differences in reactivity between geometrical isomers and between species with different ring size have been employed extensively in organic chemistry to obtain information on topology (Beak, 1992). Isotopic chirality was introduced (Cornforth *et al.*, 1969; Lutly *et al.*, 1969) for the study of mechanism in enzyme reactions; the advanced analytical procedures now available have enabled its widespread application as a mechanistic tool. The technique is particularly valuable in studies of bio-organic systems because gross structural change (such as methyl to hydrogen or ethyl, etc.) employed to introduce chirality can render a substrate completely inactive towards an enzyme or make it bind in an orientation different from that in the natural case. Isotopic change does not suffer from this problem but the price to be paid for this increased flexibility is the requirement for expensive, advanced, analytical techniques.

Stereochemical analysis

Recent advances in the study of phosphorus stereochemistry in reactions of $[^{18}O,^{17}O,^{16}O]$phosphate esters involve ^{31}P-NMR as the analytical method of choice (Hall & Inch, 1980, Knowles, 1980; Frey, 1982; Eckstein, 1983; Lowe, 1983). The method depends on two properties of oxygen isotopes bonded directly to phosphorus: the oxygen isotope ^{17}O broadens the ^{31}P resonance to such an extent that it is not therefore observed in the NMR spectrum; and the oxygen isotope ^{18}O causes a significant shift in the ^{31}P resonance to a field higher than that where the phosphorus is bonded to ^{16}O. The magnitude of the shift increases with P—O bond order and the position of the resonance therefore identifies whether the isotopic $[^{18}O]$ oxygen is in a P—O or in a P=O bond. In a typical procedure (Knowles 1980; Lowe, 1983) the stereochemistry of phosphorus in phenyl phosphate may be determined by transferring the phosphate to (S)-propan-1,2-diol with alkaline phosphatase. The transfer occurs with retention of configuration (Eqn [3.26]). Cyclization of the phosphorylated diol (to **6**—with inversion at the phosphorus) followed by methylation of the resulting phosphate ester gives *syn* (**7**) and *anti* (**8**) forms of the methyl ester; the resultant array of isotopically substituted phosphate ester molecules are displayed in Scheme 3.17.

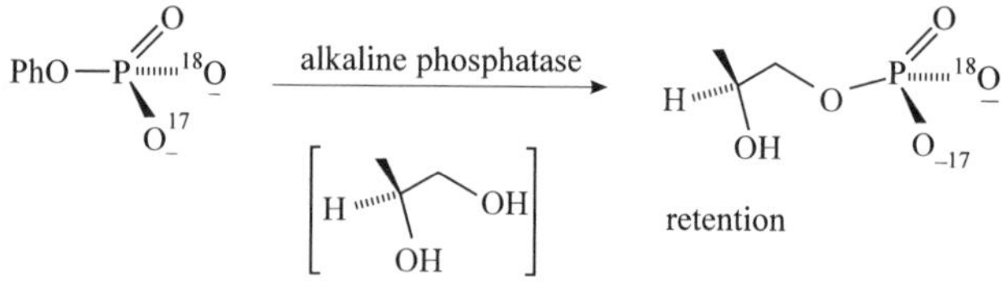

[3.26]

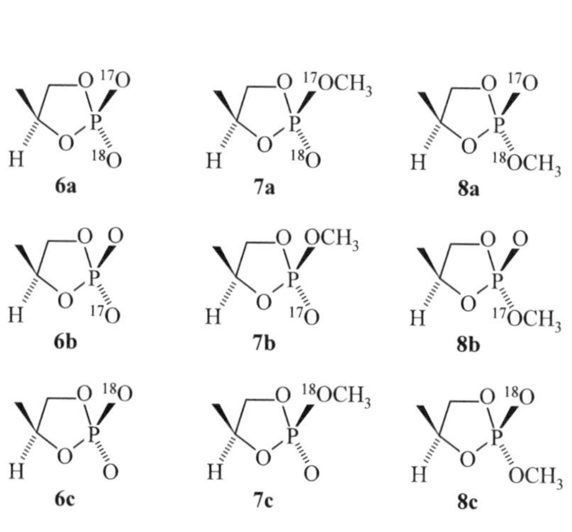

Scheme 3.17 Stereochemical analysis of the products of a phosphorylation reaction.

The species with ^{17}O does not appear in the ^{31}P-NMR spectrum but since ^{17}O is never obtained at 100% enrichment the species **7a,b** and **8a,b** have ^{31}P-NMR spectra which may be employed as internal standards to determine the chemical shifts from **7c** and **8c**. The stereochemical purity is obtained from a comparison of the peak areas, and the accuracy with which these can be estimated (associated with the repeatability of the spectra) give the confidence limits to the stereochemistry of the phosphate being analysed.

Stereochemistry at carbon resulting from asymmetric isotopic labelling with deuterium may be analysed by optical rotation but NMR methods are now available which involve direct analysis of the intact methyl group. Diastereotopic protons in $-CH_2D$ in a suitable chiral molecule have an observable chemical shift difference (Anet & Kopelwich, 1989).

Analysis of chirality at sulphur, for example in an aryl sulphate molecule, may be carried out by Fourier-transform IR (FTIR) spectroscopy (Lowe, 1991a,b). The sulphuryl group ($-SO_3^-$) is transferred to an optically active alcohol (Scheme 3.18) by heating a solution in carbon tetrachloride.

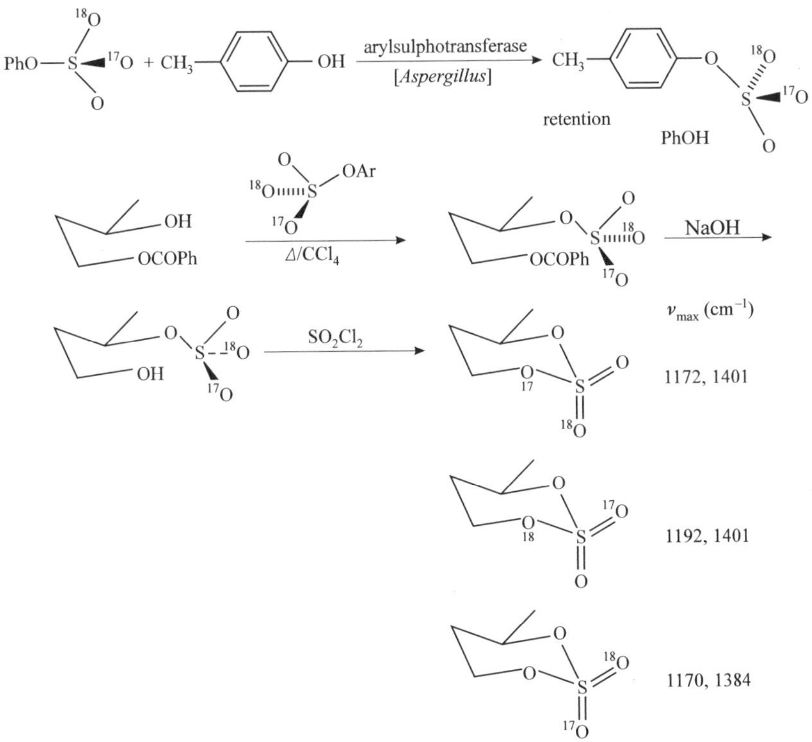

Scheme 3.18 Stereochemical analysis of a sulphuryl group ($-SO_3^-$) transfer reaction; for stereochemical clarity some charges and valencies are omitted.

The monosulphate is then deprotected and cyclized to yield a cyclic sulphate and the FTIR spectrum determined. The shift in the S=O vibration (symmetric and antisymmetric stretching frequencies of the $>SO_2$ bond at 1400 and 1200 cm^{-1}) caused by isotopic substitution of oxygen in axial and equatorial positions forms the basis for the stereochemical analysis of the chirality of the sulphur.

Mechanisms from crystallographic structure data

X-ray crystallography, augmented by the results of a small number of other techniques such as neutron diffraction, provides a database with details of angles and bond distances for a very large number of crystalline structures. The thermal ellipsoid from X-ray studies is a measure of the uncertainty with which the position of an atom is known and reveals motion in a shallow local potential well which could form the beginnings of motion along a reaction coordinate. The crystal environment of a molecule is manifestly not that in the free state and parameters such as bond angle and bond length can be correlated for a large number of similar structures, each possessing its own different environment (Burgi & Dunitz, 1983; Kirby, 1994). This approach has been applied with success to the reaction of an amino nitrogen atom with a carbonyl function (R—CO—R). Crystal structures of molecules possessing these groups have the nitrogen atom on average on a line between 100° and 110° to the bond between C and O. The angle between the R–C–R plane and the C–O bond decreases from 180° as the distance between N and C decreases; this is accompanied by a slight lengthening of the C–O bond (Scheme 3.19). The approach is likely to become attractive as access to crystallographic databases becomes easier and is accepted as part of the mechanistic chemist's armoury of tools (Burgi, 1975; Jones & Kirby, 1984).

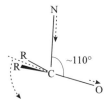

Scheme 3.19 Crystallographic data for molecules possessing R—CO—R and amino nitrogen groups.

Further reading

Bell, R. P. (1973) *The Proton in Chemistry*, 2nd edn, Chapman and Hall, London.

Billups, W. E., Houk, K. E. & Stevens, R. V. (1986) Stereochemistry and reaction mechanism, in *Investigation of Rates and Mechanisms of Reactions*, 4th edn, Part I, Bernasconi, C. F. (ed.), Wiley–Interscience, New York, p. 663.

Burgi, H. B. & Dunitz, J. D. (1983) From crystallographic studies to chemical dynamics, *Acc. Chem. Res.*, **16**, 154.

Caldin, E. F. & Gold, V. (eds) (1975) *Proton Transfer Reactions*, Chapman and Hall, London.

Chapman, N. B. & Shorter, J. (eds) (1978) *Correlation Analysis in Chemistry*, Plenum Press, New York.

Cornforth, J. W. (1969) Exploration of enzyme mechanisms by asymmetric labelling, *Q. Rev. Chem. Soc.*, **23**, 125.

Frey, P. A. (1982) Stereochemistry of enzymatic reactions of phosphates, *Tetrahedron*, **38**, 1541.

Grob, C. A. (1976) Polar effects in organic reactions, *Angew. Chem.*, **15**, 569.

Hall, C. R. & Inch, T. D. (1980) Phosphorus stereochemistry, *Tetrahedron*, **36**, 2059.

Hammett, L. P. (1935) Some relations between reaction rates and equilibrium constants, *Chem. Rev.*, **17**, 125.

Hammett, L. P. (1940) *Physical Organic Chemistry*, 1st edn, McGraw-Hill, New York.

Hansch, C., Leo, A. & Taft, R. W. (1991) A survey of Hammett substituent constants and resonance and field parameters, *Chem. Rev.*, **91**, 165.

Hine, J. (1975) *Structural Effects on Equilibria in Organic Chemistry*, John Wiley, New York.

Jencks, W. P. (1972) General acid–base catalysis of complex reactions in water, *Chem. Rev.*, **72**, 705.

Jencks, W. P. (1988) Are structure–reactivity correlations useful? *Bull. Soc. Chim. Fr.*, 218.

Johnson, C. D. (1973) *The Hammett Equation*, Cambridge University Press, Cambridge.

Kirby, A. J. (1994) Crystallographic approaches to transition state structures, *Adv. Phys. Org. Chem.*, **29**, 87.

Knowles, J. R. (1980) Enzyme-catalysed phosporyl transfer reactions, *Annu. Rev. Biochem.*, **49**, 877.

Knowles, J. R. (1990) A feast for chemists, in 'Mechanistic Enzymology', *Chem. Rev.*, **90**, 1077.

Kresge, A. J. (1973) The Brønsted relation—recent developments, *Chem. Soc. Rev.*, **2**, 475.

Lewis, E. S. (1966) Linear free energy relationships, in *Investigations of Rates and Mechanisms of Reactions*, 4th edn, Part I, Bernasconi, C. F. (ed.), Wiley–Interscience, New York, p. 871.

Lowe, G. (1983) Chiral [^{16}O,^{17}O,^{18}O]phosphate esters, *Acc. Chem. Res.*, **16**, 244.

Lowe, G. (1991a) Mechanisms of sulphate activation and transfer, *Phil. Trans. R. Soc. London, Ser. B*, **332**, 141.

Lowe, G. (1991b) The stereochemical course of sulphuryl transfer reactions, *Phosphorus, Sulphur and Silicon*, **59**, 63.

Lutly, J., Retey, J. & Arigoni, D. (1969) Preparation and detection of chiral methyl groups, *Nature (London)*, **221**, 1213.

Shorter, J. J. (1990) Hammett Memorial Lecture, *Prog. Phys. Org. Chem.*, **17**, 1.

Streitwieser, A. (1960) Stereochemical and kinetic applications of deuterium isotope effects, *Ann. N.Y. Acad. Sci.*, **84**, 576.

Verbit, L. (1970) Optically active deuterium compounds, *Prog. Phys. Org. Chem.*, **7**, 51.

Wells, P. R. (1968) *Linear Free Energy Relationships*, Academic Press, New York.

4 Kinetic and equilibrium isotope effects

4.1 Origin of isotope effects

Rotational, translational and electronic contributions to the overall energy of a molecular assembly are essentially independent of isotopic substitution and the overall change in mass thus has a negligible effect on the overall energy of a molecule. Vibrational contributions to energy, on the other hand, can be strongly dependent on isotopic substitution, and isotope effects result mainly from differences in vibrational frequencies when an atom is substituted for its isotope in a molecule. Isotope effects are measured on properties of assemblies of molecules but come close to providing a method for studying energy changes in individual bonds in a reaction.

Changing the isotopic identity of an atom in a bond causes a measurable effect on the frequency of vibration of that bond which is small except for the case of hydrogen isotopes. The C—H stretching mode has a vibrational wavenumber which changes from about $3000\,\mathrm{cm}^{-1}$ to $2200\,\mathrm{cm}^{-1}$ on substitution of D for H, whereas, for example, substitution of ^{18}O for ^{16}O causes a shift of only about $36\,\mathrm{cm}^{-1}$ for the S—O bond frequencies (Lowe, 1991a,b). Isotopic substitution does *not* change the electronic potential energy surface; its effect results entirely from changes in nuclear masses. The effect on vibrational frequencies arises because these are inversely proportional to the reduced mass. For a simple harmonic oscillator the effect of the reduced mass is given by Eqn [4.1], where ω is the frequency of vibration in wavenumbers, k is the vibrational force constant and μ is the reduced mass and c is the speed of light. The force constant is an indicator of the 'strength' of a bond and is the proportionality constant between energy and distance. For a parabolic relationship it represents the curvature. The vibrational force constant is unchanged by isotopic substitution because the nuclear electronic interaction remains essentially the same.

$$\omega = (k/\mu)^{0.5}/(2\pi c) \tag{4.1}$$

The biggest changes in reduced mass occur upon isotopic substitution for hydrogen. Because of the small atomic mass of hydrogen, bonds to hydrogen atoms usually vibrate independently of the rest of the molecule and isotopic sensitivity is usually reflected in one stretch and two bending vibrations. It is rare for more than three vibrations to be isotopically sensitive.

The isotope effect can often be associated with a particular bond vibration; an *isotopically sensitive bond* is defined as a bond where substitution by isotopes causes a measurable isotope effect.

Changing a C—H bond to C—D increases the reduced mass from 0.92 to 1.71, and consequently the vibrational frequency is decreased by a factor of 1.36. The vibrational states are quantized and the lowest level, the zero-point energy (ZPE), per mole, is related to the vibrational frequency by Eqn [4.2], where h is Planck's constant, N_0 is Avogadro's number, c is the velocity of light and ω is the frequency. Consequently the difference in zero-point energy for a C—D vibration compared with that for C—H is $5.98(\omega_D - \omega_H)$ J/mol^{-1}. For a C—H stretching frequency of 3000 cm^{-1} substitution by D reduces the stretching frequency to 2206 cm^{-1} and the C—D zero-point energy is therefore 4.75 kJ mol^{-1} lower than that for C—H.

$$ZPE = 0.5hc\omega N_0 \qquad [4.2]$$

It is conventional to refer to the isotope effect as a single parameter with the ratio of the isotopes as superscripts. Thus the kinetic isotope effect for a reaction involving fission of a carbon–hydrogen/deuterium bond (k^H/k^D) would be $k^{H/D}$.

Isotopic substitution can have an effect on both rates and equilibrium constants; the effect for the latter arises from differences in zero-point energies giving rise to only small energy differences between the isotopically substituted species. These amount to no more than about a twofold change in the majority of cases. The isotope effect on *kinetic* constants, however, can be significant. Primary kinetic isotope effects arise when the bond to the isotopically substituted atom is broken in the transition state. Smaller, secondary, kinetic isotope effects can still be seen for reactions in which this bond is not broken in the transition state. Some examples are given in Tables 4.1 and 4.2 which show that the ratios can vary from surprisingly high values down to ones close to unity.

The origin from the vibrational mode of primary kinetic isotope effects and equilibrium isotope effects can be illustrated graphically by consideration of the potential energy curves (Fig. 4.1) for the reaction represented by Eqn [4.3], where the A—H bond is broken and the H—B bond is formed. As the force constant for the vibration is independent of the isotope, it does not

$$A—H + B \longrightarrow A + H—B \qquad [4.3]$$

matter whether the hydrogen transfer occurs by proton, hydride or atom transfer. The A—H vibrations in the reactant are replaced by the B—H vibrations in the product. Isotopic replacement of H by D gives a reaction with the same potential energy diagram but the zero-point energies of reactant and product (A—D and B—D) are lower (see Fig. 4.1). The zero-point energy difference between the protio and deuterio reactant is generally similar to that in the product because there is usually not a large change in

Table 4.1 Values of maximal primary isotope effects for some standard bond fission processes.[a]

Bond a	Bond b	$k^{a/b}$	Bond a	Bond b	$k^{a/b}$
C—H	C—D	6.4	C—H	C—T	13
C—^{12}C	C—^{13}C	1.0488	C—^{12}C	C—^{14}C	1.0934
C—^{14}N	C—^{15}N	1.0443	C—^{16}O	C—^{18}O	1.0668
C—^{32}S	C—^{34}S	1.0139	C—^{35}Cl	C—^{37}Cl	1.0133
N—H	N—D	8.5	O—H	O—D	10.6
O—^{12}C	O—^{14}C	1.1184	N—^{12}C	N—^{14}C	1.1161
S—^{12}C	S—^{14}C	1.0964			

[a] Data are from Huskey (1991) and Isaacs (1995).

Table 4.2 Some kinetic hydrogen isotope effects for reactions involving C—H bond fission.[a]

Reaction	Catalyst	Isotope	Isotope effect
Bromination of acetone	acetate ion	H/D	7.0
	oxonium ion	H/D	7.7
Bromination of methylacetylacetone	acetate ion	H/D	5.8
Bromination of 2-carbethoxycyclopentanone	chloroacetate ion	H/D	3.7
Bromination of nitromethane	acetate ion	H/D	6.5
Oxidation of isopropanol	CrO_3	H/D	6.7
Elimination of Ph—CD_2—CH_2—Br	EtO$^-$	H/D	7.1
Nitration of benzene	nitronium ion	H/T	1.0

[a] Data are from Bell (1959) and Hine (1962).

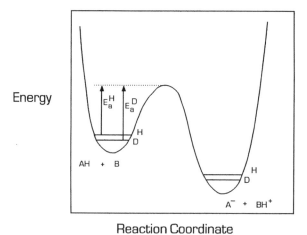

Fig. 4.1 Energy diagram to illustrate the origin of primary isotope effects from zero-point energy (ZPE).

vibrational frequencies on going from one molecule to another. Consequently, the *equilibrium* isotope effects are small: $K^{H/D}$ is less than 2.

The primary kinetic isotope effect will result from a difference in zero-point energies of the reactant state because the vibration is 'lost' in the transition state (see Fig. 4.1). If the isotope is transferred in the transition state, the vibrational motion constituting the main reaction coordinate is converted into translational motion and the force-restoring constant becomes effectively zero. Consequently there is no corresponding zero-point energy in the transition state for that motion. The difference in activation energies for the two isotopes is simply the difference in zero-point energy in the reactant state. If the reaction coordinate for a reaction involving hydrogen transfer from carbon is essentially a stretching motion, the difference in activation energies/zero-point energy is the $4.75 \, \text{kJ} \, \text{mol}^{-1}$ described earlier, giving rise to a kinetic isotope effect, $k^{H/D}$, of 6.8 at 25 °C; this is in the region of values commonly observed (cf. Table 4.1). If the reaction coordinate is essentially a lower-frequency motion, for example bending, the $C-H$ and $C-D$ frequencies could be 1400 and $1029 \, \text{cm}^{-1}$ respectively; the difference in zero-point energy of $2.22 \, \text{kJ} \, \text{mol}^{-1}$ gives a reduced $k^{H/D}$ of 2.45 at 25 °C. This can occur, for example, in intramolecular hydrogen transfer reactions.

If the bond to the isotope is not broken in the transition state but changes its vibrational frequency because of electronic changes elsewhere in the molecule, then a secondary kinetic isotope effect can arise from the difference in the two zero-point energies in the reactant and transition states (Fig. 4.2). For example, the conversion of a tetrahedral sp^3 $C-H(D)$ to a trigonal

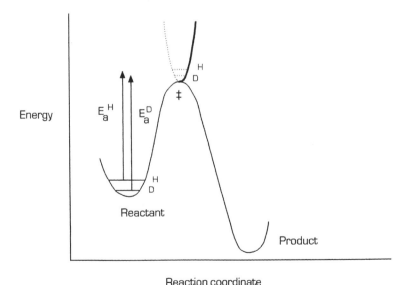

Fig. 4.2 Evolution of kinetic secondary isotope effects from vibrational degrees of freedom not lost in the transition state.

sp^2 C—H(D) is accompanied by a significant change in the in-plane bending frequency from about $1300\,cm^{-1}$ to $800\,cm^{-1}$ for the H-isotope and from $956\,cm^{-1}$ to $588\,cm^{-1}$ for the D-isotope. The conversion is thus accompanied by a change in the *difference* of the H/D zero-point energy of $2.99\,kJ\,mol^{-1}$ in the sp^3 reactant state to $2.20\,kJ\,mol^{-1}$ in the sp^2 state, which could be a transition state, an intermediate or a product. This decrease in zero-point energy differences (Fig. 4.2) would give a secondary isotope effect ($k^{H/D}$) of 1.38 at 25 °C.

The application of *fractionation factor* terminology considerably simplifies the treatment of isotope effect results. The fractionation factor (ϕ) is defined as in Eqn [4.4] in terms of the ratio of the partition function (Q) for an isotopically substituted molecule (for example AH/AD) to that of a standard molecule (SH/SD).

$$\phi_A = (Q_{AD}/Q_{AH})/(Q_{SD}/Q_{SH}) \tag{4.4}$$

The isotope effect for an equilibrium or rate constant is simply the ratio of fractionation factors for reactants and products (ϕ_{react})/($\phi_{product}$) or transition states and reactants (ϕ_{react})/(ϕ_{trans}), respectively. The deuterium fractionation factor of a hydrogen in a molecule is the tendency for deuterium to enrich in this position relative to its content in water (see under Section 4.5). An exchangeable hydrogen with a fractionation factor of 0.5 has half the deuterium content of solvent water as long as only traces of deuterium are present. Carbon-bonded hydrogens that are not readily exchangeable have fractionation factors ranging from 0.64 for acetylene to 1.23 for the hydrogen at the C1 position of glucose. The sulphydryl group is the only amino acid side chain with a fractionation factor less than unity (0.5). Carboxyl, imidazole, amino and hydroxyl groups all have fractionation factors close to unity.

4.2 Measurement of kinetic isotope effects

The earliest contributions of isotope effects to the elucidation of mechanism involved hydrogen isotope substitution because the rate constant changes were relatively large and could be measured by standard kinetic methods. Isotope effects due to substitution of other atoms (known as 'heavy'-atom isotope effects) are very small and require experimental techniques for rate constant ratios obtained through indirect but very much more accurate methods than are allowed by routine rate constant measurements. Techniques are available which enable rate constant ratios to be measured to give results with experimental errors of less than ±1%. Explicitly measured rate constants determined spectrophotometrically with computerized collection of data have been reputed to yield rate constants with a standard error of ±0.01%, with a reproducibility of ±0.3%. Since the normal errors on kinetic data determined by this method are ±5%, studies of such accuracy cannot be undertaken routinely and very great care must be given to the details of the measurements.

Competition methods are best employed for rate constant ratios of less than 5–10%; they provide a useful tool which requires less attention than the explicit approach to give results of good accuracy. When two iso-topomers have different reactivities, the mixture of the two will change its isotopic composition as the reaction progresses until at 100% reaction the composition returns to its original value. Radioactive isotopes give a specific radioactivity proportional to the ratio of radioactive to non-radioactive isotopes, which may be measured with a precision better than ±1%. The isotope effect is calculated from the change in isotopic composition and the extent of the overall reaction.

The ratio of isotopes may also be measured with a mass spectrometer and results good to ±0.1–0.2% are easily obtained for particular species; simul-taneous collection of ions from the two peaks of interest with a dual-collection instrument (isotope ratio mass spectrometer) can give precision to ±0.01%. It is advisable to employ relatively simple molecules usually obtained by unambiguous degradation from the compound in question; the peaks of these molecules should be isolated from those from possible contaminants. Thus CO_2, N_2, SO_2 or SF_6 are suitable for carbon, nitrogen and sulphur but appropriate precautions should be made to ensure that no isotope fractionation occurs during degradation; chlorine is often deter-mined as CH_3Cl. A major problem with the use of heavy-atom isotope effects is that the instruments employed to measure the isotope ratios are not widely available for everyday use and they require skilled operators and careful maintenance to achieve good results. Precision around ±0.1% can be obtained with a good routine mass spectrometer by successively scanning the two isotopic peaks in question; this is particularly good if the spectro-meter is computer-controlled (Saunders, 1986). The last method is often that of choice for deuterated compounds, using the parent peak and provided no fragmentation occurs.

An isotope effect will cause the enrichment of the product to differ from that of the reactant in the initial stages of the reaction (Eqns [4.5] and [4.6]).

$$\text{Rate of depletion of AH} = d[AH]/dt = k_H[AH][B] \tag{4.5}$$

$$\text{Rate of depletion of AD} = d[AD]/dt = k_D[AD][B] \tag{4.6}$$

Thus in unit time AH and AD will be depleted to different extents and the isotopic compositions will therefore alter as the reaction proceeds. The two equations may be combined to give Eqn [4.7], which on integration gives Eqn [4.8]. This equation can be set for time = 0 and time = t to give Eqn [4.9] and hence Eqn [4.10]:

$$k_D(d[AH]/dt)/[AH] = k_H(d[AD]/dt)/[AD] \tag{4.7}$$

$$k_D \log[AH] - k_H \log[AD] = C + t \tag{4.8}$$

$$k_D\{\log[AH]_t - \log[AH]_0\} = k_H\{\log[AD]_t - \log[AD]_0\} \tag{4.9}$$

$$k^{H/D} = \log\{[AH]_t/[AH]_0\}/\log\{[AD]_t/[AD]_0\} \tag{4.10}$$

Samples are usually isolated at 0% and 50% conversion and the isotope effect calculated from the compositions of the unused reactant. The scheme can be repeated for isotopic composition of the product when samples are isolated at other extents of conversion (such as 10% and 100%) and reproducibility can be typically ±0.05%.

4.3   Primary isotope effects

Introduction

Primary isotope effects ensue when the isotopically sensitive bond undergoes fission in the transition state of the rate-limiting step. The presence of a large deuterium isotope effect is unequivocal evidence that a bond to hydrogen is undergoing fission in the rate-limiting step of a reaction. For example, the large effect on the rate constant of replacing hydrogen by deuterium for the bromination of acetone (Table 4.2) confirms the hypothesis that enolization is the rate-limiting step, followed by fast addition of bromine.

Variation in primary isotope effects

Perusal of Table (4.3) indicates that primary hydrogen isotope effects vary from the theoretical maximum expected by the removal of a stretching motion in the transition state. The variation is sometimes due to the fission of the isotopically sensitive bond being only partially rate-limiting (see O'Leary, 1988). Originally, a simple ratio of the isotope effect to the theoretically maximal value was thought to measure the strength of the bond in the transition state relative to the full bond. Westheimer (1961) indicated that all the zero-point energy of a bond is lost in the transition state even

Table 4.3 Variation of the primary deuterium isotope effect as a function of structure in the transfer of protons to water from carbon acids.[a]

Carbon acid[b]	pK_a[c]	$k^{H/D}$
$C\underline{H}_2(COOEt)_2$	13.3	1.7
$CH_3COC\underline{H}(CH_3)COOEt$	12.7	3.8
$C\underline{H}_3NO_2$	10.2	2.9
$CH_3COC\underline{H}_2COOEt$	10.0	3.0
$(CH_3CO)_2C\underline{H}_2$	9.0	3.9
$C\underline{H}(COOEt)_3$	7.8	3.8
$(CH_3CO)_2C\underline{H}Br$	7.0	3.9
$O_2NC\underline{H}_2COOEt$	5.8	3.1
$(C\underline{H}_3)_2C{=}OH^+$	0.3	5.0

[a] Data mainly from More O'Ferrall (1975).
[b] The acceptor base is water and the donor hydrogen is underlined.
[c] pK_a of carbon acid.

though it still has a 'bond order', because according to transition-state theory a single motion along the reaction coordinate is converted to translational motion and the vibration is lost. In a linear hydrogen bond there are four vibrational modes (Scheme 4.1). Because of the relative mass of hydrogen compared with most other atoms, it is the hydrogen which effectively undergoes rapid movement relative to its bonding partners. Hydrogen transfer (between A and B), through a linear process, occurs by the relative movement of the heavy atoms corresponding to the asymmetric vibration which must cause reaction to occur. Simple transition-state theory treats the other vibrational modes as effectively existing in the transition state. If the transition-state structure is 'symmetrical' with respect to the hydrogen and its two partners, there is the expected loss of zero-point energy corresponding to the loss of stretching motion in the transition state, and a normal kinetic isotope effect will be exhibited. The changes in zero-point energy of the other vibrational modes are minimal. The symmetrical stretching vibration possesses zero-point energy but is isotopically insensitive because hydrogen is at the centre of mass and the force constants acting on it exactly balance. As a consequence there is no movement of hydrogen in this vibration and, for example, there is no isotopic shift in the symmetrical stretching frequency of a hydrogen-bonded species like HF_2^- compared with DF_2^-. Conversely, if the transition-state structure is unsymmetrical and the hydrogen lies closer to either A or B, there is still the conversion of stretching vibration into a translational motion and loss of zero-point energy in the transition state. However, there are increases in vibrational frequencies of the other motions with an accompanying *increase* in zero-point energy on going to the transition state. In an unsymmetrical transition state the force constants to hydrogen are unequal and the symmetrical stretching vibration becomes more isotopically sensitive. A vibrational zero-point energy change in the transition state can then offset that in the reactant. Consequently the total change in zero-point energies is less and a reduced kinetic isotope effect will be observed. The theoretical maximum isotope effect is by convention referred to as the *intrinsic isotope effect* and is of great importance; its application in the diagnosis of stepwise mechanisms is discussed in Section 7.1 (p. 139).

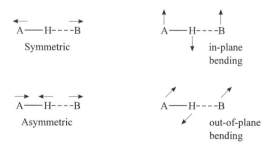

Scheme 4.1 Vibrational modes of a hydrogen bond.

More O'Ferrall (1975) collected data for the value $k^{H/D}$ for the ionization of ketones and nitroalkanes and showed that there is a maximal effect at a pK_a of the base equal to that of the acid. The symmetrical transition state thus yields a large isotope effect when the energy of the reactant is the same as that of the product. Work from the laboratories of Jencks (Cox & Jencks, 1978) (Fig. 4.3) and Kresge (Bergman *et al.*, 1978) have confirmed the validity of these notions for proton transfer between heteroatoms.

Another source of isotope variation is the quantum-mechanical 'tunnelling' effect which is due to a reaction with a 'thin' energy barrier. The effect can give rise to isotope effects larger than those normally expected (Tables 4.1 and 4.2) (Bell 1980; Klinman, 1991) and is only important for hydrogen.

4.4   Solvent isotope effects

Physical properties which are related to intermolecular interactions and vibrational frequencies often vary between H_2O and D_2O. For example, melting and boiling points and viscosities differ but dipole moments and dielectric constants are essentially the same. Reactions which are studied in H_2O and D_2O may lead to kinetic and equilibrium isotope effects. The cause of these effects is sometimes complex and can result from differences in physical parameters of the solvent system as well as direct 'chemical' effects

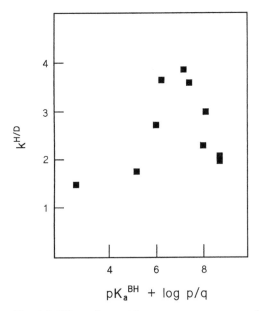

Fig. 4.3 Effect of transition-state symmetry on the primary isotope effect. Protonation of the zwitterionic intermediate by general acids (BH) in methoxyaminolysis of phenyl acetate; data from Cox & Jencks (1978).

due to changes in bonding. (Schowen, 1972; Schowen, 1978; Kresge *et al.*, 1987; Quinn & Sutton, 1991.)

Acids are 'weaker' in D_2O than in H_2O with a difference in pK_a of about 0.4–0.5, although this varies with the absolute value of the pK_a. Acid-catalysed reactions which proceed through the intermediate formation of the conjugate acid thus give *inverse* kinetic solvent isotope effects. The rate constant is greater in D_2O than in H_2O simply because the concentration of the substrate's conjugate acid is higher in D_2O.

Additionally, normal primary isotope effects may occur in water and the observed effect may be a combination of primary and secondary contributions. The use of fractionation factors ϕ enables the separation of kinetic and equilibrium effects into the individual contributions in reactants, transition states and products. When the solute AL is dissolved in solvent SL (usually water) (where L is either D or H) the value of ϕ quantifies the preference of the hydrogen site in AL for deuterium over protium relative to that for the hydrogen site in SL. Values of ϕ may be determined for various protonic species relative to water as the standard (SL) and typical values come between 0.4 and 1.3. Table 4.a1 in Appendix A.2 collates values of ϕ for a selection of solutes.

The *equilibrium isotope effect* for a reaction where the hydrogen isotope L migrates to convert AL to BL (Eqn [4.11]) is given by Eqn [4.12] which holds for solvents and solutes with single proton exchangeable sites. The equation applies generally to any isotope effect even though the isotope may not be readily exchangeable (Albery & Knowles, 1976a). For species with multiple sites it is assumed that the occupation of the other sites by D or H will not affect the deuterium preference for a particular site.

$$A-L \rightleftharpoons B-L \qquad\qquad [4.11]$$

$$K^{H/D} = \phi_A/\phi_B \qquad\qquad [4.12]$$

The relationship expressed by Eqn [4.13] is known as the Gross–Butler equation (Schowen, 1978) for equilibria and reactions such as Eqn [4.11] in mixtures of light and heavy solvents, where x is the atom fraction of deuterium. The fractionation factor may be conveniently estimated from Eqn (4.13) and the isotope effect $K^{H/x}$ for dissociation of a monobasic acid (SL) in water (L_2O). The isotope effect at $x = 1$ enables ϕ_{SL} to be estimated and ϕ_{L_3O} may be taken from Table 4.a1.

$$K^{H/x} = (1 - x + x.\phi_{SL})/(1 - x + x.\phi_{L_3O}) \qquad\qquad [4.13]$$

The application of isotope effects in solvents with varying isotopic composition involves fitting the observed relationship between $k^{H/x}$ and x to a theoretical equation. This procedure is known as the *proton inventory* technique as it effectively counts the number of protons which undergo a change from ground to transition state. The Gross–Butler equation has a term $(1 - x + x.\phi)$ for every exchangeable proton in the system and these cancel for those protons which do not change their state.

The simplest relationship between a kinetic isotope effect ($k^{H/x}$) and x is a linear equation [4.14]; the value of ϕ for a single proton which changes its state is the isotope effect, $k^{H/x}$, at $x = 1$. An example of a plot obeying Eqn [4.14] is that for the hydrolysis of dichloroacetylsalicylate ion in water (Minor & Schowen, 1973). It is compatible with a transition-state structure (1) which possesses a single proton undergoing major change from reactant to transition state (such a proton is conventionally described as 'in flight').

$$k^{H/x} = 1 - x + x.\phi \qquad\qquad [4.14]$$

An initial criticism is that the experimental results might not distinguish between the different equations represented by varying numbers of active protons. Maximum deviations between the results for different equations will occur at $x = 0.5$ and values of $(k^{H/x})_{0.5}$ may be calculated for the various formulations of the Gross–Butler equation (Albery & Davies, 1972; Albery, 1975). Values of 0.73, 0.704 and 0.695 are obtained for one, two or three protons respectively. The differences in these values are well outside the experimental error (Schowen, 1978; Quinn & Sutton, 1991) and the observed linearity of the plot (Fig. 4.4) is taken to indicate that the hydrolysis involves a single proton transfer.

The inventory technique is a potentially useful tool for elucidating enzymic transition states (however, see Chiang et al., 1995); it essentially gives information on the number of protons undergoing bond fission or formation in the transition state and therefore enables distinctions to be made between proposed mechanisms. Proton inventory of the deacetylation of acetyl-α-chymotrypsin is consistent with a single proton transfer (Pollock et al., 1973) (2); this result is not consistent with the charge-relay mechanism (3) which involves transfer of two protons and requires a quadratic equation for the dependence of $k^{H/x}$ on x. However, curvature can sometimes simply be a medium effect (Chang et al., 1996).

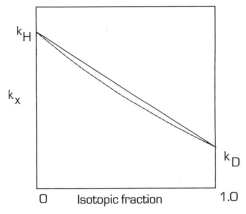

Fig. 4.4 The hydrolysis of dichloroacetylsalicylate anion, illustrating the accuracy required (and attained) for proton inventory studies.

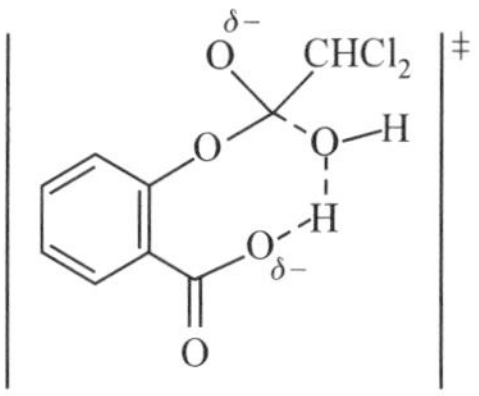

1

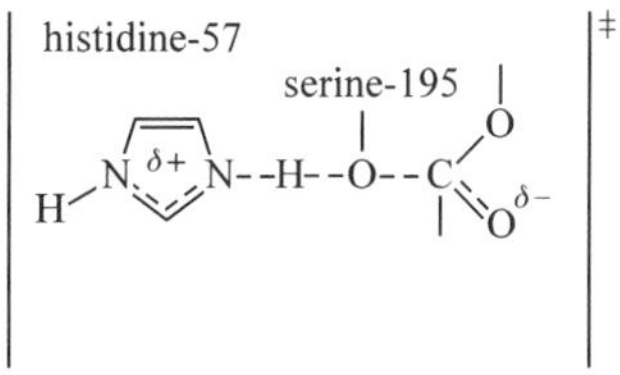

2

only one proton 'in flight' is detected

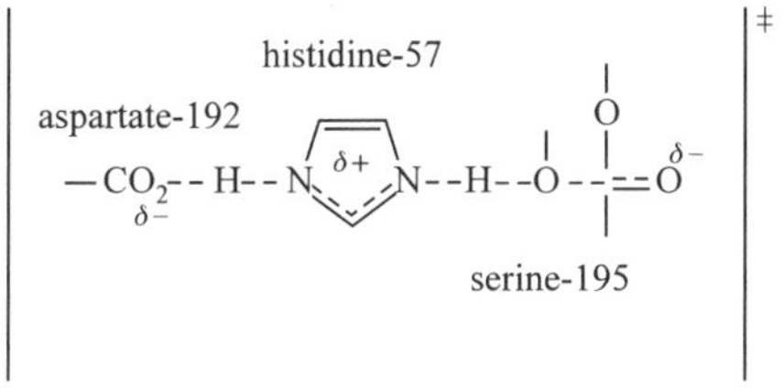

3

charge-relay mechanism has two protons 'in flight'

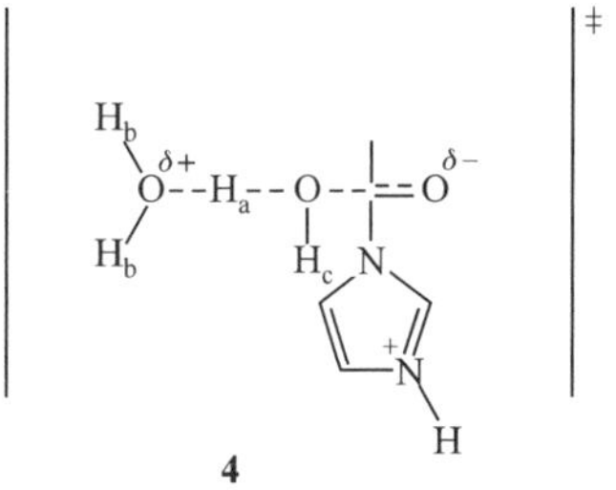

4

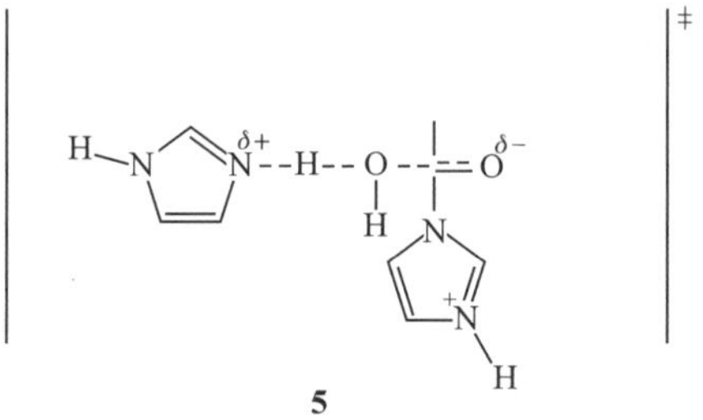

5

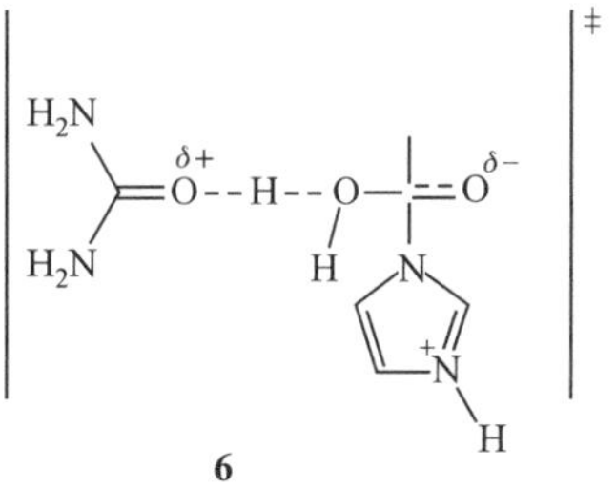

6

Trypsin and elastase also have substrate reactions with proton inventories consistent with a single proton transfer (Elrod *et al.*, 1975) but asparaginase and glutaminase, which catalyse the hydrolysis of the terminal amide bond of the respective amino acids, have proton inventories of two proton transfers (Elrod *et al.*, 1975; Quinn *et al.*, 1980). Proton inventory of the uncatalysed degradation of acetylimidazolium ion in water indicates three protons undergoing major changes from reactant to transition state (**4**) whereas the imidazole-catalysed reaction has only one proton 'in flight' (**5**). When the uncatalysed reaction is carried out in urea solutions, linear plots are obtained that are consistent with only one proton 'in flight' (Hogg *et al.*, 1977; Patterson *et al.*, 1978) (**6**). There are a very large number of exchangeable hydrogen atoms in enzymes and their aqueous solvation shells and so the interpretation of solvent kinetic isotope effects is often not unambiguous.

The solvent deuterium isotope effect for pure water and pure D_2O ($x = 1$) may be employed to solve the kinetic ambiguity whereby a nucleophile can act as either a general base or a nucleophilic catalyst. Table 4.4 illustrates the

Table 4.4 Deuterium oxide solvent isotope effects for nucleophilic attack (n) and for general base-catalysed **hydrolysis** (gb) reactions.[a]

Catalyst and reactant/substrate	Mechanism	k_{H_2O}/k_{D_2O}
Methoxylamine + *S*-ethyl trifluorothioacetate	gb	4.4
Imidazole + ethyl dichloroacetate	gb	3.2
Acetic anhydride + acetate ion	gb	1.7
Acetylimidazole + imidazole	gb	3.6
Acetoxime acetate + imidazole	gb	2.0
Dichloroethyl acetate + imidazole	gb	1.9
4-Nitrophenyl acetate + imidazole	n	1.0
4-Nitrophenyl acetate + 4-picoline	n	1.12
Phenyl acetate + methylamine	n	1.15
2,4-Dinitrophenyl acetate + acetate ion	n	1.8
Phenyl acetate + hydroxylamine	n	1.5
Phenyl acetate + imidazole	n	1.8
β-Propiolactone + imidazole	n	1.2
Phenyl acetate + piperidine	n	1.19

[a] Data mainly from Johnson (1967).

solvent isotope effects for a number of such kinetically ambiguous reactions. The expectation is that reactions involving general base catalysis and rate-limiting proton transfer should exhibit a significant primary kinetic isotope effect larger than the solvent isotope effect observed for nucleophilic catalysis which does not involve a kinetically significant proton transfer step.

The method is empirical and the range 1.5–1.9 for the value of $k^{H/D}$ divides general base from nucleophilic reactions, although there are exceptions to this criterion. The relatively low solvent deuterium isotope effects for rate-limiting proton transfer for some reactions known to follow general base catalysis, compared with those predicted (Table 4.4), has been a considerable puzzle. Proton transfer between oxygen and nitrogen atoms often occurs with a rate-limiting step which is diffusion-controlled and shows little isotopic dependence ($k^{H_2O/D_2O} \approx 1.2$) in line with the viscosity difference for the two solvents. However, a maximum in the primary isotope effects for proton transfer reactions between electronegative atoms *has* been observed as the strengths of the acid and base are raised (see Fig. 4.3). The low isotope effects observed for some of the nucleophilic reactions of Table 4.4 could be explained by the large values of ΔpK_a for these systems.

4.5   Heavy-atom isotope effects

Next to the isotope effect for hydrogen, that for carbon is the most studied, and primary isotope effects range as high as 15% with ^{14}C (Table 4.1). It is useful to introduce a slightly different view of the theory to discuss heavy-atom isotope effects. The ratios of the fractionation factors or partition functions can be considered as multiples of the isotope effect on the mass moments of inertia (MMI), on the zero-point energies (ZPE) of the normal vibrations and on the population of excited vibrational states (EXC). The kinetic isotope effect for substituting isotope a by isotope b is thus $k^{a/b} = \text{MMI} \times \text{ZPE} \times \text{EXC}$. The zero-point energy term predominates in primary isotope effects of hydrogen, but such predominance is not necessarily the case in heavy-atom isotope effects because zero-point energy values become closer to unity. The contributions of MMI, ZPE and EXC may be calculated and Table 4.5 illustrates some results for a number of reactions.

Heavy-atom isotope effects have an advantage over hydrogen isotope effects because the lower contribution of the ZPE component makes it simpler to elicit knowledge of the transition-state structure by comparison with the intrinsic isotope effect using Leffler's α; the reason for this is that, unlike the ZPE, both MMI and EXC can contribute to the transition state. For example, heavy-atom and secondary deuterium isotope effects have been applied to methyl transfer from adenosylhomocysteine to 3,4-dihydroxyacetophenone catalysed by catechol-*O*-methyltransferase (Rodgers *et al.*, 1982). The effects are used to 'bracket' Pauling bond orders in contour maps of isotope effects calculated for various bond-order values; the isotope effects ($\text{CH}_3/\text{CD}_3 = 0.83 \pm 0.05$ and $^{12}\text{CH}_3/^{13}\text{CH}_3 = 1.09 \pm 0.02$) identify the structure of the transition state within a defined area of the map (Fig. 4.5).

The limits to the possible values of the bond orders of the C—S and C—O bonds depend on the error limits on the data. The present case gives a rather large possible area for the case of the enzyme as well as ambiguity, due to two cross-over points; as experimental techniques are improved, this type of approach could become more important (see, for example, Murray & Webb, 1991).

Table 4.5 Contribution of MMI, ZPE and EXC to the heavy-atom kinetic isotope effect (KIE).[a]

Reaction	Isotopes	MMI	EXC	ZPE	KIE
$\text{CH}_3-\overset{*}{\text{C}}\overset{\text{O}}{\underset{\text{H}}{\diagup}}\ \xrightarrow{\text{OH}^-}\ \text{CH}_3-\overset{\text{O}^-}{\underset{\text{H}}{\text{C}}}-\text{OH}$	$k^{12/13}$	1.018	0.995	1.008	1.021
	$k^{16/18}$	1.021	0.997	0.995	1.012
(Hogg *et al.*, 1982)					
$-\overset{+}{\underset{\overset{\textstyle\|}{*\text{CH}_3}}{\text{S}}}-\ \xrightarrow{\text{Nu}^-}\ -\text{S}-\quad\text{CH}_3\text{Nu}$	$k^{32/34}$	1.0085	0.9934	1.0082	1.0101
	$k^{12/13}$	1.0244	0.9893	1.0566	1.0708
(Rodgers *et al.*, 1982)					

[a] The atoms starred are isotopically replaced.

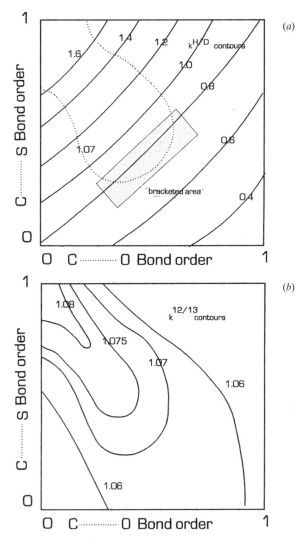

Fig. 4.5 Transition-state mapping by isotope effect; the lines plot calculated isotope effects for given C—S and C—O bond distances: (*a*) $k^{H/D}$; (*b*) $k^{12/13}$. The dotted line in (*a*) is taken from (*b*) and brackets the transition-state structure. Compiled from data in Rodgers *et al.* (1982).

4.6   Secondary isotope effects and transition-state structure

Primary isotope effects are at their best in diagnosing rate-limiting steps by comparison with the intrinsic value. Secondary isotope effects offer potential for determining the structure of a transition state from the ratio of the kinetic secondary isotope effect to that for the overall reaction, namely the equilibrium isotope effect, because the zero-point energy of the isotopically sensitive bond is not lost in the transition state (cf. Section 4.1). For example, when an organic molecule ionizes to yield a carbenium ion, any adjacent C—H bond weakens, thus reducing the contribution of vibrational modes involving this bond to the zero-point energy of the transition state. The isotope effect on atoms adjacent to bonds being formed or broken acts as a 'reporter' of the structural changes proceeding in these bonds. The concept of the reporter group is widely employed in physical organic chemistry and it depends on the property of a group, such as a chromophore, which is affected by adjacent micro-environments (cf. Section 5.2). The equilibrium secondary isotope effect for the ionization represented by Eqn [4.15] has the value $K^{H/D} = 1.29$ (Streitwieser *et al.*, 1958) and this may be employed as a standard model for complete C—O bond fission. General acid-catalysed hydrolysis of orthoesters (**7**) have isotope effects in the region of 1.05, and acid-catalysed hydrolysis of acetals 1.19 (**8**), indicating less carbenium ion character in the transition state of the orthoester (Cordes, 1970) than in acetal transition states; application to enzyme-catalysed acetal hydrolysis is described in Section 10.6 (p. 246). The secondary isotope effect has been applied to the hydrolysis of *O*-ethyl-*S*-phenyl-thiobenzaldehyde acetal hydrolysis (Eqn [4.16]). Acid-catalysed hydrolysis of these acetals has $k^{H/D} = 1.038 \pm 0.008$ (Faroz & Cordes, 1979), indicating little C—S fission; more C—S bond fission is indicated for the uncatalysed (or water-catalysed) reaction by an isotope effect of 1.13 ± 0.02. Phenyl formates exhibit isotope effects of the order of 1.12–1.28 consistent with a transition state resembling a tetrahedral intermediate (do Amaral *et al.*, 1979; Eqn [4.17])

$$\text{(D)} \qquad K^{H/D} \qquad \text{(D)}$$
$$\text{Ph}_2\text{CHOH} + \text{H}^+ \rightleftharpoons \text{Ph}_2\text{CH}^+ + \text{H}_2\text{O} \qquad\qquad [4.15]$$

7

8

$$\text{(D)} \qquad -\text{PhSH}, -\text{EtOH} \qquad \text{(D)}$$
$$\text{Ph-CH(SPh)(OEt)} \xrightarrow{\quad\text{H}_2\text{O/H}^+\quad} \text{PhCHO} \qquad\qquad [4.16]$$

Further reading

Caldin, E. F. & Gold, V. (eds) (1975) *Proton Transfer Reactions*, Chapman and Hall, London.

Cleland, W. W. (1987) Secondary isotope effects on enzymatic reactions, in *Isotope Effects in Organic Chemistry*, Buncel, E & Lee, C. C. (eds), Elsevier, Amsterdam, Vol. 7, Chapter 2.

Cleland, W. W., O'Leary, M. H. & Northrup, D. B. (eds) (1977) *Isotope Effects in Enzyme Catalyzed Reactions*, University Park Press, Baltimore.

Collins, C. J. (1964) Isotopes and organic reaction mechanisms, *Adv. Phys. Org. Chem.*, **2**, 1.

Cook, P. F. (ed.) (1991) *Enzyme Mechanism from Isotope Effects*, CRC Press, Boca Raton.

Hogg, J. L. (1986) Secondary hydrogen isotope effects, in *Investigation of Rates and Mechanisms of Reactions*, 4th edn, Part 1, Bernasconi, C. F. (ed.), Wiley–Interscience, New York, p. 201.

Isaacs, N. S. (1995) *Physical Organic Chemistry*, 1st edn, Longman Scientific, Harlow.

Klinman, J. P. (1978) Primary hydrogen isotope effects, in *Transition States of Biochemical Processes*, Schowen, R. L. & Gandour, R. D. (eds), Plenum Press, New York, p. 165.

Melander, L. & Saunders, W. H. (1980) *Reaction Rates of Isotopic Molecules*, John Wiley, New York.

Schowen, R. L. (1972) Mechanistic deductions from solvent isotope effects, *Prog. Phys. Org. Chem.*, **9**, 275.

Streitwieser, A. (1960) Stereochemical and kinetic applications of deuterium isotope effects, *Ann. N.Y. Acad. Sci.*, **84**, 576.

Suhnel, J. & Schowen, R. L. (1991) Theoretical basis for primary and secondary hydrogen isotope effects, in Cook, P. F. (1991).

Wiberg, K. B. (1955) The deuterium isotope effect, *Chem. Rev.*, **55**, 713.

5 Transition states from external effects

5.1 Introduction

The energies of ground/reactant, transition and product states may be affected by changes in the structure of the molecule itself. Consideration of the whole reaction system reveals that external variables such as temperature, pressure or dielectric constant of the solvent also affect the energies of the states and, in principle, the relationship between the change in a parameter and its effect should be predictable from the state structures. Since external properties refer to the bulk solvent, the structure–reactivity correlations for external parameters are more tenuously connected with transition-state structure than those relating to the structure of the discrete reactant molecules. The 'micro-force' referring to the property in the immediate vicinity of the reactant is only partially related to the bulk property, and knowledge of the effect on reactivity of external, bulk, properties is relatively weak. Since bio-organic reactions usually involve bringing a substrate into a preformed micro-solvation region (for example an enzyme active site), the effect of external parameters is of relevance and importance. The external effects considered in this chapter—variation of temperature, pressure and medium—are properties associated with the whole system and their application is therefore to diagnose pathways rather than to measure detailed transition-state structure.

5.2 Solvent effects

Reactivity is significantly affected by the micro-environment of the substrate in the solvent or within the complex (Abraham, 1974; Reichardt, 1979, 1994; Israelachvili, 1987). Solvent variation may significantly alter the energies of both the transition state and the ground state and may completely alter the mechanism or even the nature of the reaction; the term 'solvent effect' covers all effects due to changes in medium including solute additives which do not involve themselves covalently with the reaction. Solvent effects may be used as tools for the elucidation of enzyme reaction mechanisms but are limited because the active site provides the micro-solvent (Knowles, 1990) for the transition state of the rate-limiting step and cannot easily be changed.

Variation of the bulk solvent generally has relatively little effect on the stability of the enzyme–substrate complex, but it may be important if the solvent is involved in the reaction being catalysed or there is a large ground-state effect. However, it can affect the stability of the substrate and hence its free energy of transfer to the enzyme.

The effect of solvent change has been employed extensively to study mechanisms of non-enzymic reactions. These studies provide underpinning for the overall understanding of bio-organic mechanisms, which often involve transport of a substrate from bulk solvent into an active site that provides alternative solvation. There is current interest in reactions catalysed by enzymes in solvent media possessing large fractions, even up to 100%, of non-aqueous solvent (Koskinen & Klibanov, 1996).

Solvent is composed of an assembly of molecules, each having a random energy within an energy range; the interaction of solute with solvent is usually dynamic, involving a rapid exchange of partners. The incorporation of a solute into a solvent changes the nature of the assembly of solvent molecules and it is only in dilute solution that any solute–solute or solute–solvent effects on the solvent may be disregarded. This problem is dealt with formally by the use of the concept of *activity*, which becomes equal to concentration at infinite dilution. The nature of the bulk solvent changes as the concentration of the solute increases. A number of polar solvents, including water, partially conserve the structure of the solid phase. Water can be considered to consist of an assembly of clusters (Frank & Wai, 1957; Jencks, 1969) of hydrogen-bonded regions which are continually forming and breaking. Such residual structures can be markedly changed by the addition of interacting solute molecules. However, solvents such as carbon tetrachloride or alkanes have very little inter-solvent inter-action, and the solvent has little structure.

Hydrophobic interactions occur between relatively apolar groups or molecules in water and represent the favourable tendency for such groups and molecules to associate. The detailed understanding of the driving force for these interactions is still controversial but a popular molecular inter-pretation is based on the unusual thermodynamic parameters observed for the hydration of apolar moieties in water. The unusually large and positive entropies of association in water are often attributed to the structuring of water molecules in the hydration sphere of the hydrophobic solute. Associa-tion of apolar molecules is then accompanied by a release of this 'structured' hydration water into the bulk phase; this gain in entropy is assumed to be the driving force for hydrophobic interactions. Alternative views emphasize the importance of the small size of water molecules, their tendency to retain as many hydrogen bonds as possible, and the London dispersive forces between the apolar groups (Provalov & Gill, 1989; Muller, 1990; Blokzijl & Engberts, 1993).

Kinetics of reactions in solutions with concentrations of the scale used in preparative work can introduce a time-variable parameter due to substantial structure changes in the medium resulting from the change in solute/

reactant as it is converted to product under these conditions. Such variation has an essentially uncontrolled effect on kinetics, product ratios and stereochemistry, and it is therefore imperative that mechanistic studies be carried out with dilute solutions and that data from 'preparative' experiments be regarded with caution. It is important that conditions are kept constant throughout the reaction and, moreover, to enable comparison, throughout a series of experiments. It is essential to ensure that ionic strength is kept constant by ensuring that it is large (often between 0.1 and 0.5 M) relative to changes in molarity occurring as a result of reaction. Buffering is also essential even in non-protogenic reactions. It is necessary to ensure that the additives to keep both ionic strength and pH constant do not interfere with the reaction or that the interference is controlled—for example by measuring the effect on the rate constant of change in concentration of the additive. In some studies, such as in micelle work, it is also necessary to ensure that the concentration of a particular ion remains constant.

The study of solvent effects often involves comparison of the kinetics in solution with those for the same reaction in the gas phase and a study of microscopic solvation in clusters. There is no direct comparison possible between macroscopic solvent parameters and microscopic details of reaction processes, but often quite good correlations exist between rate constants and macroscopic parameters such as the dielectric constant of the solvent (Abraham *et al.*, 1988). There has been reasonable success in theoretically modelling reactions in solution and in the gas phase where the solvent is treated as a limited number of molecules within a given theoretical space. An excellent example of this approach is for the attack of chloride ion on methyl chloride in water and in the gas phase, where the results give good agreement with experimental values (Jorgensen, 1989). Reactions which are studied in mixed solvents have additional difficulties because the microscopic environment can be very different from the bulk phase. For example, hydrophilic substituents appear to sequester water molecules preferentially from water–organic solvent mixtures so that 'solvent sorting' occurs and different solutes/reactants may have varied micro-environments (Capon & Page, 1971).

Lone pairs which may act as general bases or nucleophiles will usually be 'solvated' by hydrogen bonding either from water or intramolecularly from within the enzyme. Reagent groups will normally require 'desolvation' before bond making or breaking can occur. This process is energetically expensive and yet it is an essential part of the normal activation energy, which may be compensated by favourable interactions between the substrate and enzyme (Page, 1984a,b).

Changes in solvent composition may cause variation in rate constant due to *differences* in ground-state and transition-state energies; these could alter the substrate specificity of enzymes (Bell *et al.*, 1974). Subtilisin catalyses the transesterification reaction of *N*-acetyl-L-serine ethyl ester, eightfold faster in dichloromethane than the corresponding reaction with *N*-acetyl-L-phenylalanine ethyl ester, whereas in *t*-butylamine these relative rates are inverted (Wescott & Klibanov, 1993). The parameter k_{cat}/K_m is the second-order rate constant for an enzyme-catalysed reaction (see Section 10.1) and measures the free-energy difference between transition state $TS^{\ddagger}$ of the rate-limiting step and ground-state E and S (Eqn [5.1]). Except in ideal, dilute solutions, many measurements are not directly related to the experimental concentrations of solutes. To correct for this, 'effective concentrations' or activities are used which are obtained from the product of the concentration and the activity coefficient, γ. If γ_S, γ_E and $\gamma_{\ddagger}$ represent the activity coefficients of the substrate, enzyme and transition state, respectively, and are defined relative to a common value of unity in purely aqueous solution, then k_{cat}/K_m for the enzyme-catalysed reaction in a given solvent mixture is related to its value k_{cat}^0/K_m^0 in pure aqueous solution by Eqn [5.2]. For sparingly soluble substrates, γ_S may be obtained from the ratio of the solubilities, S and S^0, measured respectively in the presence and absence of organic solvent (Eqn [5.3]).

$$E + \text{substrate} \xrightarrow{\ k_{cat}/K_m\ } TS^{\ddagger} \to E + \text{product} \tag{5.1}$$

$$k_{cat}/K_m = \frac{k_{cat}^0}{K_m^0} \cdot \frac{\gamma_S \cdot \gamma_E}{\gamma_{\ddagger}} \tag{5.2}$$

$$\gamma_S = S^0/S \tag{5.3}$$

Solvation effects on the enzyme may be determined if solubility and kinetic data are obtained for *two* substrates, 1 and 2. In a given solvent system γ_E may be eliminated because it is the same for both reactions. The effect of the solvent on the ratio of the two transition-state activity coefficients is given by Eqn [5.4] where k is the value of k_{cat}/K_m with subscripts 1 and 2 referring to the substrate identity and the superscript (0) referring to the value in purely aqueous solution. The right-hand side of Eqn [5.4] contains only measurable quantities (rate parameters and solubilities) and the transition-state ratio for the two substrates 1 and 2, $\gamma_2^{\ddagger}/\gamma_1^{\ddagger}$, may be compared with the ground-state ratio, γ_2/γ_1, to provide useful information concerning solvation.

$$\frac{\gamma_{\ddagger}^2}{\gamma_{\ddagger}^1} = \frac{k_1 S_1}{k_2 S_2} \cdot \frac{k_2^0 S_2^0}{k_1^0 S_1^0} \tag{5.4}$$

The rate of the α-chymotrypsin-catalysed hydrolysis of 5-nitrophenyl acetate and N-acetyl-L-tryptophan methyl ester in organic solvent mixtures decreases with increasing amounts of dioxan or propan-2-ol (Bell *et al.*, 1974). Measurement of the solubilities of the substrates in the solvent mixtures indicates that the *difference* in reactivity of the two substrates according to solvent composition is largely a *ground-state effect*. In general, the different specificities shown by α-chymotrypsin may be due to ground-state solvation effects (reflected in the different solubilities of the substrates) rather than to effects on the transition state. Desolvation of the substrate may be responsible for the fact that specific substrates bearing an acylamino side-chain are more reactive by several powers of ten than their analogues containing a free amino group. This has sometimes been explained in terms of hydrogen-bonding effects but the differences between the hydrogen-bond energies of the amino and acylamino groups with the carbonyl oxygen atom of the enzyme, or with the solvent, are quite inadequate to explain the large rate differences. It seems more likely that the increase in reactivity often observed when the amino group is converted into acylamino arises more from the relatively weak solvation of the latter group in the ground state than from favourable transition-state interactions, though of course the net result will depend upon the sum of these two effects.

It should be emphasized that structural specificity attributed to the desolvation of the substrate in enzyme-catalysed reactions results from a *retardation* effect on the *less* reactive substrates. It has been suggested (Cohen *et al.*, 1970) that a part of the rate enhancement associated with enzyme-catalysed reactions is due to the preliminary formation of a complex in which the site of reaction is made susceptible to attack by the removal of solvating water molecules. However, since the overall rate of reaction depends on the free-energy difference between ground and transition states, any desolvation of the substrate, other than that occurring in comparable non-enzymic reactions and necessary for approach of the catalytic groups, will raise the transition-state energy and decrease the rate unless the unfavourable energy change is compensated by favourable binding energy between substrate and enzyme.

The nature of the solvent can affect the enantioselectivity shown by an enzyme-catalysed reaction of a racemic mixture (Klibanov, 1990). As changes in solvent must change the ground-state energy of the enantiomers equally, any changes in discrimination induced by solvent changes must be a transition-state effect.

5.3 Reactions in strongly acidic and strongly basic media

Although the conditions normally employed in studies in strongly acidic and basic media are far outside the limits normally encountered in bio-organic systems, such studies are directly relevant to proton transfer, which is central to all bio-organic—and indeed to most organic—reactions. Many substrates are weak organic bases and acids and yet their bio-organic reactions often involve protonation and deprotonation steps. It is therefore of relevance to understand the protonation and deprotonation behaviour of these species. As the acidity of the medium is increased, the rate constants for acid-catalysed reactions increase; but then they often level off and start to decline in the region where the water content in an acid–water mixture becomes small (Fig. 5.1). The cause of this effect is the decrease in water activity and the change in structure of the solvent compared with that of the

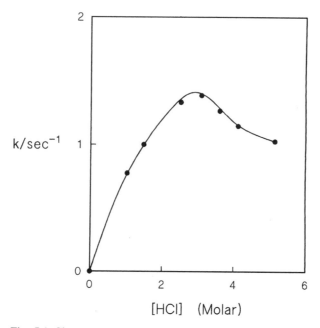

Fig. 5.1 Change in the observed rate constant for the hydrolysis of acetamide as the concentration of acid catalyst is increased.

more dilute solutions of acid, i.e. it is essentially a medium effect (Bagno *et
al.*, 1987). The problem with quantifying reactions in strongly acidic solutions is that there are at least two variables—the effective acid concentration
and the medium change with increasing concentration. Similar effects are
observed for base-catalysed reactions in strongly basic solutions.

Reactions at high acidities may be treated by the use of *acidity functions*,
which are empirical indicators of both acidity and medium changes. These
acidity functions are defined by the effect the acid solution has on the
equilibrium between a weak base (A^-) and its conjugate acid (HA). In
strongly acidic solutions the degree of protonation of weak bases changes
in an unpredictable way. This is true not only for, say, an increase in the
molarity from 2 M to 4 M being different from a change 4 M to 8 M, but also in
that, say, 5 M $HClO_4$ acts differently from 5 M HCl or 5 M HNO_3. The pK_a of
an acid (HA) is given by Eqn [5.5], which may be transformed to Eqn [5.6],
where H_0 is the original acidity function of Hammett and Deyrup (1932).

$$pK_a = -\log(a_H.a_A/a_{HA}) = \log(a_{HA}/a_A) - \log a_H$$
$$(= \log([HA]/[A^-]) + pH \text{ for dilute solutions}) \qquad [5.5]$$
$$= \log([HA]/[A^-]) - \log(a_H.\gamma_A/\gamma_{HA})$$

$$\log[A^-]/[HA] = H_0 - pK_a$$
$$H_0 = -\log h_0 \qquad [5.6]$$

The acidity function, H_0, reflects the ratio of the concentration of a weak base to that of its conjugate acid, and its pK_a; the *ratio* of the activity coefficients γ_{A^-}/γ_{HA} is assumed to be independent of structure for a series of structurally related acids (HA).

The determination of H_0 through an evaluation of pK_a and the concentrations of HA and A^- is often difficult when HA itself is a very strong acid, but is usually obtained by using a series of structurally related acids, HA_1, $HA_2, \ldots, HA_n$, of increasing acidity. It is experimentally very difficult to measure the pK_a of an acid by pH-titration methods when it lies close to or beyond values represented by the extremes of the pH scale (< 2 and > 12) because at these extremes the solvent is being titrated as well as solute. Lower and lower H_0 values may be determined by use of progressively stronger HA_n species, each with a range of observable changes (e.g. in an absorption spectrum) in a region of acidity which overlaps with that of its neighbour. It should be noted that the pK_a of the acid might itself vary over the range of compositions studied to determine H_0 and the methods employed to determine the ratio $[HA]/[A^-]$ (usually spectroscopic) involve parameters (such as extinction coefficient) which might be dependent on solvent composition.

Acidity functions have been employed to distinguish between an acid-catalysed solvolysis mechanism involving solvent in the rate-limiting step (an A-2 mechanism) and one where the rate-limiting step is the unimolecular decomposition of the protonated substrate followed by fast water addition (A-1). Let us consider the reaction where the species R_1OR_2 undergoes specific acid-catalysed hydrolysis (Eqn [5.7]).

A-2 mechanism

$$R_1OR_2 + H^+ \underset{K_a}{\overset{}{\rightleftharpoons}} R_1\overset{+}{\underset{H}{O}}R_2 \xrightarrow[\text{rate-limiting step}]{[H_2O]k_2} R_2OH + R_1OH_2^+ \qquad [5.7a]$$

A-1 mechanism

$$R_1OR_2 + H^+ \underset{K_a}{\overset{}{\rightleftharpoons}} R_1\overset{+}{\underset{H}{O}}R_2 \xrightarrow[\text{rate-limiting step}]{k_2} R_2OH + R_1^+ \xrightarrow{\text{fast}} R_1OH \qquad [5.7b]$$

Zucker & Hammett (1939) proposed that reactions involving the transformation of a protonated substrate $[R_1OH^+R_2]$ via the A-1 mechanism should give a linear plot of $\log k$ versus $-H_0$. A plot of $\log k$ linear in $\log [H^+]$, where $[H^+]$ is the stoichiometric concentration of hydrogen ions, is consistent with a mechanism involving bond formation with the solvent in the rate-limiting step. Equations [5.8] and [5.9] relate to this diagnosis, where γ_S, γ_{HB}, $\gamma_{\pm}$ and γ_B are the activity coefficients respectively of solvent, conjugate acid of substrate, transition state and substrate.

A-1 mechanism

$$k_{obs} = (k_2/K_a).h_0.(\gamma_S\gamma_{HB}/\gamma_{\pm}\gamma_B)$$

$$H_0 = -\log h_0 \qquad [5.8]$$

$$\log k_{obs} = -H_0 + \log(k_2/K_a)$$

A-2 mechanism

$$k_{obs} = \{k_2/K_a\}[H_3O^+]a_{H_2O}\cdot\gamma_{H^+}\cdot\gamma_S/\gamma_{\ddagger}$$

$$\log k_{obs} = \log[H_3O^+] + \log k_2/K_a + \log(a_{H_2O}\cdot\frac{\gamma_{H^+}\cdot\gamma_S}{\gamma_{\ddagger}}) \tag{5.9}$$

The Zucker & Hammett treatment makes the assumption that the ratio of the activity coefficients in Eqns [5.8] and [5.9] is unity. Although the approach gives either linear or non-linear plots these generally have non-unit slopes which indicate that the assumption about the activity coefficients is probably not valid. The method was modified by Bunnett (1961) by including an empirical parameter ω (Eqn [5.10]), the value of which is related to the number of solvent molecules involved. Thus an ω value less than or equal to zero corresponds to the absence of water participation in bond formation in the rate-limiting step, and $\omega > +1.2$ indicates the involvement of solvent.

$$\log k_{obs} + H_0 = \omega \log a_{H_2O} + \text{constant} \tag{5.10}$$

Application of the method to the specific acid-catalysed hydrolysis of esters, amides, ethers and cyclic ethers gives linear plots of $\log k + H_0$ against $\log a_{H_2O}$ with slopes consistent with the A-2 mechanism. Acetal, ketal, orthoester and glycoside hydrolyses catalysed by acid have ω-values consistent with A-1 mechanisms involving no bonding water in the rate-limiting step.

Reactions in strongly basic media also exhibit activity effects and a 'basicity function' (H_-) is obtained, analogous to the H_0 function. The basicity function may be employed to study the solvent participation in base catalysed reactions whereby a plot of $\log k$ versus H_- should be linear if the solvent is not involved and a plot of $\log k$ versus $\log$ [lyate ion] (for example, water has the lyate ion OH^- and for methanol it is CH_3O^-) should be linear if the solvent is involved. More O'Ferrall & Ridd (1963) and Allison *et al.* (1958) studied the methanolysis of chloroform in basic methanol solutions and found that the dependence of $\log k$ on H_- is consistent with a B-1 process not involving methoxide ion in the rate-limiting step (Eqn [5.11]). Equation [5.12] illustrates a mechanism requiring methoxide ion in the rate-determining step.

B-1 mechanism

$$CHCl_3 \rightleftharpoons {}^-CCl_3 \xrightarrow[\;-Cl^-\;]{\text{rate-limiting step}} :CCl_2 \xrightarrow[\text{MeOH}]{\text{fast}} HCOOMe \tag{5.11}$$

B-2 mechanism

$$CH_3\bar{O} \;\curvearrowright H{-}CH_2{-}Cl \xrightarrow{\text{rate-limiting step}} Cl^- + CH_3OH + H_2C: \xrightarrow{\text{fast}} HCOOMe \tag{5.12}$$

The methoxide ion-catalysed elimination of phenylethyl chloride, known to involve methoxide ion in the rate-limiting step, exhibits a linear dependence between $\log k$ and $\log [MeO^-]$.

The equations given above for the application of acidity and basicity functions are somewhat complicated but overall they are essentially linear free-energy relationships between the reaction under investigation and a reference reaction (Bagno *et al.*, 1987, 1993, 1995).

5.4 Solvent ionizing power

The extent to which a solvent stabilizes ions can be measured by comparing the equilibrium constant for a standard ionization in that solvent with the value in a standard solvent (Abraham, 1974; Bentley & Llewellyn, 1990; Ta-Shma & Rappoport, 1991). Unfortunately no cation-forming equilibrium is known where the positive charge is localized solely on the central carbon and which would be convenient to use as a standard for S_N1-type processes. In place of this, Grunwald & Winstein (1948) sought a standard reaction where it may be assumed that the transition state has almost complete carbenium ion character and the rate constant k therefore measures the ionizing power (Y, defined in Eqn [5.13]) for the solvent; the solvolysis of *t*-butyl chloride is such a standard reaction and the standard solvent is 80% $EtOH/H_2O$:

$$Y = \log k_s / \log k_{standard\ solvent} \qquad [5.13]$$

When the solvolysis reaction in question is studied, a plot of $\log k_s$ versus Y (Appendix A.2, Table 5.a1) should be linear (Eqn [5.14]). Here m is a measure of the carbenium ion character of the transition state relative to that of the standard reaction, where m is unity (Scheme 5.1).

$$\log k_s = m \cdot Y + \log k_{standard\ solvent} \qquad [5.14]$$

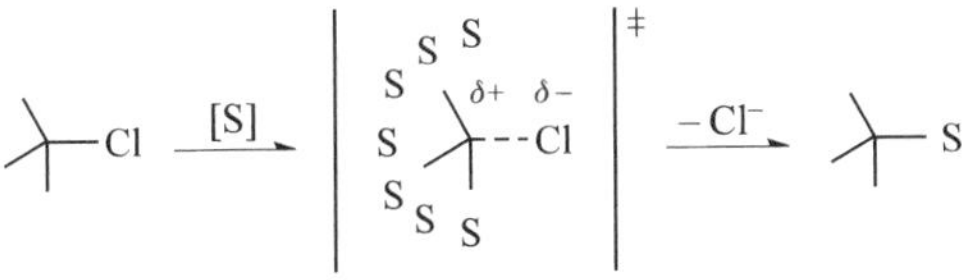

Scheme 5.1 Standard reaction defining the Grunwald and Winstein solvent parameter (Y); S = solvent.

The value of m (0.343) for the solvolysis of ethyl bromide in 80% ethanol/water set against $m = 0.941$ for the solvolysis of t-butyl bromide suggests little charge separation in the transition state for the former reaction. Values of m for a selection of reactions are given in Table 5.1.

The cyclization reaction (Eqn [5.15]) carried out in various solvents has a value of $m = 0.13$, which is consistent with little charge expression in the transition state of the rate-limiting step (Scott *et al.*, 1971).

Table 5.1 Values of m for some solvolysis reactions.[a]

Substrate	Solvent	m
EtBr	EtOH/H_2O	0.343
Bu^{t}Cl	EtOH/H_2O	(1.000)
Bu^{t}Br	EtOH/H_2O (0 °C)	1.02
Bu^{t}Br	EtOH/H_2O (25 °C)	0.94
Bu^{t}Br	CH_3COOH/HCOOH	0.95
HOCH$_2$CH$_2$Br	EtOH/H_2O	0.23
PhCH$_2$OTs	acetone/H_2O	0.65
CH$_3$COOCOCH$_3$	acetone/H_2	0.58
CH$_3$SO$_2$Cl	dioxane/H_2O	0.47
PhCHClCH$_3$	EtOH/H_2O	1.00
PhCHClCH$_3$	CH$_3$COOH/H_2O	1.14
PhCHClCH$_3$	dioxane/H_2O	1.14
PhC(CH$_3$)$_2$CH$_2$Cl	EtOH/H_2O	0.83
Ph$_2$CHCl	EtOH/H_2O	0.74
Ph$_2$CHCl	CH$_3$COOH/H_2O	1.56
Ph$_3$CF	EtOH/H_2O	0.89
Ph$_3$CF	acetone/H_2O	1.58

[a] Parameters taken from Leffler & Grunwald (1963) and Hine (1962).

$$\text{Ph} \quad \overset{\text{H}}{\underset{\text{O}}{\text{N}}} \quad \text{Br} \quad \xrightarrow{\text{OH}^-} \quad \text{Ph} \underset{\text{O}}{\overset{\text{N}}{\diagup}} \quad + \; H_2O + Br^- \qquad [5.15]$$

Difficulties in the interpretation of the Grunwald–Winstein m-values arise because the solvent itself is almost certainly involved in the mechanism of the standard reaction. Internal return could interfere with the value of the rate constant if decomposition of the carbenium ion became rate-limiting as a result of a change in the solvent required to estimate m. The first problem was recognized by Schadt *et al.* (1976), who suggested that nucleophilic assistance to ionization was occurring. The solvolysis of 1-adamantyl chloride **1** → **2** has been employed as a standard reaction where it is 'impossible' for solvent to assist the nucleophilic substitution by attack backside to the C—Cl bond (Eqn [5.16]). The Y-values defined by this new standard appear to depend on the nature of the leaving group (Br—,TsO— etc.); further difficulty is the geometry of the putative carbenium ion, which is constrained by the adamantane architecture. These considerations indicate that the transition-state structure of the 1-adamantyl halide solvolysis may not be a satisfactory model for the formation of a hypothetical carbenium ion.

$$\qquad [5.16]$$

1   **2**

Other measures of solvent ionizing power are obtained from spectral studies of charged 'reporter' molecules; certain transitions in the electronic spectrum, particularly charge transfer transitions, are well known to suffer medium effects. These systems, including the Grunwald–Winstein approach, suffer from the problem that the system being investigated is normally very unlike that of the standard process.

Kosower's Z-parameter ($Z = h\nu$) (Kosower, 1958) is derived from the effect of the solvent on the charge transfer spectrum of the pyridinium iodide **3**. The electronic transition involves transfer of an initial state to an excited state where the two ions have partially neutralized each other, and therefore provides a measure of solvent polarity. Similar parameters have been obtained for other reporter molecules such as **4** which are advocated by Reichardt (1965, 1979, 1994), thus generating the so-called 'E_T scale'.

Applications of the effect of solvent on electronic spectra in the study of enzyme mechanisms are fairly sparse. Nevertheless the information derived is quite useful and an interesting example was reported by Kallos & Avatis (1966), who measured the difference spectrum between 4-nitrobenzenesulphonyl-α-chymotrypsin and tosyl-α-chymotrypsin; they demonstrated (by comparison with model chromophores in standard media) that the subsequently discovered 'tosyl pocket' (see Chapter 10) provided a microenvironment corresponding to that of cyclohexane.

5.5   Variation of pressure and temperature

Volume of activation

Pressure can cause a rate or equilibrium constant to vary if there is a change in partial molar volume (V) on going from ground to transition or product states. Increasing pressure on a reaction which exhibits a decrease (ΔV) in partial molar volume will enhance a rate or equilibrium constant (Fig. 5.2). Since predictions can often be made about the volume change expected for a given mechanism, the effect of pressure (Eqn [5.17]) is a valuable diagnostic tool. Increasing pressure increases boiling and melting points; increases the solubility of gases in liquids but decreases that of solids in liquids; increases

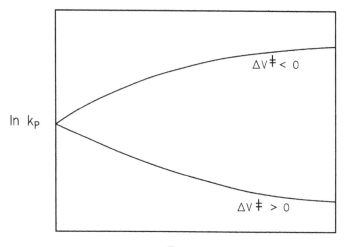

Fig. 5.2 Dependence of rate constant on pressure for reactions involving positive and negative $\Delta V^{\ddagger}$ values.

viscosity (by a factor of *ca* 2 per kbar (10^5 kPa)), and compresses liquids (by *ca* 10% per kbar).

$$\Delta V = \Sigma V_{\text{products}} - \Sigma V_{\text{reactants}} = -RT(\text{d ln } K\text{d}P)_T$$
$$\Delta V^{\ddagger} = V^{\ddagger} - \Sigma V_{\text{reactants}} \qquad\qquad [5.17]$$
$$= -RT(\text{d ln } k_P/\text{d}P)_T$$

In its simplest terms an association between reagents will decrease the volume of activation due to the reduced requirement for free space. Ion formation is accompanied by a volume decrease due to solvent electrostriction (where solvent becomes tightly associated with the ion). Neutralization of charge would cause an opposite effect (Fig. 5.2). Bimolecular associations typically have $\Delta V^{\ddagger}$ of -20 to $-30\,\text{cm}^3\,\text{mol}^{-1}$ (see Table 5.2) and reaction rates need to be studied up to 10 kbar (10^6 kPa) to ensure the accurate determination of volumes of activation.

Table 5.2 Typical contributions to volumes of activation.[a]

Process	Contribution to $\Delta V^{\ddagger}$ ($\text{cm}^3\,\text{mol}^{-1}$)
Molecular association	-20 to -30
Ion formation	-20
Charge localization	-5
Molecular dissociation	10
Ion recombination	20
Charge dispersal	5
Diffusion-controlled reactions	20

[a] Those processes involving charge give contributions which are solvent-dependent.

The application of pressure effects in elucidating mechanism requires the volume of activation to be divided into components (ΔV_R) due to the interaction of the reacting molecules with each other and (ΔV_S) due to the interaction of the reactant molecules with the solvent and the solvent molecules with each other. When two molecules combine, a van der Waals' separation is converted into a partial valence bond in the transition state. Crude calculations based on knowledge of the cross-section area of the bond (ca $10\,\text{Å}^2$) and the change in distance of $3.6\,\text{Å}$ to $1.6\,\text{Å}$ give a value of $-12\,\text{cm}^{-3}\,\text{mol}^{-1}$ for the volume change for a bond formation. It has been argued that destruction of empty space within a ring too small to accommodate solvent will give rise to a volume change, as exemplified in the hydrolysis of butyrolactone (Eqn [5.18]) compared with the hydrolysis of ethyl acetate (Eqn [5.19]). However, explanations of the difference between $\Delta V^{\ddagger}$ for lactones and esters call on the difference in dipole moments, which results in different solvation shells or different rate-limiting steps.

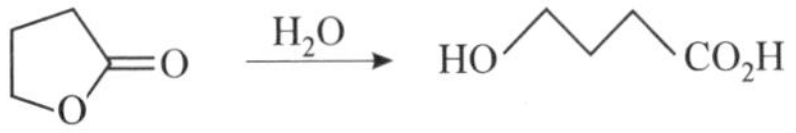

$$ \xrightarrow{\text{H}_2\text{O}} \quad \text{HO}\diagup\diagdown\diagup\text{CO}_2\text{H} \qquad \Delta V^{\ddagger} = -5.5\,\text{cm}^{-3}\,\text{mol}^{-1} \qquad [5.18] $$

$$ \text{CH}_3\text{COOEt} + \text{H}_2\text{O} \longrightarrow \text{CH}_3\text{COOH} + \text{EtOH} \qquad \Delta V^{\ddagger} = 0\,\text{cm}^{-3}\,\text{mol}^{-1} \qquad [5.19] $$

Formation and destruction of electrostatic charge and hydrogen bonds would appear to be the main source of the ΔV_S component. Volumes of mixing of liquids which do not form hydrogen bonds are small—of the order of 1 to $2\,\text{cm}^{-3}\,\text{mol}^{-1}$. The bimolecular attack of hydroxide ion on esters, and the nucleophilic attack of hydroxide ion on saturated carbon, possess volumes of activation of the order of -7 to $-10\,\text{cm}^{-3}\,\text{mol}^{-1}$ (Scheme 5.2). The electrostriction effects presumably cancel, because charge is not destroyed during reaction although it suffers a certain amount of delocalization in the transition state.

$$ \text{ClCH}_2\text{CO}_2^- + \text{OH}^- \longrightarrow \text{HOCH}_2\text{CO}_2^- + \text{Cl}^- \qquad \Delta V^{\ddagger} = -8.2\,\text{cm}^{-3}\,\text{mol}^{-1} $$

$$ \text{(cyclopropane-O)} + \text{OH}^- \longrightarrow \text{HO}\diagup\diagdown\diagup\text{O}^- \qquad \Delta V^{\ddagger} = -7\,\text{cm}^{-3}\,\text{mol}^{-1} $$

$$ \text{CH}_3\text{COOCH}_3 + \text{OH}^- \longrightarrow \text{CH}_3\text{CO}_2^- + \text{CH}_3\text{OH} \qquad \Delta V^{\ddagger} = -9.4\,\text{cm}^{-3}\,\text{mol}^{-1} $$

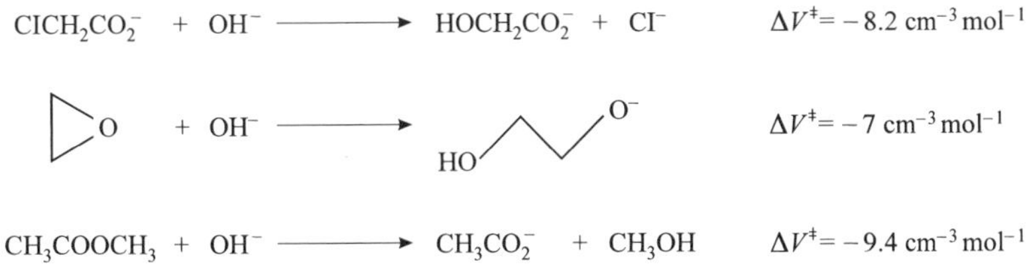

Scheme 5.2 Volumes of activation ($\text{cm}^{-3}\,\text{mol}^{-1}$) for nucleophilic attack.

Reactions where ions are involved but not destroyed involve only small volume changes (Scheme 5.3); however, the formation of ions gives a very large decrease in volume (Eqn [5.20]).

Volumes of activation are useful in diagnosing the role of solvent in solvolysis reactions, and complement acidity and basicity function techniques. Earlier equations for acid-catalysed solvolyses (Eqns [5.7a] and [5.7b]) indicate two types of process, one with the solvent participating in the transition state of the rate-limiting step (A-2) and one involving solvent attacking an intermediate formed in the rate-limiting step (A-1). With the assumption that the protonation step involves little volume change (see Scheme 5.3), the A-1 mechanism should give rise to an increase in volume and the A-2 mechanism should result in a decrease.

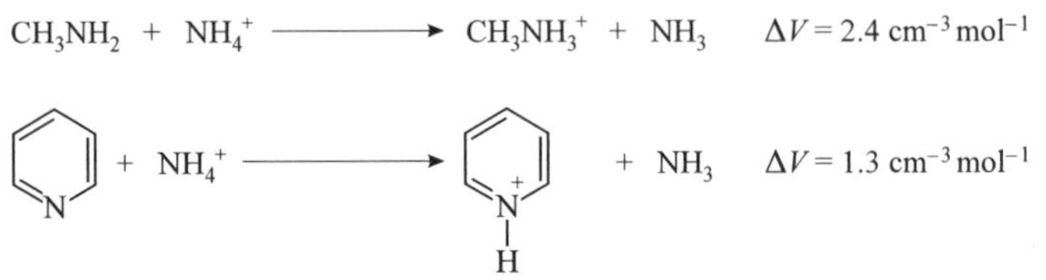

$$CH_3NH_2 + NH_4^+ \longrightarrow CH_3NH_3^+ + NH_3 \qquad \Delta V = 2.4 \text{ cm}^{-3}\text{mol}^{-1}$$

Scheme 5.3 Volumes of reaction ($\text{cm}^{-3}\text{mol}^{-1}$) for non-ionogenic reactions.

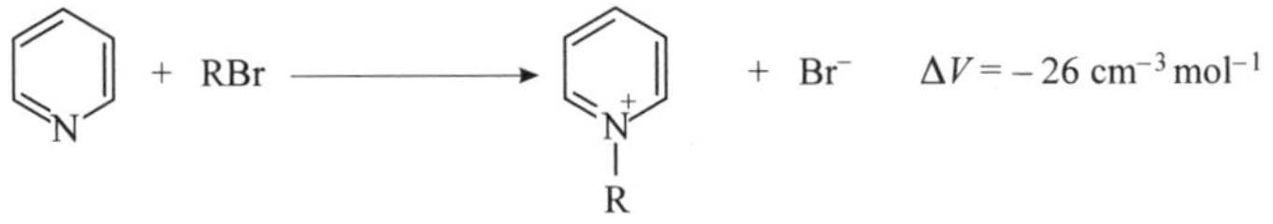

The acid-catalysed hydrolyses of esters and amides usually involve A-2 processes and this is consistent with the decrease in volume ($-9.4 \text{ cm}^{-3}\text{mol}^{-1}$) observed for both methyl acetate and acetamide hydrolyses. Ethers also hydrolyse in acid with a decrease in volume consistent with the A-2 process, but acetals and orthoesters exhibit an increase or a very slight decrease which confirms results from other methods that indicate the mechanisms to be A-1.

The alkaline hydrolysis of chloroform could take either a B-1 (Eqn [5.21]) or a B-2 (Eqn [5.22]) pathway (analogous to A-1 and A-2) and may be diagnosed as B-1 by observation of a positive volume of activation. This is consistent with the results for the analogous methoxide ion-catalysed methanolysis reaction (Eqn [5.11]), which has been shown to involve a B-1 mechanism by use of basicity functions.

B-1 mechanism

$$CHCl_3 + OH^- \longrightarrow CCl_3^- \xrightarrow[\text{Cl}^-]{\text{rate-limiting step}} : CCl_2 \xrightarrow{\text{fast}} HCOOH \qquad [5.21]$$

B-2 mechanism

$$CHCl_3 + OH^- \xrightarrow[\text{-Cl}^-]{\text{rate-limiting step}} {}^-CCl_3 \xrightarrow{\text{fast}} H_2\overset{+}{O}{-}\overset{-}{C}Cl_2 \longrightarrow HCOOH \qquad [5.22]$$

Mechanistic information about transition-state structure can be obtained from volumes of activation if they can be normalized by comparison with the corresponding volume change from reactants to products. For example, $\Delta V^{\ddagger}$ values for Diels–Alder reactions are typically -25 to $-45\,\text{cm}^3\,\text{mol}^{-1}$ and are very close to the corresponding equilibrium ΔV values consistent with a product-like transition-state structure. The ratio of $\Delta V^{\ddagger}/\Delta V$ sometimes even exceeds unity for those reactions when there are polar substituents attached to the diene and dienophile, indicative of contributions from solvation as well as association.

Entropy of activation

The treatment of reaction rates by transition-state theory gives rise to pseudo-thermodynamic activation parameters—free energy, enthalpy and entropy of activation. The effect of change in temperature on a rate constant can be used to elucidate the entropy difference between reactants and transition state. Entropy changes imply changes in degrees of freedom, which in turn may be used to elucidate information about the mechanism. Simple bimolecular mechanisms involving the bringing together of reagents should be associated with a decrease in entropy simply because of the loss of translational and rotational degrees of freedom. Reactions involving the creation of charge likewise give rise to negative entropies of activation because of electrostriction of solvent molecules. The reverse of these processes, dissociation and charge neutralization, are associated with positive entropies of activation. These are the largest effects, but smaller entropy changes result from the restriction of internal degrees of freedom such as occur in cyclization reactions.

Entropy changes for reactions can be calculated theoretically for the gas phase but not for solution. Entropies of activation may be used to discriminate between pathways rather than to elucidate detailed transition-state structures. More information can be gained if the entropies of activation can be normalized by comparison with a corresponding entropy of reaction. Finally, it is important to remember that, like most other parameters, entropies of activation represent the *difference* between reactant state and transition state. Variation in the values as a result of structural changes in a reaction are just as likely to be a reactant/ground-state effect as a transition state effect.

Entropies of activation may be determined from temperature effects by use of the transition-state Eqns [5.23] and [5.24], where κ is Boltzmann's constant and E_a is the enthalpy of activation:

$$k = (\kappa T/h)\exp(\Delta S^{\ddagger}/R)\exp(-E_a/RT) \qquad [5.23]$$

$$\ln k = \ln(\kappa T/h) + (\Delta S^{\ddagger}/R) - E_a/RT \qquad [5.24]$$

The theoretical equation [5.24] is not linear in $1/T$ but measurements are usually taken over a relatively small range of temperature so that a plot of $\ln k$ against $1/T$ is linear because changes in $\ln(\kappa T/h)$ are relatively small. The entropy is obtained from the intercept $(\ln(\kappa T/h) + \Delta S^{\ddagger}/R)$; the entropy obtained is temperature-dependent and the mid-point temperature of the range must therefore be quoted. The rate constant k is in units of s^{-1} and the order of the reaction is taken into account by employing standard states of the reactants at unit molarity; the standard state of the solvent is often taken as unity but it is important to state this explicitly as it will affect the value of an entropy of activation. Schaleger & Long (1963) showed that only 3% error in the rate constants will translate into an error of $\pm 10.8\,\mathrm{J\,K^{-1}\,mol^{-1}}$ for two rate constants measured at 25 °C and 35 °C; it is therefore imperative that the temperature is controlled precisely in the kinetic measurements, and that the temperatures should be accurately known.

The electrostriction effect is well exemplified by the dissociation of neutral species such as carboxylic acids where the ionization constant is associated with some $84\,\mathrm{J^{-1}\,K^{-1}\,mol^{-1}}$ decrease in entropy. The ions must be considerably solvated because the dissociation would be expected to involve a component due to gain in degrees of freedom associated with an increase in entropy. Reactions of ions with neutral molecules should have entropy effects largely reflecting the association process. Thus hydroxide ion attack

Table 5.3 Entropies of activation for some hydrolytic reactions.[a]

Reaction	$\Delta S^{\ddagger}(\mathrm{J\,K^{-1}\,mol^{-1}})$[b]	Reaction	$\Delta S^{\ddagger}(\mathrm{J\,K^{-1}\,mol^{-1}})$[b]
$HCOOEt + OH^-$	-75	$(CH_3O)_2CH_2 + H_3O^+$	28.5
$CH_3COOEt + OH^-$	-126	$(CH_3O)_2CHCH_3 + H_3O^+$	54.4
$C_2H_5COOEt + OH^-$	-109	$(EtO)_3CH + H_3O^+$	25.1
$CH_3COOPh + OH^-$	-92	$CH_3COOEt + H_3O^+$	-109
$CF_3COOPh + H_2O$	-188	$ClCH_2COOEt + H_3O^+$	-142
$CH_3COOCOCH_3 + H_2O$	-176	$CF_3COO^tBu + H_2O$	62
$CH_3COOH + CH_3OH + H^+$	-134	$CF_3COOCH_3 + H_2O$	-136

$CH_3Cl + H_2O$	-51.7	S_N2
$Pr^iCl + H_2O$	-34	$\downarrow$
$Bu^tCl + H_2O$	60	S_N1

Claisen/Cope

-38 (loss of internal rotation)

Neighbouring-group participation

26.9

[a] Data from Schaleger & Long (1963), Isaacs (1996) and Martin & Scott (1967).

[b] Entropy values are temperature-dependent and are normally quoted for 25 °C; since only moderate ranges of temperature are employed, Arrhenius plots are normally linear. The standard state is unit molarity.

on esters is associated with a large decrease in entropy (Table 5.3). The electrostrictive component constituting an increase in entropy due to the delocalization of the charge on the hydroxide ion is not sufficient to counterbalance the decrease due to the loss of rotational and translational entropy.

Alkaline hydrolysis of esters which can ionize and follow eliminative mechanisms can be distinguished from the bimolecular type of process by entropy arguments. For example the 1-hydroxynaphthalene-2-sulphonate ester (Scheme 5.4) is hydrolysed in base and two mechanisms may operate; the kinetic ambiguity may be resolved by observing that the entropy of activation for the apparent second-order rate constant involving neutral ester and hydroxide ion is $+45\,\mathrm{J\,K^{-1}\,mol^{-1}}$. The positive entropy is not consistent with a bimolecular displacement (k_{OH} in the scheme) but fits the ionization of the phenolic group (involving hydroxide ion exchange for phenoxide ion and associated with little entropy change) and the eliminative step (associated with positive entropy) to give a sulphene-like intermediate which reacts in a fast step with water or hydroxide ion yielding product. The kinetically equivalent attack of water on the ionized phenolic ester should also give an overall negative $\Delta S^{\ddagger}$.

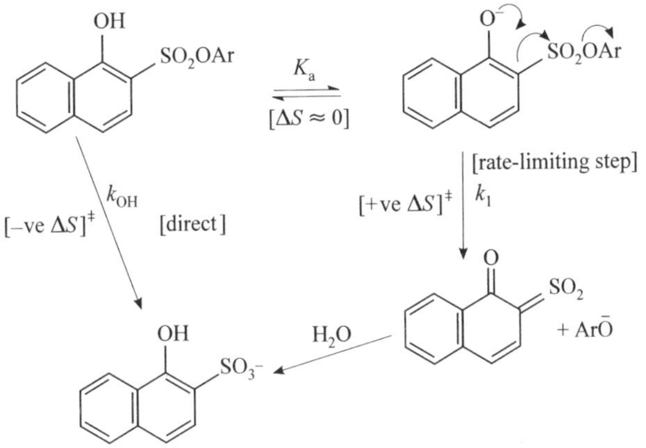

Scheme 5.4 Alkaline hydrolysis of aryl 1-hydroxynaphthalene-2-sulphonate esters.

The entropy changes observed for the acid hydrolysis of the esters quoted in Table 5.3 indicate the involvement of water in the transition state of the rate-limiting step. However, the proton-catalysed hydrolysis of *t*-butyl ester has a positive, or slightly less than zero, entropy of activation consistent with an A-1 mechanism.

The alkaline hydrolysis of formate esters is some hundred-fold faster than that of esters of other aliphatic carboxylic acids. The original explanation for this was thought to involve the effect of the greater electron-donating power of alkyl groups compared with that of hydrogen. The less negative entropy of activation observed for the hydrolysis of the formate ester was then thought to be due to the free rotation of the alkyl group being restricted in going from the trigonal reactant to the tetrahedral transition state (Schaleger & Long, 1963). A similar restriction is not felt for the hydrogen in formate esters (Scheme 5.5). However, this argument does not fit the observation that $S^0_{(\text{internal rotation})}$ of the $CH_3—C(sp^3)$ is similar to that for $CH_3—C(sp^2)$ (Page, 1973). The dominant effect is probably the difference in solvation on formation of the tetrahedral intermediate in the formate and acetate cases.

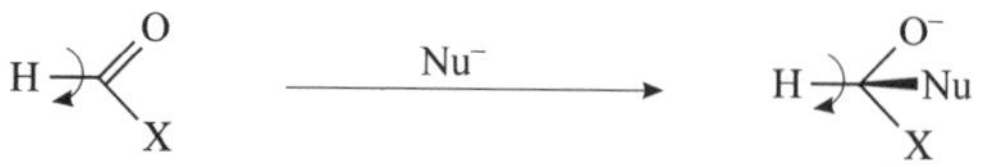

[rotation of the C—H bond is not hindered in reactant or product]

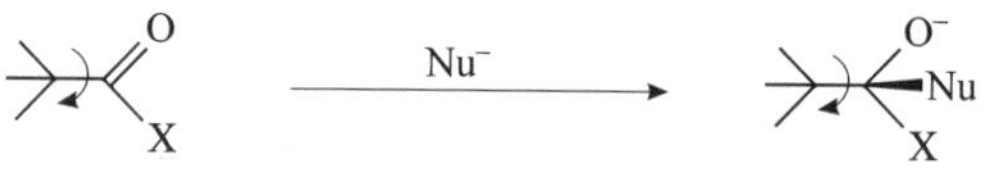

less hindered   more hindered

[rotation is apparently hindered in the product and transition state relative to reactant]

Scheme 5.5 Suggested entropic advantage in formate ester hydrolysis (but see text).

Electrostrictive effects on reactions are exemplified by the Menschutkin reaction for the nucleophilic attack of pyridines on alkyl halides, which typically has large decreases in entropy of activation ($167\,\mathrm{J\,K^{-1}\,mol^{-1}}$ for the reaction of pyridine with methyl iodide). These can be compared with the small decreases for anion attack on alkyl halides (ethoxide ion on methyl bromide has a decrease of $25\,\mathrm{J\,K^{-1}\,mol^{-1}}$) where a small increase in entropy due to release of electrostriction by delocalization is outweighed by the loss of degrees of freedom on bringing the reactants together in the transition state.

Further reading

Asano, T. & le Noble, W. J. (1978) Activation and reaction volumes in solution (I), *Chem. Rev.*, **78**, 407.

Bagno, A., Scorrano, G. & More O'Ferrall, R. A. (1987) Linear free energy relationships for acidic media, *Rev. Chem. Ind.*, **7**, 313.

Bowden, K. (1966) Acidity functions for strongly basic solutions, *Chem. Rev.*, **60**, 119.

Cox, R. A. & Yates, K. (1983) Acidity functions—an update, *Can. J. Chem.*, **61**, 2225.

Cox, R. A. (1987) Organic reactions in sulphuric acid: the excess acidity method, *Acc. Chem. Res.*, **20**, 27.

Gal, J. F. & Maria, P. C. (1990) Correlation analysis of acidity and basicity, from solution to gas phase, *Prog. Phys. Org. Chem.*, **17**, 159.

Hammett, L. P. (1970) *Physical Organic Chemistry*, 2nd edn, Academic Press, New York, Chapter 9, p. 263.

Hine, J. (1962) *Physical Organic Chemistry*, McGraw-Hill, New York.

Isaacs, N. S. (1981) *Liquid Phase High Pressure Chemistry*, Wiley, New York.

Isaacs, N. S. (1995) *Physical Organic Chemistry*, 2nd edn, Longman Scientific, Harlow.

Israelachvili, J. (1987) Solvation forces and liquid structures, as probed by direct force measurements, *Acc. Chem. Res.*, **20**, 415.

Kamlet, M. J., Abboud, J. M. & Taft, R. W. (1981) An examination of linear solvation energy relationships, *Prog. Phys. Org. Chem.*, **13**, 485.

Koskinen, A. M. P. & Klibanov, A. M. (eds) (1996) *Enzymatic Reactions in Organic Media*, Blackie, London.

Leffler, J. E. & Grunwald, E. (1963) *Rates and Equilibria of Organic Reactions*, Wiley, New York.

Long, F. A. (1960) Acid–base studies with deuterium oxide, *Ann. N.Y. Acad. Sci.*, **84**, 596.

Page, M. I. (1984) in *The Chemistry of Enzyme Action*, Page, M. I. (ed.), Amsterdam, Elsevier, pp. 1–54.

Reichardt, C. (1965) Empirical parameters of the polarity of solvents, *Angew. Chem. Int. Ed. (Engl.)*, **4**, 29.

Reichardt, C. (1979) *Solvent Effects in Organic Chemistry*, Verlag Chemie, Weinheim.

Reichardt, C. (1994) Solvatochromic dyes as solvent polarity indicators, *Chem. Rev.* **94**, 2319.

Rochester, C. H. (1970) *Acidity Functions*, Academic Press, New York.

Satchell, D. P. N. (1957) The use of the acidity function H_0 as a tool for the study of reaction mechanisms in mixed solvents, *J. Chem. Soc.*, 2878.

Schaleger, L. L. & Long, F. A. (1963) Entropies of activation and mechanism of reactions in solution, *Adv. Phys. Org. Chem.*, **1**, 1.

van Eldik, R. & Jonas, J. (eds) (1987) *High Pressure Chemistry and Biochemistry*, Reidel, Dordrect.

van Eldik, R., Asano, T. & le Noble, W. J. (1989) Activation and reaction volumes in solution (II), *Chem. Rev.*, **89**, 549.

Wescott, C. R. and Klibanov, A. M. (1993) Solvent variation inverts substrate specificity of an enzyme, *J. Am. Chem. Soc.*, **115**, 1629.

Whalley, E. (1964) Use of volumes of activation for determination of reaction mechanisms, *Adv. Phys. Org. Chem.*, **2**, 93.

6 Transition-state structures— anomalies

6.1 Introduction

Experimental science depends above all on the accuracy of its measurements but simple error is often a major problem; such an apparently mundane fact often sets a limit on techniques which purport to distinguish between two mechanisms. For example, in stereochemical analysis the observation of 'complete inversion' cannot exclude a percentage of racemization *less* than the error on the stereochemical measurement; the consequence is that it is impossible to distinguish between a concerted process and a stepwise process with a tight 'ion pair', using this convenient experimental tool. For similar reasons care must be taken when interpreting linearity in Brønsted relationships or the plots of solvent kinetic isotope effects against the fraction of D_2O. Probably one of the most serious problems associated with studies of mechanism is overconfidence in the power of one particular technique or theory.

Methods which are employed to study such an ephemeral structure as that of the transition state have been shown to give rise to a number of anomalies, apart from the limits imposed by experimental error, which affect the conclusions. These anomalies and difficulties are best gathered together; this chapter discusses the most important of them.

6.2 Reactivity–selectivity relationship

It is often accepted as a general principle (Buncel *et al.*, 1982) that an increase in reactivity is accompanied by a decrease in selectivity, because the transition-state structure is implicitly assumed to become closer to that of the reactant state as the energy barrier decreases. This idea has some truth for a hypothetical A-to-B reaction model but it often does not hold for real reactions. Rate constants are a measure of reactivity and relative rate constants are an index of selectivity as measured by the proportionality factor in linear free energy relationships such as Hammett's ρ or Brønsted's β. A corollary of the reactivity–selectivity relationship is therefore that free-energy correlations should be curved, with slope decreasing with increasing reactivity, even though the reaction in question is not suffering a change in

rate-determining step. Bearing in mind that the slope of a free-energy correlation is a function of transition-state structure, the following cases may be
observed: (a) a linear free-energy relationship where the transition-state
structure measured by the effect is constant; (b) non-linear free-energy
relationships with (i) a changing transition-state structure, (ii) a change in
rate-limiting step and (iii) a change in mechanism. The reactivity–selectivity
relationship refers to cases (a) and (b)(i).

The effect of variation in structure—for example changing the leaving
group in the series fluoride, chloride, bromide, iodide and azide ion—has
been used extensively in mechanistic studies; however, results should be
viewed only in a confirmatory sense (Lowry & Richardson, 1987) because
the essentially gross structural change could easily cause mechanistic
changes. The relative rates of methylation of 3- and 4-substituted pyridines
are remarkably insensitive to the reactivity of the alkylating agent. The
Brønsted β_{nuc} of about 0.3 is insensitive to the nature of the leaving group
or reaction conditions over a rate range of nearly 10^9 By contrast
2-substituted pyridines do appear to follow the reactivity–selectivity principle, presumably due to the importance of steric effects (Arnett & Reich,
1980). Changing the structure at the reaction site by small increments,
such as by variation of substituent in aryl esters or in pyridine nucleophiles,
should not cause significant changes in mechanism and the effects should
therefore be applicable to diagnosis.

The linearity of the majority of Hammett- and Brønsted-type plots
appears paradoxical because their constant slopes would indicate that the
transition-state structure is energy-independent; variation in structure might
be expected to occur due to the change in energy brought about by the
change in substituent and thus to yield curved relationships. The relationship between reactivity and selectivity (Pross, 1977) is based on such a
simple model (Leffler, 1953, Chapter 3) that most reactions do not fit in
with expectation because they often involve complex bonding changes and
the contribution of solvation changes to the free energy of activation is often
dominant.

Hammond's postulate is that if two states occur consecutively in a reaction and have nearly the same energy content their interconversions will
involve only a small reorganization of the molecular structure. In its simplest interpretation this means that the structures of these two states are
almost identical. The applicability of the Hammond postulate (Hammond,
1955) to transition-state structure can be rationalized by a model of intersecting potential energy curves employed by Bell, Evans and Polanyi (Fig.
6.1). Making a reaction step more exothermic, for example, by making the
product more stable, will shift the position of the intersection towards reactant and make the transition-state structure more 'reactant-like' (Fig. 6.1).
The Hammond postulate relates the *structures* of reactive species to their
corresponding *potential energy*. Kinetic measurements and their application
to elucidating structure using linear free-energy relationships reflect Gibbs'
free-energy changes (see Chapter 1). The reason why entropy differences in

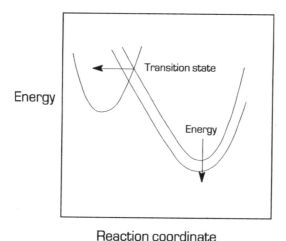

Reaction coordinate

Fig. 6.1 Potential energy diagram illustrating how changing energies of reactant and product can change transition-state structure.

these processes can often be neglected is that the entropy changes cancel in the measured and reference reaction, or they are proportional to the enthalpy changes. Changes in transition-state structure are often deduced from changes in the 'selectivity parameter' of linear free-energy relationships—the Hammett ρ- or Brønsted β-values. The application of the reactivity–selectivity principle would then suggest that the selectivity coefficients should vary according to the Hammond postulate, i.e. they should *decrease* as reactivity *increases* within a reaction series. This is manifestly untrue for many reactions studied over a wide range of reactivity. However, there is curvature in rate–equilibrium relationships which is not attributable to a change in rate-limiting step (case b(ii)) or to a change in mechanism (case b(iii)) (see p.117). For example, a 10^4-fold *increase* in the rate of E2 elimination of HBr from alkyl bromides, brought about by a more electron-withdrawing group at C2, is accompanied by an *increase* in the sensitivity to the strength of the catalysing base—the Brønsted β-value increases from 0.39 to 0.67 (Hudson & Klopman, 1964).

Many reactions occur by several bonding changes, and hence it is not surprising that the transition-state structure is dependent on several variables. For example, 1,2-elimination (Eqn [6.1]) involves C—H bond fission and proton transfer to a base, C—Lg bond fission (where Lg is the leaving group) and C=C bond formation. These steps can occur separately in a stepwise mechanism by the formation of a carbenium ion intermediate (E1) or a carbanion (E1cb) or in a concerted process (E2). These processes may be illustrated on a Jencks–More O'Ferrall diagram (Fig. 6.2) which draws the two bond fission steps orthogonal to each other so that reactants are placed in the bottom left-hand corner and the products in the top right-hand corner, whereas the putative reactive intermediates are placed in the remaining opposite corners (Fig. 6.2). As described in Chapter 3, the degree of proton transfer to the base may be deduced from the Brønsted β-value obtained by varying the strength of the base; the degree of carbenium ion or carbanion character may be elucidated by Hammett ρ-values from substituents at C1 and C2, whereas the degree of bond fission to the leaving group could be determined from a Brønsted β_{lg}-value obtained by changing the nature of the nucleofuge. Provided that the selectivity coefficients are normalized, changes in transition-state structure are often predictable by the Hammond postulate for the two stepwise processes E1 and E1cb, along the edges of the diagram. For the *concerted*, E2, mechanism, however, changes in transition state structure and selectivity parameters are more difficult to predict.

$$B + \underset{[2]}{\overset{H}{\underset{|}{-C}}}\underset{[1]}{\overset{|}{-C}}-Lg \longrightarrow BH^{+} + Lg^{-} + \overset{\diagdown}{\underset{\diagup}{C}}=\overset{\diagup}{\underset{\diagdown}{C}} \qquad [6.1]$$

The Hammond postulate predicts that stabilization of the reactants or products moves the transition-state structure *away* from the corner which is becoming more stable—a so-called 'parallel to the reaction coordinate' effect. However, stabilization of the putative intermediate corners (whether the intermediate is actually formed or not) moves the transition-state structure *towards* the corner becoming more stable. The direction of the movement is perpendicular to the reaction coordinate and is called a 'perpendicular' effect (or sometimes a 'Thornton' (1967) or an 'anti-

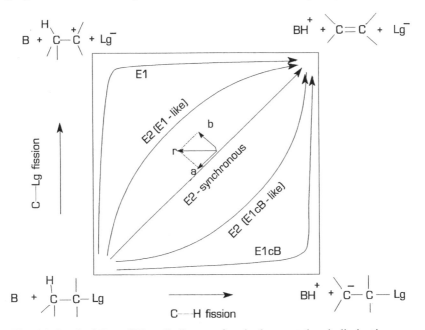

Fig. 6.2 Jencks–More O'Ferrall diagram for the base-catalysed elimination reaction (Eqn [6.1]).

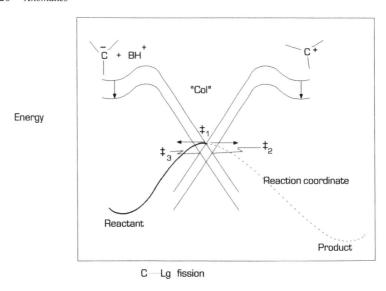

C—Lg fission

Fig. 6.3 Representation of the 'col' in the elimination reaction (Eqn [6.1]). The intersections $\ddagger_2$ and $\ddagger_3$ represent the transition states when the carbenium ion and carbanion respectively become more stable.

Hammond' effect). For example, structural changes in the reactant which would be expected to stabilize an intermediate carbenium ion, were it formed, or stabilize the leaving group (Lg) will make the transition-state structure of the E2 concerted process more E1-like, i.e. with more C—Lg bond fission and less C—H bond fission.

When structural changes are predicted to cause both parallel (vector **a**) and perpendicular (vector **b**) effects, the change in transition-state structure is the resultant (**r**) of the two vectors (shown in Fig. 6.2). The two opposing effects result simply from the difference in position of overlap of parabolic minima and maxima; Fig. 6.1 can be contrasted with Fig. 6.3 where energy is plotted against C—Lg distance. The transition-state structure $\ddagger_1$ is moved towards more C—Lg bond fission, $\ddagger_2$, by making the carbenium ion/leaving group more stable. The transition-state structure $\ddagger_3$ has less C—Lg bond fission when structural changes make the carbanion/conjugate acid (BH^+) more stable, i.e. the transition state becomes more E1cb-like.

Stabilizing Lg^- would tend to make the transition-state structure for the almost synchronous E2 process have less C—Lg bond fission by the parallel effect (vector **a**) but more by the perpendicular effect (vector **b**) (Fig. 6.2). The net result (**r**) could be no change in the observed β_{lg} despite a decrease in the degree of proton transfer and of the Brønsted β-value for the base-catalysed elimination.

If the transition-state structure with which comparisons are being made is imbalanced with, say, an 'early' E1-like mechanism, then the **a** and **b** vectors may be unequal with a large parallel effect dominating β_{lg} so that there will be a net decrease in this parameter whereas the decrease in β for the base is smaller.

Prediction of the transition-state structure changes described above are embodied in Thornton's 'rules' (1967) which resolve the movement into components along and perpendicular to the reaction coordinate.

6.3 Microscopic medium effects

Free-energy relationships sometimes exhibit scatter-type deviations from linearity which are outside experimental error. These may be attributable to microscopic medium effects caused by a difference between the standard equilibrium and the reaction being studied. In the reaction of substituted pyridines with an alkyl function (Eqn [6.2]) the reactivity is compared with the equilibrium constant for ionization of pyridinium molecules under the same conditions (Arnett & Reich, 1980).

[6.2]

The cause of the scatter effect is probably the non-systematic influence of the substituent on the microscopic environment of the transition state. The linear free-energy relationship between product state xpyH$^+$ (Eqn [6.2]) and the transition state will be modulated by second-order non-systematic variation because the microscopic environment of the reaction centre in the standard reaction (xpyH$^+$) will differ slightly from that (xpyCH$_3^+$) in the reaction under investigation, and this will give rise to small specific substituent effects.

The microscopic medium effect should be reduced if the model reaction chosen as standard resembles the reaction in question more closely than the ionization (Fig. 6.4). Figure 6.4 indicates that free-energy relationships with small numbers of points should be treated with suitable caution if useful conclusions are to be made; it should be noted that the example given has more data points than most! Confidence in the parameters such as slope and curvature will increase with the number of points in the correlation; such consideration becomes crucial if small variations in β or ρ or curvatures in plots are under investigation.

Free-energy relationships for proton transfer reactions often exhibit little scatter because the proton being transferred at the transition state has a microscopic environment similar to that of the completely transferred

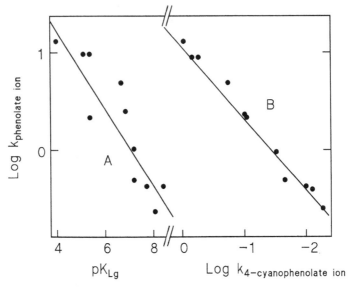

Fig. 6.4 Reduction of the scatter in a Brønsted-type plot by choice of the appropriate standard reaction (B). The graphs are drawn from data in Ba-Saif *et al.* (1991) for phenoxide ion attack on substituted phenyl acetate esters ($PhO^- + CH_3CO_2Ar \longrightarrow CH_3CO_2Ph + ArO^-$). A, plot of the second-order rate constant (units are $M^{-1} sec^{-1}$ at 25 °C) against the pK_a of the leaving phenoxide ion; B, plot against a standard reaction having a similar transition state, namely 4-cyanophenoxide ion attack on the same esters.

proton used in the standard ionization equilibrium (Scheme 6.1). There are exceptions, however, of which the transfer of a proton to hydroxide ion is notable. The microscopic environment of the variant portions (boxed in Scheme 6.1) is approximately the same for both transition state and product state (HB^+). In addition to the influence of microscopic medium effects and experimental error, scatter is also caused by a statistical effect where a multiplicity of reaction centres is involved (Bell, 1959, p. 159). The reactivity of a base or nucleophile possessing q identical sites (which could be protonated) compared with the ionization of its conjugate acid possessing p identical sites (from which equivalent protons could be 'lost') requires a statistical correction to the simple Brønsted-type relationship for proton transfer to a variant base or nucleophilic reaction, as given in Eqn [6.3].

$$\log(k_{base}/p) = \beta pK_a - \beta \log(q/p) + C' \tag{6.3}$$

$$\log(k_{acid}/q) = \alpha \log(p/q) + \alpha pK_a + C \tag{6.4}$$

A reaction where a general acid donates a proton in the rate-limiting step requires the corrected relationship of Eqn [6.4]. For example, triethylamine

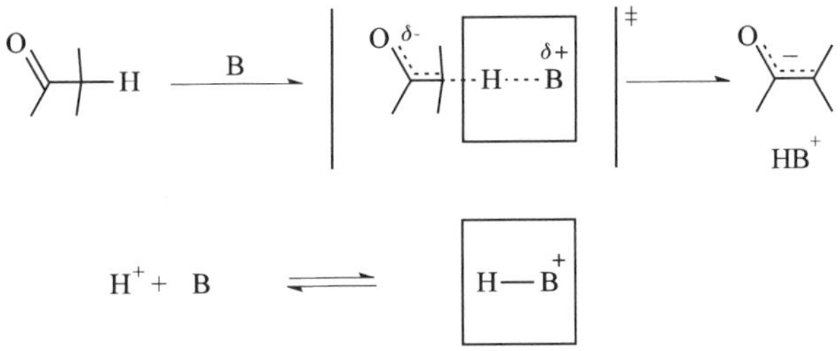

Scheme 6.1 Partial cancellation of microscopic medium effects in proton transfer reactions.

and diaminoethane dication need corrections because the former has $p = 1$ and $q = 1$ whereas the latter has $p = 6$ and $q = 1$. The treatment of the results is not yet satisfactory for acid–base pairs with uncertain structures (such as H_3O^+ and HO^-) or species with multiple centres of differing reactivity; hydronium and hydroxide ions almost always show anomalous reactivities in corrected Brønsted correlations, which are partly due to solvation effects. The most reliable Brønsted-type correlations come from those reactions in which the reagents have a common structure (such as all pyridines or all phenoxide ions) so that p and q are constant throughout.

Resonance effects may also give rise to differences, and alcohols and phenols sometimes show separate correlations even though their pK_a values may overlap. The second-order rate constant for the hydroxide ion-catalysed hydrolysis of phenyl acetate is fourfold less than that for trifluoroethyl acetate, even though the latter has a more strongly basic and, hence expected, poorer leaving group (by about 2.4 pK_a units; Barton *et al.*, 1994).

6.4 Curvature in free-energy relationships

Curvature is most reliably detected for proton transfer because the microscopic medium effects are minimal for these reactions. Proton transfer between electronegative atoms A and B usually involves diffusion steps (k_1 and k_3) and an ionization step (k_2) within the encounter complex (Eqn [6.5]). The energy diagram for this reaction is illustrated in Fig. 6.5.

$$AH + B \underset{k_{-1}}{\overset{k_1}{\rightleftharpoons}} [AH \cdots B \underset{k_{-2}}{\overset{k_2}{\rightleftharpoons}} A^- \cdots BH^+] \underset{k_{-3}}{\overset{k_3}{\rightleftharpoons}} A^- + BH^+ \qquad [6.5]$$

Equation (6.5) gives rise to the rate laws (Eqn [6.6]) for forward (k_f) and reverse (k_r) reactions:

$$k_f = k_1 k_2 k_3 / N \quad \text{and} \quad k_r = k_{-1} k_{-2} k_{-3} / N \qquad [6.6]$$

where $N = k_{-1}(k_{-2} + k_3) + k_2 k_3$. When the rate constants for proton transfer within the encounter complex are larger than that of complex formation or decomposition (i.e. $k_2 > k_{-1}$ or $k_{-2} > k_3$), Eqn [6.6] reduces to Eqn [6.7], where $K_2 = k_{-2}/k_2$; a Brønsted plot is predicted (Fig. 6.6), consisting of two intersecting straight lines and a sharp break-point at $K_2 = 1$ (the 'Eigen' line). Simple application of the rules given in Chapter 2 indicates that this is consistent with a change in the rate-limiting step at a ΔpK of HB^+ equal to that of HA ($\Delta pK = 0$). The plot is exemplified by the proton transfer between heteroatom acids and bases as given in the example in Fig. 6.6 (Eigen, 1964).

$$k_f = k_1/(1 + K_2 k_{-1}/k_3) \quad \text{and} \quad k_r = k_{-3}/(1 + k_3/k_{-1}K_2) \tag{6.7}$$

The rate-limiting step in proton transfer between electronegative atoms is either k_1, the diffusion-controlled encounter of the acid–base pair, for thermodynamically favourable transfers or k_3, the dissociation of the acid–base pair, for thermodynamically unfavourable transfers. The rate-limiting step is rarely the proton transfer step itself. Eigen plots are often observed in reactions in which proton transfer occurs to or from reactive intermediates. For example, the aminolysis of penicillin is general base-catalysed and the Brønsted plot for the hydrazinolysis reaction (Scheme 6.2), catalysed by a series of bases of varying pK_a, shows a sharp break at around pK_a 8.5. This is consistent with a stepwise process of proton transfer from the tetrahedral intermediate, $T^{\pm}$, which has a calculated pK_a of 8.5 (Morris & Page, 1980a; see also Appendix A.3). However, proton transfers involving carbon do not usually exhibit sharp breaks in the Brønsted plot in the region of $\Delta pK = 0$.

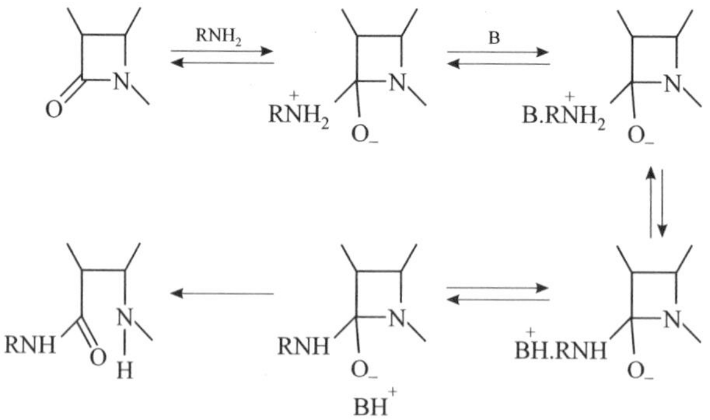

Scheme 6.2 Aminolysis of the β-lactam of penicillins.

The deprotonation of acetylacetone is a typical example where the reaction may be followed easily by observing the uptake of iodine which occurs in a fast step following the rate-controlling ionization (Eqn [6.8]).

$$(CH_3CO)_2CH_2 \underset{\text{(rate-limiting step)}}{\overset{B}{\rightleftharpoons}} (CH_3CO)_2CH^- \overset{I_2 \text{ (fast step)}}{\rightleftharpoons} (CH_3CO)_2CHI \tag{6.8}$$

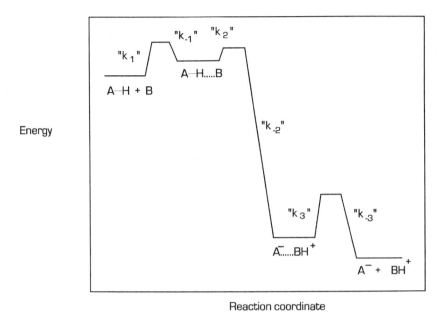

Fig. 6.5 Energy diagram for the proton transfer reaction (Eqn [6.5]). The energy barriers are identified by their corresponding rate constants.

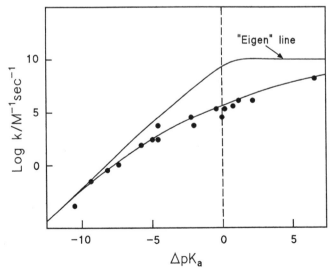

Fig. 6.6 Transfer of proton between acid donor and base acceptor: upper curve, 'Eigen' line for diffusion-limited reactions; lower curve, proton transfer from acetylacetone. The figure is compiled from data in Eigen (1964). In this example the acid (HA) is kept constant and the base (B) is varied.

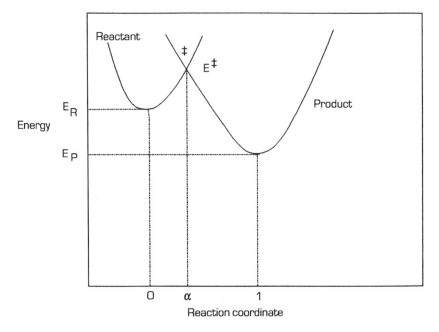

Fig. 6.7 Illustration of the derivation of the Marcus equation from intersecting parabolic curves.

The free-energy relationship (Fig. 6.6) of log k_f versus the pK_a of the conjugate acid of the base (B) does *not* have a sharp break between the linear segments, although the mechanism follows Eqn [6.5]. The reason for this is that the k_{-2} and k_2 steps within the encounter complex are slow compared with diffusion (the k_1 and k_3 steps) and the 'second-order' curvature results from the reaction occurring within the encounter complex.

Curvature may also arise as described in the following argument, from non-linearity of potential energy surfaces in the region where they intersect (Fig. 6.7). A reasonable assumption (over a small range of pK_a variation) is that the two intersecting curves are parabolic (the reactant parabola is Eqn [6.9a] and the product parabola is Eqn [6.9b]; α is Leffler's parameter) and the Marcus (1975) equation [6.12] may then be obtained.

$$E^{\ddagger} = a \cdot \alpha^2 + E_R \tag{6.9a}$$

$$E^{\ddagger} = b(1 - \alpha^2) + E_P \tag{6.9b}$$

In the symmetrical reaction ($E_R = E_P$) Eqns [6.9a] and [6.9b] give $\alpha = 0.5$ and $a = b = 4\Delta E_0^{\ddagger}$. These values may be substituted to give Eqn [6.10], which may be transformed to Gibbs' free energy to give Eqn [6.11], where $\Delta G_0^{\ddagger}(\Delta G_0^{\ddagger} \geq \Delta G^{\circ})$ is the intrinsic barrier for the ergoneutral reaction (at $\Delta G^{\circ} = 0$).

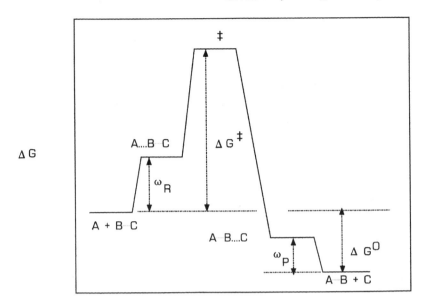

Reaction coordinate

Fig. 6.8 The origin of the work terms, ω, in elementary processes.

$$\alpha = 0.5 + (E_R - E_P)/8\Delta E_0^{\ddagger} \qquad\qquad [6.10]$$

$$\alpha = \mathrm{d}\Delta G^{\ddagger}/\mathrm{d}\Delta G^{\circ} = 0.5 + \Delta G^{\circ}/8\Delta G_0^{\ddagger} \qquad\qquad [6.11]$$

Equation [6.11] indicates that the slope of a free-energy relationship, as given by the Leffler parameter, should vary.

The component processes in the simple reaction $A + B\!-\!C \rightleftharpoons A\!-\!B + C$ can be described by so-called 'work terms' (ω) which represent the energies required to bring the reactants together and to separate the products and may be represented graphically as in Fig. 6.8. Integration of Eqn [6.11] gives Eqn [6.12] and inclusion of work terms yields Eqn [6.13], from which the parameters may be obtained (see below) for $\Delta G_0^{\ddagger}$ and ω^R values (Table 6.1).

$$\Delta G^{\ddagger} = \Delta G_0^{\ddagger} + \Delta G^{\circ}/2 + (\Delta G^{\circ})^2/16\Delta G_0^{\ddagger} \qquad\qquad [6.12]$$

$$\Delta G^{\ddagger} = (\Delta G_0^{\ddagger} + \omega_R) + (\Delta G^{\circ} + \omega_R - \omega_P)/2 + (\Delta G^{\circ} + \omega_R - \omega_P)^2/16\Delta G_0^{\ddagger} \qquad [6.13]$$

If the reaction is made thermodynamically unfavourable ($\Delta G^{\circ} > 0$) then α becomes more than 0.5 and the transition-state structure will become more product-like. The transition-state structure, as indicated by α, will change with the free energy of the reaction according to Eqn [6.14], which is obtained by differentiating Eqn [6.11].

$$\mathrm{d}\alpha/\mathrm{d}\Delta G^{\circ} = 1/8\Delta G_0^{\ddagger} \qquad\qquad [6.14]$$

Table 6.1 Intrinsic barriers ($\Delta G_0^{\ddagger}$) and work terms (ω^R) for some proton transfer reactions.[a]

Reaction[b]	$\Delta G_0^{\ddagger}$ (kJ mol^{-1})	ω^R (kJ mol^{-1})
Protonation of aromatic nuclei in acid	42	42
$CH_3CH(OH)_2 + AH \rightleftharpoons CH_3CHO + H_2O$	21	54
$R_2CHCOR + B \rightleftharpoons R_2CCR(O)^- + BH^+$	33	25
$RCHN_2 + AH + H_2O \rightleftharpoons RCH_2OH + N_2 + AH$	4–21	33–60
$B + AH \rightleftharpoons BH + A$	8.2	13

[a] Data from Kresge, 1973, 1975a.
[b] Base (B) and acid (AH) are 'normal' nitrogen or oxygen species.

Equation [6.14] is a quantitative expression of the Hammond postulate. Substitution into Eqn [6.11] indicates that $\alpha = 0.5$ for identity reactions ($\Delta G^\circ = 0$).

It is unlikely that the intrinsic kinetic barrier is constant (Lewis & Hu, 1984); the variation of α with ΔG° becomes Eqn [6.15] and α is no longer restricted to 0.5 for the identity reactions.

$$\alpha = d(\Delta G^{\ddagger})/d(\Delta G^\circ) = 0.5(1 + \Delta G^\circ/8\Delta G_0^{\ddagger})$$
$$+ (1 - (\Delta G^\circ)^2/16(\Delta G_0^{\ddagger})^2)\{d(\Delta G_0^{\ddagger})/d(\Delta G^\circ)\} \qquad [6.15]$$

Equation [6.16] is a quadratic form of Eqn [6.15] consistent with curved rate equilibrium relationships. Deduction of the parameters a, b and c from Eqn [6.16] allows the work terms ω_R, ω_P and the intrinsic kinetic barrier, $\Delta G_0^{\ddagger}$ to be calculated. The interpretation of the work terms (Table 6.1) is not obvious when they are larger than that required for the simple encounter reaction. The usual assumption is that they reflect the energy required for the formation of the correct geometry, and for solvation effects.

$$\log k = a + b \log K + c(\log K)^2 \qquad [6.16]$$

6.5   Brønsted anomalies

The equilibrium constant for a proton transfer reaction from a series of proton donors (AH) to base (B) (Eqn [6.17]) has a β-value of unity for the Brønsted dependence on the pK_a of the conjugate acids (AH). This is simply because the equilibrium constant $[HB^+][A^-]/[AH][B]$ is equal to K_{AH}/K_{HB}, where K_{AH} is the ionization constant of the variant proton donor and K_{HB} is the ionization constant of the conjugate acid of the base (HB).

$$A-H + B \underset{\beta_r}{\overset{\beta_f}{\rightleftharpoons}} A^- + H-B^+ \qquad [6.17]$$

$$\beta_f - \beta_r = \beta_{eq} = 1 \qquad [6.18]$$

It follows from Eqn [6.18] that the *numerical* value of β_r or β_f would not be expected to exceed unity. In the proton transfer reaction between nitro-alkane donors and hydroxide ion, the value of β_f (1.54) far exceeds unity when the acidity of the nitroalkane is varied (Eqn [6.19]) (Bordwell *et al.*, 1969; Bordwell & Boyle, 1971). This astounding result naturally caused considerable discussion in the early 1970s and its study provided substantial insight into the mechanism of proton transfer reactions.

$$ArCH_2-NO_2 + OH^- \underset{(\beta_{eq}=1)}{\overset{\beta_f=1.54}{\rightleftharpoons}} ArCH = N^+ \begin{smallmatrix} O^- \\ \\ O^- \end{smallmatrix} + H_2O \tag{6.19}$$

The value of β_f compared with that of β_{eq} indicates that the transition-state structure has a negative charge on carbon *greater* than that in the product, which is not in accord with a smooth transition of structure from reactant to product. Reactions involve a number of molecular processes which include bond formation, bond fission, solvent reorganization, formation and fission of hydrogen bonds associated with the reaction centre and changes in delocalization and hybridization. When these processes make unequal progress at the transition state, the reaction is considered to be imbalanced (see Chapter 3; Jencks & Jencks, 1977; Jencks, 1985) and anomalies occur such as that observed by Bordwell.

For example, the deprotonation of nitroalkanes could be separated into carbanion formation (proton removal) and geometrical rearrangement from tetrahedral to trigonal carbon to allow delocalization (Scheme 6.3). If delocalization lagged behind proton transfer, the carbon would bear greater negative charge in the transition state than in the product, giving rise to $\beta_f > \beta_{eq}$.

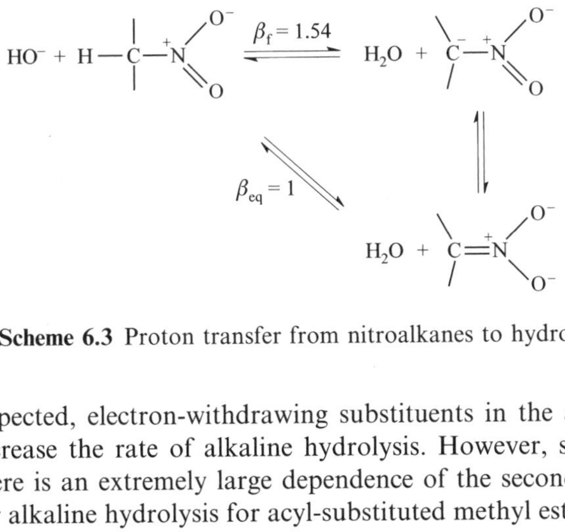

Scheme 6.3 Proton transfer from nitroalkanes to hydroxide ion.

As expected, electron-withdrawing substituents in the acyl group of an ester increase the rate of alkaline hydrolysis. However, slightly unexpectedly, there is an extremely large dependence of the second-order rate constant for alkaline hydrolysis for acyl-substituted methyl esters on the pK_a of the product carboxylic acid, generating a Brønsted β_{lg} of -1.3 (Barton *et al.*, 1994). The transition-state structure in ester hydrolysis is effectively more negatively charged compared with the neutral ester than is the carboxylate anion compared with its undissociated acid. Presumably the ionized tetrahedral intermediate (1) has more localized negative charge than the corresponding carboxylate anion (2), which gives its stability a greater dependence upon substituents.

1   2

The study of substituent effects on bond formation and bond fission in non-proton transfer reactions requires knowledge of the value of β_{eq} which is unity only for proton transfer reactions. It is sometimes found that β_{lg} exceeds β_{eq} in non-proton transfer reactions and it is often assumed that the effect is due to bond fission being in advance of solvent reorganization. If such is the case in, for example, departure of aryl oxide ions, it is conceivable that the lack of solvation causes the charge on the oxygen to be greater than in its solvated product state (Cox & Gibson, 1975; Bell & Sorensen, 1976; Arora *et al.*, 1979; Jencks & Gilbert, 1979; Thea *et al.*, 1979; Jencks *et al.*, 1982).

6.6   Do experimental parameters directly measure transition-state structures?

The polar substituent effect has come under criticism as a structure-determining tool. It relates to changes in charge or dipole in the system which derive from the difference in electronic structure between ground and transition state. A perspicacious study (Pross & Shaik, 1989) draws together many aspects of the polar substituent effect:

(a) Each rate process has a reference equilibrium process.
(b) The polar effect (slope of a linear free-energy relationship) represents a single transition state.
(c) The relationship between charge development and bond order is not necessarily linear.
(d) Solvation changes, including hydrogen bonding, may not be in step with bonding changes.
(e) Bonding change is not necessarily a single process—indeed, there are often two or more bonds undergoing change in any reaction.
(f) Each bonding change involved in the reaction may not be synchronized.

The above factors indicate that great care must be taken in interpreting polar substituent effects in terms of simple bond order. The shorthand

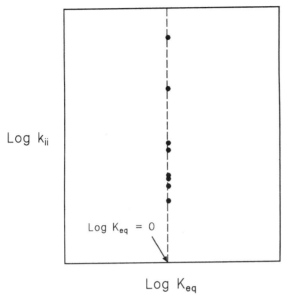

Fig. 6.9 Free-energy relationship with equilibrium constant for the identity rate constants (k_{ii}) for the attack of substituted phenoxide ions on substituted phenyl acetates. The figure is compiled from data in Ba-Saif *et al.* (1991).

notation used in illustrating reaction mechanisms, which we described in Chapter 1, neglects most of the above factors and bonding is tacitly thought of as the most important energy process in a reaction. A deliberately naïve application of bonding and polar substituent effects was used by Pross (1983) to give a dramatically wrong picture of mechanism. We emulate this demonstration by the rate–equilibrium plot of the identity exchange reaction between substituted phenoxide ions and substituted phenyl acetates (Fig. 6.9), where the infinite slope would appear to indicate infinite build-up of charge in the transition state! The microscopic reverse of an identity reaction is the same as the forward process. The explanation of the paradox is that the equilibrium constant is invariant for an identity reaction ($K = 1$), whereas the intrinsic rate constants are variant. No conclusions may be drawn from this type of free-energy relationship except that a more informative analysis should be sought.

A further example illustrates that changing the solvent can have enormous energy effects: the gas-phase nucleophilic displacement at a saturated carbon centre has no energy of activation, whereas in solution there is a substantial barrier due to the effect of the solvent (Fig. 6.10) (Pellerite & Brauman, 1980; Caldwell *et al.*, 1984; Chandrasekhar & Jorgensen, 1985).

Considerable effort has been expended in discussion of polar substituent effects as direct measures of bonding in the transition state; whereas these effects indicate the difference in electronic structure between ground and

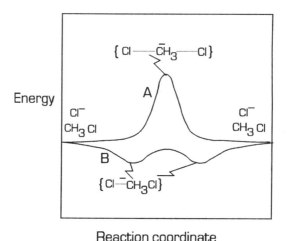

Fig. 6.10 Schematic potential energy profiles for the identity reaction of Cl^- with CH_3Cl in the gas phase (B) and in the aqueous phase (A). The figure is compiled from data in Chandrasekhar & Jorgensen (1985).

transition state, they do not indicate the bond order of individual bonds. Other techniques have not been subjected to such rigorous investigation, probably because polar substituent effects themselves are experimentally the most accessible technique available to all laboratories and do not require expensive equipment or specialist labour. Nevertheless, the same restrictions apply to the majority of experimental techniques for investigating mechanism. Because there is no other experimental access to the energy of the transition state, the techniques depend on knowledge of rate constants; they refer to *states* and thus are subject (to a greater or lesser extent) to the factors discussed previously.

Further reading

Bell, R. P. (1959) *The Proton in Chemistry*, Methuen, London.

Eigen, M. (1964) Proton transfer, acid base catalysis and enzymatic hydrolysis, *Angew. Chem., Int. Ed,. (Engl.)*, **3**, 1.

Farcasiu, D. (1975) The use and misuse of the Hammond postulate, *J. Chem. Ed.*, **52**, 76.

Giese, B. (1977) The basis and limitations of the reactivity–selectivity principle, *Angew. Chem., Int. Ed,. (Engl.)*, **16**, 125.

Hammond, G. S. (1955) A correlation of reaction rates, *J. Am. Chem. Soc.*, **77**, 334.

Jencks, D. A. & Jencks, W. P. (1977) On the characterization of transition states by structure–reactivity coefficients, *J. Am. Chem. Soc.*, **99**, 7948.

Jencks, W. P. (1985) A primer for the Bema Hapothle. An empirical approach to the characterisation of changing transition states, *Chem. Rev.*, **85**, 511.

Jencks, W. P., Brant, S. R., Gandler, J. R., Fendrich, G. & Nakamura, C. (1982) Non-linear Brønsted correlations. The roles of resonance, solvation, and changes in transition state structure, *J. Am. Chem. Soc.*, **104**, 7045.

Johnson, C. D. (1980) The reactivity–selectivity principle: fact or fiction, *Tetrahedron Rev.*, **36**, 3061.

Johnson, C. D. & Schofield, K. (1973) A criticism of the use of the Hammett equation in structure–reactivity correlations, *J. Am. Chem. Soc.*, **95**, 270.

Kamlet, M. & Taft, R. W. (1985) Linear solvation free energy relationships. Local empirical rules or fundamental laws of chemistry? A reply to the chemometricians, *Acta Chem. Scand. Ser. B*, **39**, 611.

Klots, C. E. (1988) The reaction coordinate and its limitations: an experimental perspective, *Acc. Chem. Res.*, **21**, 16.

Kresge, A. J. (1973) The Brønsted relationship—recent developments, *Chem. Soc. Rev.*, **2**, 475.

Kresge, A. J. (1975a) What makes proton transfer fast? *Acc. Chem. Res.*, **8**, 354.

Kresge, A. J. (1975b) The Brønsted relationship: significance of the exponent, in *Proton Transfer Reactions*, Caldin, E. F. & Gold, V. (eds), Chapman and Hall, London, p. 179.

Lowry, T. H. & Richardson, K. S. (1987) *Mechanism and Theory in Organic Chemistry*, 3rd edn, Harper and Row, New York, p. 596.

McLennan, D. J. (1978) Hammett ρ values—are they an index of transition state character? *Tetrahedron Rev.*, **34**, 2331.

Pross, A. (1977) The reactivity–selectivity principle and its mechanistic applications, *Adv. Phys. Org. Chem.*, **14**, 69.

Pross, A. (1983) On the breakdown of rate equilibrium relationships, *Tetrahedron Lett.*, **24**, 835.

Pross, A. & Shaik, S. S. (1989) Brønsted coefficients. Do they measure transition state structure? *New J. Chem,* **13**, 427.

Sjostrom, M. & Wold, S. (1981) Linear free energy relationships. Local empirical rules or fundamental laws of chemistry? *Acta Chem. Scand. Ser. B*, **35**, 537.

Wold, S. & Sjostrom, M. (1986) Linear free energy relationships. Local empirical rules or fundamental laws of chemistry. A reply to Kamlet and Taft, *Acta Chem. Scand. Ser. B*, **40**, 270.

7 Bio-organic group transfer reactions

7.1 Relative timing of bond formation and bond fission

Introduction

As reactions very rarely involve fewer than two major bond changes, the question of their relative timing in mechanisms is very important. This is especially so in group transfer reactions, which probably constitute the majority of bio-organic reactions. A concerted mechanism is a process where all bond changes occur in a single step without the intervention of an intermediate (see Chapter 1).

The classical concerted reaction is nucleophilic aliphatic substitution—the S_N2 reaction—where bond formation and bond fission steps are in unison. Other well-known examples of concerted reactions include pericyclic reactions and base-catalysed eliminations (Scheme 7.1).

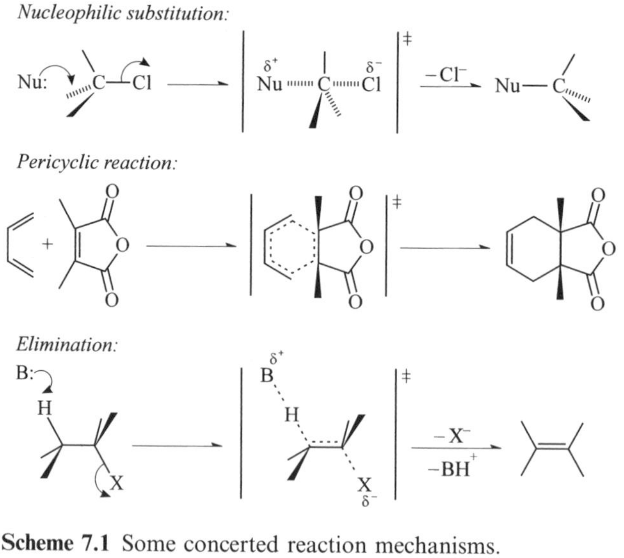

Scheme 7.1 Some concerted reaction mechanisms.

A concerted mechanism can assist the progress of a reaction because energy which is being released in bond *formation* can be directly employed to compensate for unfavourable bond *fission*, provided there is coupling between the changes. A stepwise process can lead to the formation of unstable intermediates such as relatively high-energy radicals or ions, and the concerted mechanism is an obvious route whereby these energetically unfavourable intermediates can be avoided. The concerted process, however, involves more nuclear motion in a particular step than does a stepwise mechanism, and the principle of least nuclear motion (Hine, 1966, 1977; Sinnott, 1988) is a factor which can militate against its occurrence.

There is an element of fashion in the acceptance of the concerted process in the classical cases, which has depended on its logic and appeal coupled with the attractiveness of using series of curved arrows to simulate electron flow (Bordwell, 1970). Improved mechanistic tools and knowledge (Sneen, 1973) led to significant understanding of alkyl group transfer, and a number of aliphatic nucleophilic substitution reactions originally thought to be concerted are now known to be stepwise. Conversely, there are now cases where concertedness has been shown to operate in reactions hitherto thought to be stepwise (acyl group transfer). Investigations of concertedness have been subject to considerable scrutiny and critique (Bordwell, 1970; Williams, 1989).

An intermediate structure is not a real entity in the limit where its decomposition has no energy barrier and it is therefore not in equilibrium with reactants and products. The question naturally arises as to 'when a barrier is not a barrier'. Irrespective of the availability of experimental techniques to detect a very unstable intermediate, an entity is not considered as an intermediate if its lifetime is less than h/kT, i.e. *ca* 1.7×10^{-13} s at 25 °C. The stepwise paths in Scheme 7.2 would become concerted when the intermediates CAB and A became too reactive to exist for any longer than this time.

$$A - B \xrightarrow{+C} [C - A - B] \xrightarrow{-B} A - C$$
$$A - B \xrightarrow{-B} [A] \xrightarrow{+C} A - C$$

Scheme 7.2 Mechanisms with intermediates.

This special type of process is called an 'enforced concerted' mechanism (Jencks, 1980) because a concerted mechanism is forced to ensue as the intermediate becomes progressively more reactive. The transition-state structure would correspond to that for the formation of the putative intermediate which approximates the structures represented by the top left or bottom right corners of the reaction map (Fig. 3.5 or Fig. 3.6).

Demonstration of a concerted mechanism

Concertedness is about the *number* of transition states along a reaction path as well as the nature of the structure of a transition state. In order to demonstrate that a reaction has a concerted mechanism, it is necessary to employ a tool that counts the number of transition states; proof that the stepwise pathways are *not* operating would be good evidence for concerted-ness in the simplest model, such as one of those in Scheme 7.2.

At present there are two major techniques available to disprove a stepwise path; one of these is to do with free-energy relationships whereby a stepwise route can give rise to a non-linear plot caused by a change in rate-limiting step (see Section 2.10 and Chapter 6). The other is the application of isotope effects (see Chapter 4) to indicate whether a change in rate-limiting step occurs. Both methods possess the potential of being able to *predict* a change in rate-limiting step for a proposed mechanism or indicate that two bonds are undergoing substantial change in the rate-limiting step. Although the *absence* of an anticipated change in rate-limiting step is not *positive* evidence for a concerted mechanism, comparisons with similar systems—showing such changes—does give confidence in the technique. Moreover, there is no positive evidence to be expected for a concerted mechanism in the same way that one can positively demonstrate an intermediate. Indeed, at one time concerted mechanisms were denoted as 'no mechanism' reactions because of the absence of any positive evidence; it is perhaps useful to note here that the term *'positive* evidence' is questionable—it is only possible to *exclude* hypotheses, and such phrases as 'direct evidence *for*' really mean 'direct evidence *consistent with*'.

Free-energy relationships—the method of quasi-symmetrical reactions

In a simple two-step process occurring through the formation of an inter-mediate (Eqn [7.1]), either k_1 can be rate-limiting when $k_2 > k_{-1}$ or the second step can, when the first is readily reversible and $k_{-1} > k_2$.

$$A-Lg \underset{k_{-1}}{\overset{k_1[Nu]}{\rightleftharpoons}} Nu-A-Lg \xrightarrow{k_2} Nu-A+Lg \qquad [7.1]$$

A common technique used to deduce the presence of the intermediate is the observation of a non-linear free-energy relationship brought about by a change in rate-limiting step. Usually the transition-state structures for the two steps, k_1 and k_2, are significantly different and their energies vary differently in response to substituent change, giving rise to different susceptibility parameters ρ, β, etc.

The reaction of substituted pyridines (xpy) with isoquinoline sulphate ($isq^+ - SO_3^-$) could involve a concerted (Eqn [7.2a]) or a stepwise route (Eqn [7.2b]) with SO_3 as an intermediate. Equation [7.2b] is a pre-association mechanism (see Section 8.3); it requires formation of an encounter complex between nucleophile and substrate before the latter can dissociate. There is good kinetic evidence that the SO_3 involved in a putative stepwise route does not become free (Hopkins & Williams, 1982b) and would react within the encounter complex faster than it would diffuse out. Plotting the logarithm of the second-order rate constant against the pK_a^{xpy} of the attacking pyridine nucleophile gives a linear correlation (Fig. 7.1) (Bourne *et al.*, 1983).

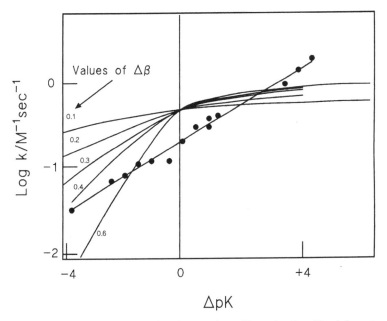

$$\overset{+}{\text{isq}}SO_3^- \xrightarrow{\text{xpy}} \left[\begin{array}{c} \delta- \\ O \\ \| \\ \overset{\delta+}{N}\cdots\overset{\delta+}{S}\cdots\overset{\delta+}{N} \\ X \\ \delta-O \quad O\delta- \end{array} \right]^{\ddagger} \xrightarrow{-\text{isq}} \overset{+}{\text{xpy}}SO_3^- \qquad [7.2a]$$

$$\text{xpy} + \text{isq}^+ - SO_3^- \underset{}{\overset{\text{diffusion}}{\rightleftharpoons}} [\text{xpy.isq}^+ - SO_3^-] \underset{}{\overset{\text{dissociation}}{\rightleftharpoons}} [\text{xpy.isq.SO}_3]$$

$$\underset{}{\overset{\text{association}}{\rightleftharpoons}}[\text{xpy}^+ - SO_3^-.\text{isq}] \underset{}{\overset{\text{diffusion}}{\rightleftharpoons}} \text{xpy}^+ - SO_3^- + \text{isq} \qquad [7.2b]$$

Fig. 7.1 Brønsted-type plot for the reaction of isoquinoline-N-sulphonate with substituted pyridines (plot compiled from data in Bourne *et al.*, 1983). Lines (offset in a vertical direction) are calculated from Eqn [7.3] for a series of $\Delta\beta$-values (figures on lines) for attack of xpy on SO_3 in the encounter complex for the stepwise path (Eqn [7.2]).

It is to be expected that the reactivity of isoquinoline and substituted pyridine will be the same or very close to each other when the pK_a^{xpy} of the conjugate acid of the substituted pyridine is the same as that of the isoquinoline. In the series of pyridine nucleophiles there should be a change in rate-limiting step for a pyridine nucleophile which has a pK_a^{xpy} equal to that of isoquinoline. When the pK_a^{xpy} of the attacking substituted pyridine is less than that of isoquinoline ($k_2 < k_{-1}$), k_2 is rate-limiting (see Eqn [7.1]); when pK_a^{xpy} exceeds that of the isoquinoline ($k_2 > k_{-1}$), k_1 becomes rate-limiting. Since k_1 is unlikely to have a significant substituent effect (because it does not involve chemical reaction of the substituted pyridine), the β_{nuc} value for

k_1 should be close to zero. A family of Brønsted lines may be calculated for a two-step reaction from Eqn [7.3]:

$$k_{nuc}/k_0 = 10^{\beta_1 \Delta pK}/(1 + 10^{-\Delta \beta.\Delta pK})$$ [7.3]

where $\Delta\beta = \beta_2 - \beta_{-1}$; $\Delta pK = pK_{nuc} - pK_{lg}$; $\log k_1 = \beta_1 \Delta pK + C_1$; $\log k_2 = \beta_2 \Delta pK + C_2$; $\log k_{-1} = \beta_{-1} \Delta pK + C_{-1}$; and $k_0 = 0.5k_1$. A linear plot results when $\Delta\beta = 0$ and may therefore be used to exclude a stepwise mechanism. In order for the experiments to be diagnostic, the Brønsted plot has to involve pK_a values well above and below the pK_a ($\Delta pK = 0$) where change in the rate-limiting step is predicted for the putative stepwise mechanism. It is necessary to determine a maximum $\Delta\beta$ from the error in fitting the points to a curved plot. Even if the plot is experimentally linear it is possible to determine an upper limit to $\Delta\beta$ from the standard deviation in $\Delta\beta$ in force-fitting the data to a curved plot. Thus, if the reaction were putatively stepwise, the 'effective charge' (see Chapter 3) on the attacking nitrogen in the two transition states would vary by less than the standard deviation in $\Delta\beta$ in an overall range of variation equal to β_{eq}. It is possible that there would be too little room on the energy surface within the area bounded by the error limits of $\Delta\beta$ for an intermediate to exist if it is to have a significant barrier for its collapse; thus even reactions with $\Delta\beta \neq 0$ could in principle have concerted mechanisms. The effective charge being measured on the intermediate would be less than that on one of its transition states by an amount equal to the error in $\Delta\beta$ (Skoog & Jencks, 1984; Dietze & Jencks, 1986; Bourne *et al.*, 1988). One can envisage a gradation of mechanisms running from a concerted reaction ($\Delta\beta = 0$) through to a full-blown stepwise process as shown in Fig. 7.2.

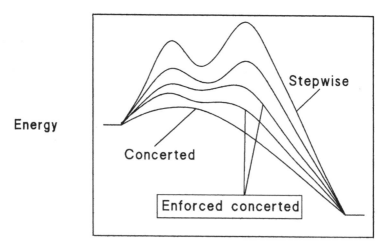

Reaction Coordinate

Fig. 7.2 Two-dimensional potential energy diagrams for reactions illustrating the change from stepwise, through 'enforced-concerted', to concerted mechanisms as a function of energy changes in the putative intermediate.

Other examples of this diagnostic method involving free-energy correlations of quasi-symmetrical reactions are given in the later sections of this chapter covering acyl group transfer mechanisms.

It is important to note that a linear free-energy plot in itself is *not* diagnostic of a concerted process, although a correlation involving two lines with decreasing slope at the intersection does indicate a stepwise process with an intermediate (see Chapter 2). A linear plot merely demonstrates a single structure for the transition state over the range of parameters employed. The indication of concertedness requires that the change in rate-limiting step be predictable, and this is only possible at the moment for quasi-symmetrical reactions. The change should also give rise to a significant difference in the susceptibility parameters for the two steps.

The method of quasi-symmetrical reactions is general and can be applied to all reactions, provided that smooth changes in structure can be made.

Isotope effects and the diagnosis of concerted mechanisms

Kinetic isotope effects have proved to be useful tools in the study of stepwise and concerted mechanisms (Saunders, 1976; Melander & Saunders, 1980; Belasco *et al.*, 1983). In these studies the ratio k_2/k_{-1} of a stepwise process (e.g. Scheme 7.3) has been given a special term (R), the partitioning factor or commitment (O'Leary, 1988). If substitution of an isotope (such as deuterium for protium) causes the k_{-1} step to vary (relative to k_2) then the partitioning factor of Scheme 7.3 will be altered; if k_2 is the 'isotopically sensitive' step, the isotope effect on the overall rate constant should also be affected in a predictable manner. The study of isotope effects for both H- and C-isotopically sensitive bonds is called the 'double isotope fractionation' technique and it provides a very useful, although experimentally demanding, tool for the diagnosis of stepwise versus concerted mechanisms. In principle the double isotope fractionation technique can apply to the effect of any pair of isotopes.

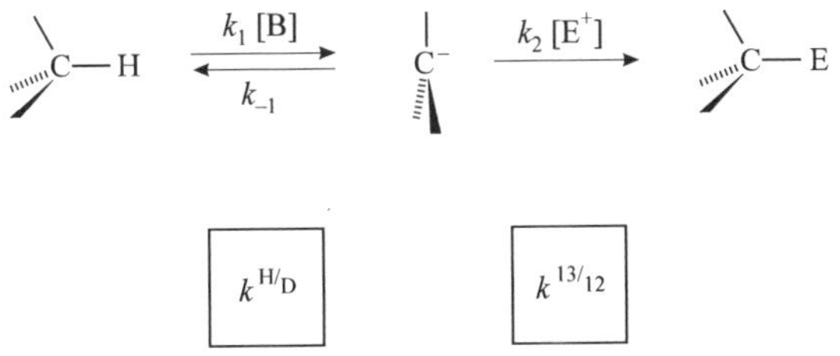

Scheme 7.3 Effect of isotopic substitution in a two-step process.

The principle is illustrated in Scheme 7.3 for a hypothetical reaction where carbanion formation precedes bond formation between carbon and electrophile in a stepwise process:

Case 1. For the stepwise process when k_2 is rate-limiting, there will be a small ^{13}C isotope effect (largely MMI and EXC — see Section 4.5) and a secondary ^{2}H effect.

Case 2. When neither k_1 or k_2 is rate-limiting in the stepwise process, substitution of deuterium for hydrogen will alter the partitioning factor (R) and thus cause a variation in the ^{13}C isotope effect.

Case 3. When k_1 is rate-limiting in the stepwise process, there will be a primary ^{2}H kinetic isotope effect and no ^{13}C isotope effect ($k^{12/13} = 1$ within the error limits).

Case 4. In a concerted mechanism, both primary ^{13}C and ^{2}H isotope effects will be exhibited and each will be independent of the isotope substitution of the other.

The aspartate deaminase reaction (Scheme 7.4) (Botting & Gani, 1989) has been studied by the double isotope technique. A ^{15}N isotope effect of 1.0246 ± 0.0013 and 1.0405 ± 0.0015 respectively for methylaspartic acid and aspartic acid indicates that C—N bond fission is at least partially rate-limiting for the latter substrate and that the enzyme mechanism occurs in a stepwise process. However, the equality of deuterium and hydrogen in the ^{15}N-substituted methylaspartic acid (^{15}N$(V/K)_D = {}^{15}$N$(V/K)_H$) indicates that C—N and C—H fission is both concerted and rate-limiting. The term V/K is the ratio V_{max}/K_m (see Section 10.1); it is the most readily determined rate parameter for enzymes and measures the energy difference between the free enzyme plus substrate and the transition state.

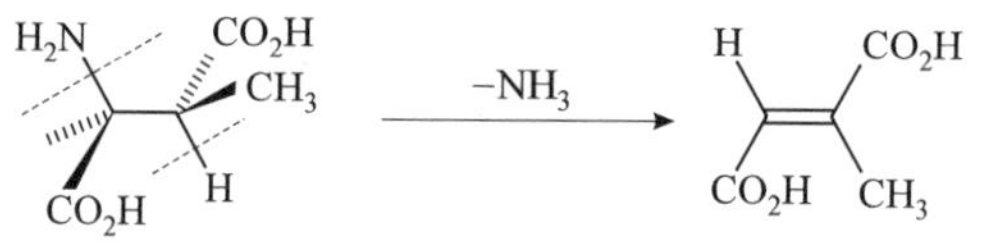

Scheme 7.4 Aspartate deaminase reaction for methylaspartic acid.

Secondary heavy-atom isotope effects can be employed to study the nature of the bonding in the transition state and hence to discover whether the fission of a particular bond is rate-limiting in a putative stepwise process. The secondary ^{15}N isotope effect for the deprotonation equilibrium of 4-nitrophenol is 1.0023 ± 0.0001 (Hengge & Cleland, 1990); the alkaline hydrolysis of 4-nitrophenyl diethylphosphate has an isotope effect of 1.0007 ± 0.0001 indicative of partial fission of the P—O bond in the rate-limiting transition state consistent with a concerted process. Further ^{15}N work (Hengge, 1992) has shown that the attack of phenoxide ion on 4-nitrophenyl acetate has an isotope effect of 1.0010 ± 0.0002, indicating fission of the C—OAr bond in the transition state which cannot arise from a stepwise process (see also Caldwell *et al.*, 1991).

A full isotopic study of a reaction might provide evidence for both bond fission *and* formation in the transition state of the rate-limiting step. An example of this is the elimination reaction of 2-phenylethyltrimethylammonium ion in base to give styrene, which proceeds via a stepwise mechanism. The isotope effects ($k^{H/D} = 3.2$ and $k^{14/15} = 1.0133 \pm 0.0002$) for the bonds undergoing fission (Scheme 7.5) (Smith & Bourns, 1970) indicate that both are broken in the transition state of the rate-limiting step. Examination of the two-step mechanism (Eqn [7.4]) indicates that the overall rate constant k_{obs} is given by Eqn [7.5a] and the ^{15}N isotope effect is therefore given by Eqn [7.5b] (note that $k^{H/D}_{obs}$ will approximate to $k^{H/D}_1$); the return step will always be $k_{-1}[HOEt]$ because the solvent is ethanol and is only diluted partially with deuterium when the reaction occurs. The intrinsic isotope effect for the fission of $^{15}N{-}C$ lies between 1.04 and 1.07 so that the observed value is less than the full quota; it is therefore possible that an explanation of the results is that neither k_1 nor k_2 is rate-limiting. Because the carbanion intermediate is never formed in significant concentrations, the overall rate equation will be second-order even if k_2 is rate-limiting (in the latter case $k_{-1}[EtOH]$ is invariant).

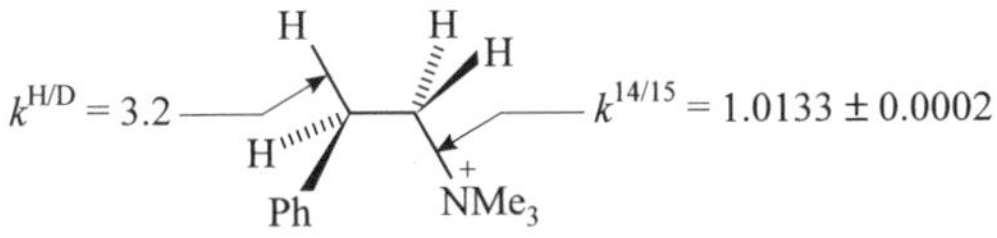

$$k^{H/D} = 3.2 \qquad \qquad k^{14/15} = 1.0133 \pm 0.0002$$

Scheme 7.5 Isotope effects in elimination of ammonium species.

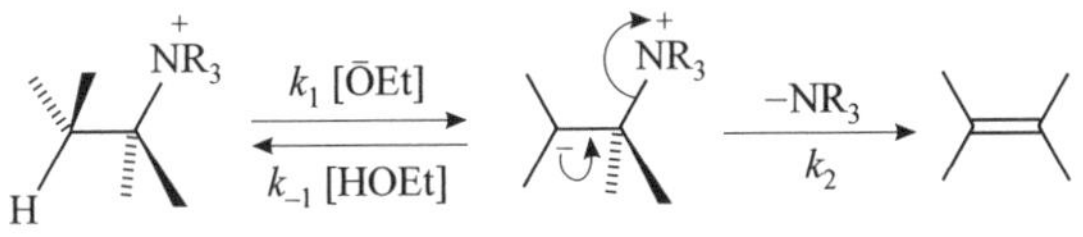

$$\underset{k_{-1}[HOEt]}{\overset{k_1[\bar{O}Et]}{\rightleftarrows}} \qquad \overset{-NR_3}{\underset{k_2}{\longrightarrow}} \qquad \qquad [7.4]$$

$$k_{obs} = k_1 k_2 [OEt]/\{k_{-1}[HOEt] + k_2\} \qquad [7.5a]$$

$$k^{14/15} = (k_2^{14}/k_2^{15})(k_{-1}^H[HOEt] + k_2^{15})/(k_{-1}^H[HOEt] + k_2^{14}) \qquad [7.5b]$$

Application of isotope effects to distinguish between stepwise and concerted mechanisms requires that we know reasonably precisely the *intrinsic* isotope effect in order to judge the 'commitment' or 'partitioning factor'. The magnitude of an isotope effect depends on the size of the 'intrinsic' effect and the degree to which the isotopically sensitive step limits the rate of the overall reaction. Malate synthase is shown to involve a stepwise path (Scheme 7.6) rather than a concerted route (Clark *et al.*, 1988) by study of the ^{13}C isotope effect for the aldehyde carbon of the glyoxylic acid as a function of the deuterium isotope change on the methyl of the acetyl coenzyme A. The interpretation of the isotope effects is complicated in this case by the dehydration of the glyoxylate hydrate prior to reaction with acetyl coenzyme A which itself has an isotope effect.

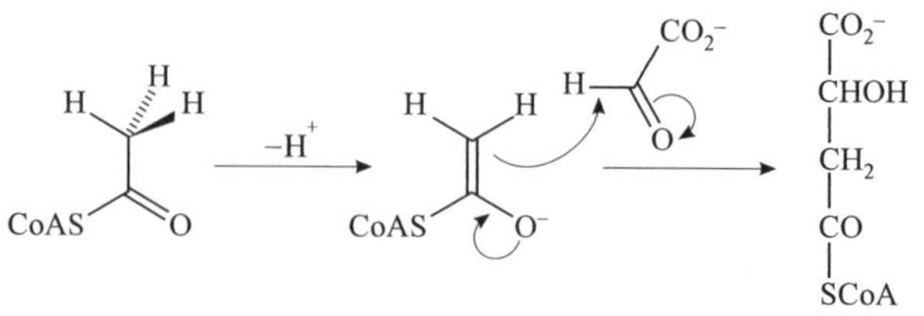

Scheme 7.6 Malate synthase involves a stepwise mechanism.

Stereochemistry and concerted mechanisms

Knowledge of the stereochemistry of reaction of chiral substrates is manifestly important in our understanding of enzymic reaction mechanisms; it enables the topology of the reactive groups at the active site to be mapped in the transition state if the reaction is concerted.

Racemization will indicate that a reaction has passed through a symmetrical intermediate which has lost its chirality. The observation of *inversion* is not conclusive evidence for a concerted process because it could result from intervention of an intermediate with a lifetime less than the time it takes to diffuse away from the reaction complex. An example is the solvolysis of 1-phenyl-2,2-dimethylpropyl-4'-toluenesulphonate in acetic acid which gives an acetate product which is only 90% racemized. This is readily explained on the basis that the reaction involves a short-lived carbenium ion, where the leaving group thus protects one side of the carbenium ion to attack by the solvent. The logical conclusion of this result is that even if 100% inversion were to be observed the experiment would still not be able to distinguish unequivocally between a concerted mechanism and a stepwise one involving a carbenium ion. It is conceptually difficult to distinguish between a mechanism involving a transition state with weak bonding between the central atom and the entering and leaving atoms, and one where there is no bonding (as in the S_N1 process) (Jencks, 1981).

Racemization could occur prior to a chiral reaction path: this phenomenon may be responsible for a number of as yet undiscovered errors in the literature. The methoxide ion causes the racemization of chiral methyl phosphinate esters (Hudson & Green, 1963) and reaction of nucleophiles with such species could be accompanied by racemization caused by substitution occurring at a slower rate than racemization (Eqn [7.6]).

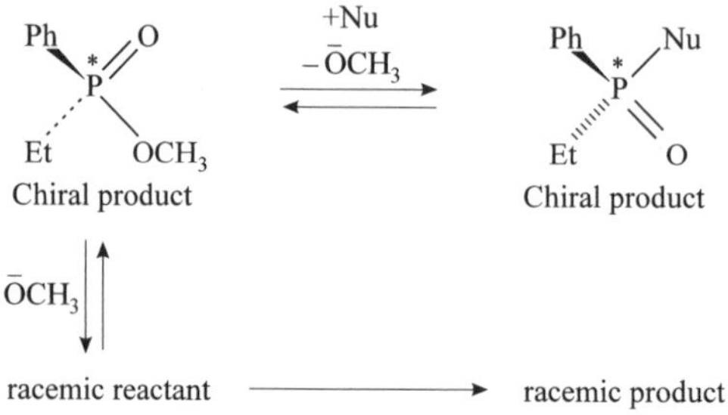

[7.6]

A classic application is the reaction of 2-octyl iodide in a solution of radioactive iodide ions (Eqn [7.7]), where the racemization occurs at twice the rate of incorporation of the labelled iodine into the molecule; this is consistent with inversion of configuration at the 2-position (Hughes *et al.*, 1935).

$$[7.7]$$

Retention of configuration or maintenance of chiral integrity through a reaction has been observed in a number of reactions at carbon and phosphorus. In some cases this has been explained on the basis of two sequential reactions, each occurring with inversion; the neutral hydrolysis of chloroacetate ion is an example (Eqn [7.8]). The overall retention of configuration is the result of a two-step process, each step occurring by S_N2 inversion. It could be proposed that any even number of reactions each involving inversion could explain retention, but Occam's razor would severely limit such an explanation for more than two such reactions. Front-side attack (see Scheme 1.6) can also explain retention of configuration.

$$[7.8]$$

Chlorination of alcohols through the chlorosulphite or chlorocarbonate esters leads to retention of configuration at the alkyl carbon which is thought to result from tight ion pairs reacting internally at a faster rate than the components can diffuse away. Inversion occurs in the presence of free chloride ion (Scheme 7.7) which results when, for example, the sulphite half-ester is produced by reaction of $SOCl_2$ with the alcohol in pyridine. Reaction between the carbenium ion and chlorosulphite or chlorocarbonate ion in the tight ion pair is faster even than the rotational lifetime of the carbenium ion in the complex.

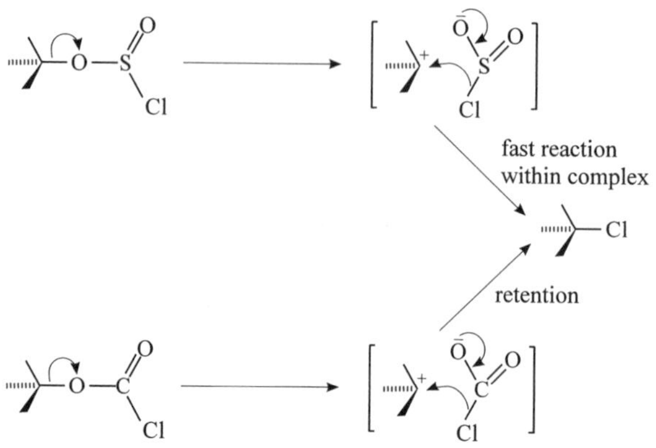

Scheme 7.7 Retention of configuration in nucleophilic aliphatic substitutions.

Retention also occurs in the bromination of phenyl alkylmethanols by hydrobromic acid because bromide ion is formed in a tight ion pair after protonation of the alcohol (Eqn [7.9]).

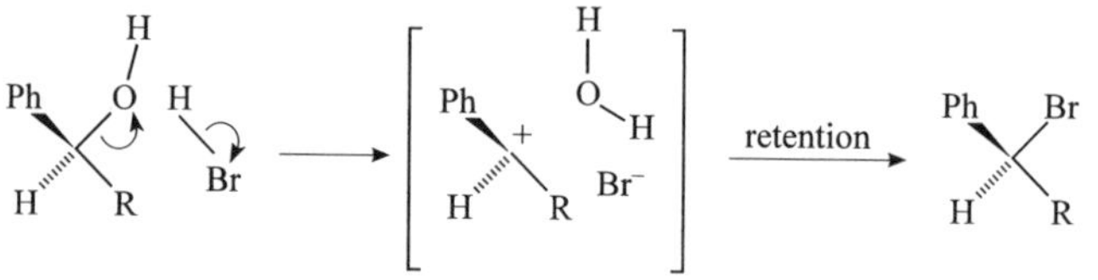

[7.9]

Retention of configuration in the rearrangement of a chiral molecule indicates that the group bearing the chiral centre does not dissociate when migration occurs to another centre in the reactant. For example, the Stevens and Wittig reactions (Scheme 7.8) involve retention of configuration at the carbon centre; CIDNP experiments indicate that such rearrangements involve caged radical pairs. The radical pair reacts before there is time for the constituents of the cage to diffuse to bulk solvent, and retention of configuration results (Scheme 7.8).

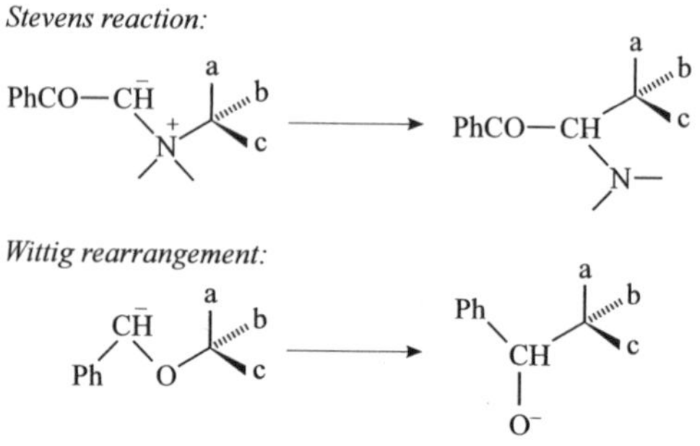

Stevens reaction:

Wittig rearrangement:

Scheme 7.8 The Stevens and Wittig reactions.

7.2 Acyl group transfer

Introduction

The transfer of an acyl group between acceptor and donor atoms forms the basis of many synthetic routes in organic chemistry, one of which is exemplified by the Claisen condensation (Eqn [7.10]). This family of reactions has application in the synthesis of heterocycles and condensation polymerization (Scheme 7.9) and it has manifest importance in biochemistry, where examples of acyl group transfer include the formation of acetyl coenzyme A from pyruvic acid via the intervention of lipoic acid (Eqn [7.11]), the various reactions involved in aminoacyl-tRNA synthesis and aldol-type condensations mediated by thiamine pyrophosphate (Scheme 7.10) (Westheimer, 1987).

$$\mathrm{RCO{-}OEt} \xrightarrow[-\mathrm{EtO^-}]{\mathrm{R'\bar{C}HCOOEt}} \mathrm{RCO{-}CHR'CO{-}OEt} \qquad [7.10]$$

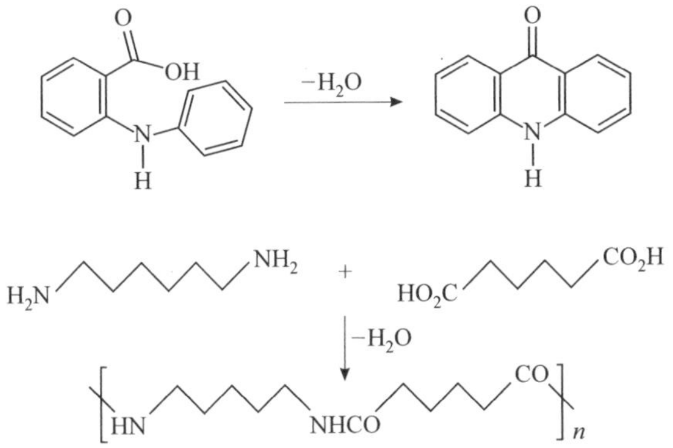

Scheme 7.9 Two industrially important acyl group transfer reactions.

Protein biosynthesis:

$\mathrm{RCO{-}OH + ATP \longrightarrow RCO{-}OPO_3^-{-}adenosine}$

$\mathrm{RCO{-}OPO_3^-{-}adenosine \xrightarrow{HO{-}tRNA} RCO{-}O{-}tRNA}$

$\mathrm{RCO{-}O{-}tRNA \xrightarrow{RNH_2} RCO{-}NHR}$

Action of thiamine:

$$\mathrm{RCHO + E^+ \xrightarrow{\text{thiamine as catalyst}} RCO{-}E + H^+}$$

$\mathrm{E^+ = RCHO, CO_2, H^+}$

Scheme 7.10 Acyl group transfer in biochemistry.

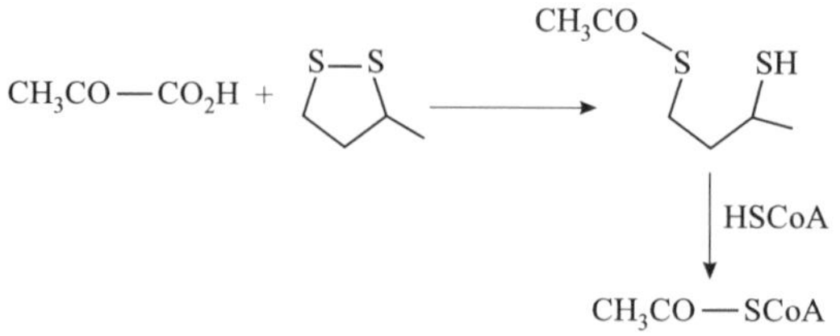

$$CH_3CO-CO_2H \;+\; \overset{S-S}{\underset{}{\bigsqcup}} \longrightarrow \overset{CH_3CO}{\underset{}{\bigsqcup}} \overset{S\quad SH}{}$$

[7.11]

$$\Big\downarrow HSCoA$$

$$CH_3CO-SCoA$$

Acyl group transfer reactions form the basis of catalyses by proteolytic enzymes which range from blood clotting, through metabolism of proteins, to key steps in the development of the AIDS virus (see Chapter 10).

Mechanisms

Mechanisms available for transfer of the acyl group can be discussed in the context of group transfer and are available for the transfer of a *general* group; they bear a strong relationship to other group transfers such as alkyl substitution. If A is defined as the acyl group and X is the leaving group substituted by the incoming nucleophilic group, Nu, or electrophilic group, E, then there are three basic types of mechanism (Scheme 7.11). The scheme for radical mechanisms also includes types similar to those for the heterolytic processes, for example involving associative paths.

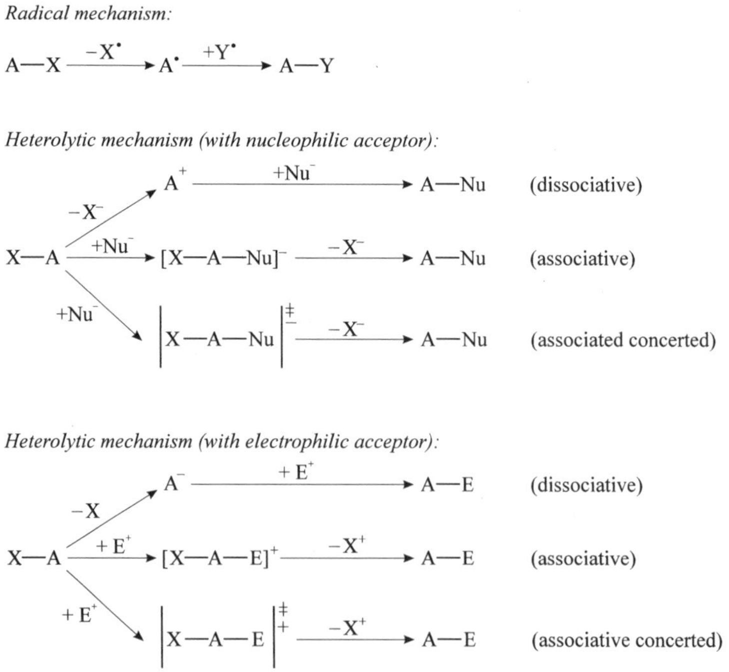

Radical mechanism:

$$A-X \xrightarrow{\;-X^{\bullet}\;} A^{\bullet} \xrightarrow{\;+Y^{\bullet}\;} A-Y$$

Heterolytic mechanism (with nucleophilic acceptor):

$$X-A \begin{cases} \xrightarrow{-X^-} A^+ \xrightarrow{\;+Nu^-\;} A-Nu \quad \text{(dissociative)} \\[2mm] \xrightarrow{+Nu^-} [X-A-Nu]^- \xrightarrow{\;-X^-\;} A-Nu \quad \text{(associative)} \\[2mm] \xrightarrow{+Nu^-} \left|X-A-Nu\right|^{\ddagger} \xrightarrow{\;-X^-\;} A-Nu \quad \text{(associated concerted)} \end{cases}$$

Heterolytic mechanism (with electrophilic acceptor):

$$X-A \begin{cases} \xrightarrow{-X} A^- \xrightarrow{\;+E^+\;} A-E \quad \text{(dissociative)} \\[2mm] \xrightarrow{+E^+} [X-A-E]^+ \xrightarrow{\;-X^+\;} A-E \quad \text{(associative)} \\[2mm] \xrightarrow{+E^+} \left|X-A-E\right|^{\ddagger}_{+} \xrightarrow{\;-X^+\;} A-E \quad \text{(associative concerted)} \end{cases}$$

Scheme 7.11 Radical and heterolytic mechanisms for acyl group transfer.

What is an acyl group?

Normally the term 'acyl' refers to a group derived from a carboxylic acid although this function should strictly always be called the 'carbonyl' group. It is often useful to compare and contrast transfer of groups derived from other acids, notably the phosphates, and it is then most convenient to have a general term; 'acyl' may then be used to denote a whole range of groups (A) derived from acids (A—OH). The methyl group (derived from the 'acid' CH$_3$OH) can in principle be considered as an acyl function on the above definition, but normally the hydroxyl acid (A—OH) has to be relatively strong, with a pK_a below *ca* 4 before the A group can be called 'acyl'. Even the strength of the acid is not definitive, because the picryl group (where the acid has a pK_a of less than 1) is not normally considered to be an acyl function. Normally only acids with extra oxygen attached to the A function are considered to give rise to acyl groups. Table 7.1 lists groups which are commonly accepted as acyl functions; the sulphenyl group is an exception to these 'rules'.

Table 7.1 (a) Some examples of acyl functions and their nomenclature.[a]

Acyl group A	Name	Acyl group (A)	Name
RCO—	carbonyl	$^{2-}$O$_3$P—	phosphoryl
RSO$_2$—	sulphonyl	—O$_3$S—	sulphuryl
RSO—	sulphinyl	R$_2$PO—	phosphinyl
RS—	sulphenyl	(RO)$_2$PO—	dialkylphosphoryl

(b) The following groups X, derived from the species X—OH, are not considered as acyl groups even though the acid is as strong as, or stronger than, some carboxylic acids:

X	X—OH	pK_a	X	X—OH	pK_a
		0.96	(CF$_3$)$_3$C—	(CF$_3$)$_3$C—OH	5.4
		0.79	(CF$_3$)$_2$C=N—	(CF$_3$)$_2$C=N—OH	6.0

[a] Thatcher & Kluger (1989) and Kice (1980) give further nomenclature of phosphoryl and sulphyl groups, respectively.

Carbonyl group transfer

Radical mechanisms

The paucity of examples of this process for acyl group transfer may be due to the emphasis on studies of heterolytic reactions in general. Radical mechanisms have been shown to occur (Applequist & Kaplan, 1965; Applequist & Klug, 1978; Pfenninger *et al.*, 1980; Giese, 1986; Crich & Fortt, 1989; Bacaloglu *et al.*, 1990; Boger & Mathvink, 1992); Eqns [7.12]–[7.15] exemplify the process, which can include both dissociative and associative types.

$$RCHO \xrightarrow{\;-H^\bullet\;} RCO^\bullet \xrightarrow{\;CCl_4\;} RCOCl + {}^\bullet CCl_3 \qquad [7.12]$$

$$RCOR \xrightarrow{\;{}^\bullet CH_3\;} R_2\overset{\displaystyle O^\bullet}{\underset{\displaystyle CH_3}{C}} \longrightarrow RCOCH_3 + R^\bullet \qquad [7.13]$$

$$RCO-SePh \xrightarrow{\;Bu_3Sn^\bullet\;} RCO^\bullet \qquad [7.14]$$

$$RCO-SePh + CH_2{=}CH-Cl \xrightarrow{\;Bu_3Sn^\bullet(via\ RCO)\;} RCO-CH_2CH_2-Cl \qquad [7.15]$$

Heterolytic associative mechanisms (see also Section 8.3)

The associative mechanism for carbonyl group transfer reactions involves the formation of a tetrahedral intermediate, in which the bond to the incoming group is made *before* that to the leaving group is broken. The conversion of trigonal carbon to a tetrahedral configuration occurs because:

(a) the carbonyl carbon–oxygen π-bond is usually weaker than the bond to the leaving group $(C-Lg)$;
(b) four-coordinate carbon presents no unusual steric or electronic problems;
(c) solvation of the negative charge density on the oxygen helps to stabilize the tetrahedral intermediate;
(d) there are usually significant barriers to expulsion of the incoming and leaving groups, giving rise to an intermediate with a significant lifetime.

Most carbonyl group transfer reactions appear to occur through an associative pathway, even with reactive acylating agents. For example, nucleophilic substitution at the β-lactam of penicillins occurs through the formation of a tetrahedral intermediate (Scheme 7.12). The stepwise process (Page, 1987) is consistent with (i) a Brønsted β_{nuc} of *ca* 1.0 compatible with full charge development/removal on the attacking nucleophile in the transition state, (ii) a non-linear Brønsted plot for the general base-catalysed aminolysis reaction against the pK_a of the catalytic base—indicative of a stepwise process for proton transfer—and (iii) a non-linear dependence on hydroxide ion concentration of the second-order rate constant for the hydroxide ion-catalysed aminolysis reaction—indicative of a change in rate-limiting step and hence the presence of an intermediate.

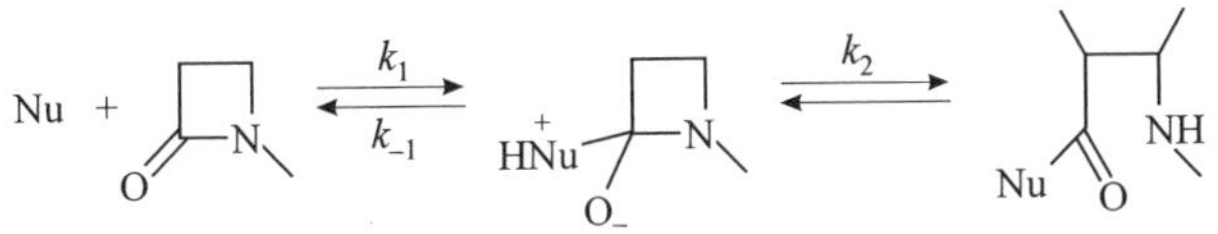

Scheme 7.12 Nucleophilic substitution at β-lactams is stepwise.

The observation of exchange between the ester carbonyl oxygen and ^{18}O-labelled water during the hydrolysis of esters (Bender, 1951) is consistent with an associative mechanism (see Chapter 2, Scheme 2.9); the proton transfer step required to make the oxygens in the tetrahedral intermediate equivalent, is fast compared with the other steps.

Detection of the 'tetrahedral' intermediate is generally difficult because of its reactivity, and many laboratories have searched for stable derivatives. For example, tetrodotoxin (**1**) was shown to possess a tetrahedral arrangement (Goto *et al.*, 1965) that is stable relative to ring-opened hydroxy ester. If the potential leaving group is very unstable (such as H^-) then the tetrahedral intermediate will be relatively stable and may accumulate, as does the adduct from water and trichloroacetaldehyde.

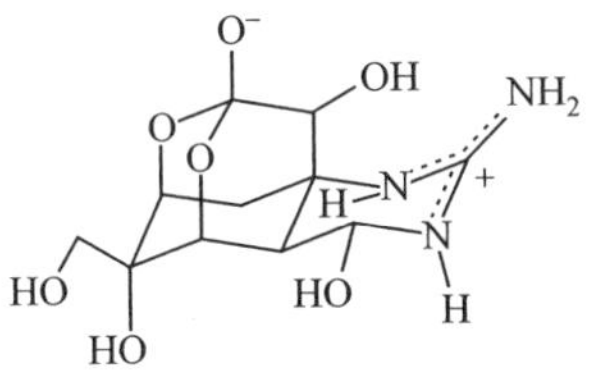

1

Observable *reactive* tetrahedral intermediates which may be generated by very fast reactions (Eqns [7.16] and [7.17]) have been thoroughly investigated (Capon *et al.*, 1981; McClelland & Santry, 1983).

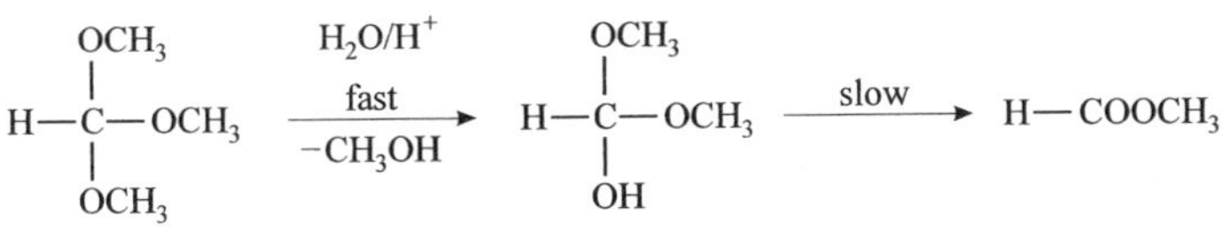

[7.16]

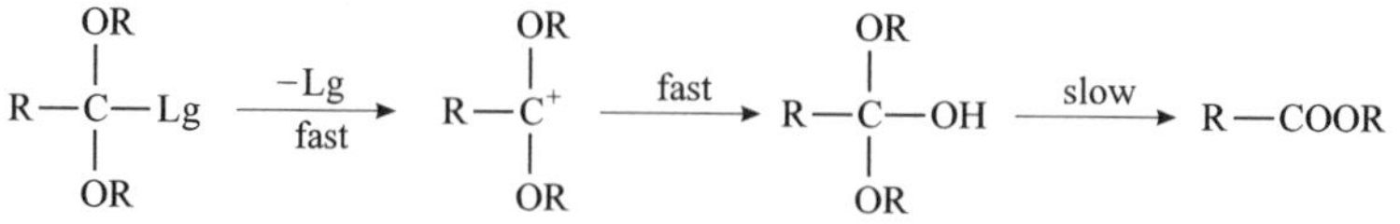

$$R-\underset{\underset{OR}{|}}{\overset{\overset{OR}{|}}{C}}-Lg \xrightarrow[\text{fast}]{-Lg} R-\underset{\underset{OR}{|}}{\overset{\overset{OR}{|}}{C^+}} \xrightarrow{\text{fast}} R-\underset{\underset{OR}{|}}{\overset{\overset{OR}{|}}{C}}-OH \xrightarrow{\text{slow}} R-COOR \qquad [7.17]$$

Heterolytic dissociative mechanisms

The acylium ion (RCO^+) has long been accepted as an intermediate in the gas phase (Bender & Chen, 1963; Olah *et al.*, 1967; Williams & Douglas, 1975; Pau *et al.*, 1978; Kim & Caserio, 1981, 1982) and is likely to be very reactive in nucleophilic solvents. The intervention of an acylium ion intermediate in solution reactions (Scheme 7.13) depends on the factors stabilizing it and on the stability of the leaving group.

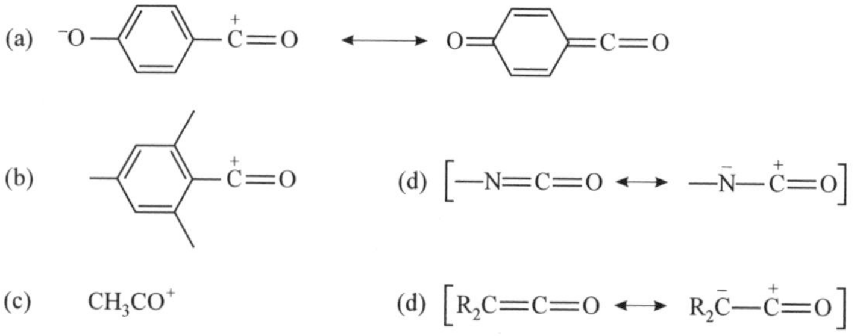

(a)

(b)

(c) CH_3CO^+

(d) $\left[-N{=}C{=}O \longleftrightarrow -\bar{N}-\overset{+}{C}{=}O\right]$

(d) $\left[R_2C{=}C{=}O \longleftrightarrow R_2\bar{C}-\overset{+}{C}{=}O\right]$

Scheme 7.13 Some acylium ions observed as intermediates in solution: (a) Cevasco *et al.* (1985); (b) Bender & Chen (1963); (c) Olah *et al.* (1967); (d) Williams & Douglas (1975).

The introduction of 2,6-dimethyl substituents into benzoic acid derivatives hinders nucleophilic attack on the carbonyl group and consequently enables 2,6-dimethylbenzoate esters to undergo acid-catalysed hydrolysis by a dissociative pathway. The dissociative mechanism can also be accelerated and the dramatic difference in reactivity between the conjugate base of 2,6-dimethyl-4-hydroxybenzoate ester and that of the ester without the 4-phenoxide ion substituent could be ascribed to the carbonyl function being forced into the conformation where the leaving group is 'antiperiplanar' to the lone pairs of the oxyanion in the substituted ester (Eqn [7.18]); this enables the elimination reaction to compete successfully with the BAc2-type attack normally associated with ester hydrolysis (Thea *et al.*, 1983, 1985).

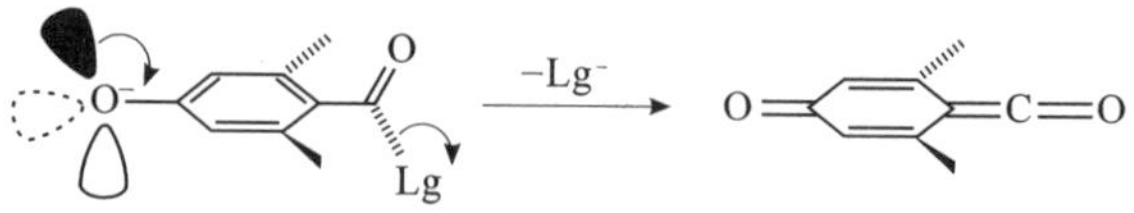

$$\qquad [7.18]$$

The concerted mechanism

The concerted mechanism for ester hydrolysis was formally expressed by Dewar (1948) but little subsequent work was done to verify this mechanism experimentally, probably as a result of Myron Bender's seminal paper

(Bender, 1951) on the stepwise mechanism; it was only in the 1980s that new evidence became available consistent with the concerted mechanism (Williams, 1989). A possible transition state for this mechanism has square planar geometry which would become tetrahedral, depending on the advancement of bond formation and fission to entering and leaving groups respectively. The reaction of substituted pyridines with the N-methoxycarbonylisoquinolinium ion to yield N-methoxycarbonylpyridinium ions (Eqn [7.19]) has a linear Brønsted correlation over a range of pK_a values greater than *and* less than that of isoquinoline (Chrystiuk & Williams, 1987). This result is not consistent with a stepwise process, which would require a change in the rate-limiting step to occur at the pK_a of isoquinoline and give rise to a breakpoint there. The data can be analysed according to Eqn [7.3] and the value $\Delta\beta = $ zero indicates that there is no charge difference between the nitrogen on the substituted pyridine for both transition states of the putative stepwise process. This means that the transition states are identical; this can only be so if the reaction is concerted. Reaction of substituted phenoxide ions with 4-nitrophenyl acetate has likewise been shown to involve a concerted pathway (Ba-Saif *et al.*, 1987). Guthrie (1991) used thermodynamic techniques to demonstrate that the adduct from methyl acetate and hydroxide ion decomposes with a half-life of about 10^{-7}s. The adduct between phenoxide ions has a lifetime estimated to be too short for it to exist for a significant period, and the reaction is thus considered to be concerted, in agreement with the results from the quasi-symmetrical reaction method.

[7.19]

The stereochemistry of nucleophilic attack on the carbonyl group has been addressed by X-ray crystallographic studies of stable species containing both nucleophile and carbonyl which show that the nucleophile—C bond is at a mean angle of between 105° and 110° to the carbonyl bond (2) (Scheme 7.14; Burgi & Dunitz, 1983) (see Section 3.7). The orbital description of the addition step (3) involves the lone pair forming on the carbonyl oxygen (a) with an orbital almost antiperiplanar to the entering pair of electrons, and that forming on the leaving atom (b) also antiperiplanar to the entering pair of electrons (the antiperiplanar lone pair hypothesis, ALPH; see Chapter 1). The microscopic reverse of this step, namely decomposition of the tetrahedral adduct, therefore possesses lone pairs on the entering atom and a CO oxygen antiperiplanar to the leaving bond. Many studies have been carried out to test the importance of this stereochemistry to the decomposition of the tetrahedral adduct (Deslongchamps, 1983; Kirby, 1983, 1996). Most of the stereochemical results for nucleophilic attack at a carbonyl centre can also be explained by the principle of least nuclear motion (Sinnott, 1988) but both least motion and ALPH apply with some exceptions to relatively small energy differences.

Scheme 7.14 Stereochemistry of nucleophilic attack on carbonyl carbon.

Transfer of the phosphyl group

Introduction

The valencies of phosphorus give rise to a number of acidic species and the acyl functions derived from these are often referred to as 'phosphyl' (being derived from *phosph*orus ac*yl*). The most important biological phosphyl groups are the phosphoryl group ($-PO_3^{2-}$), the phosphodiester group ($RO-PO_2^-$), and the neutral phosphoryl group (($R-O)_2PO-$); and considerable mechanistic work has centred around the group transfer of these and related species.

Dissociative mechanism

The metaphosphate ion (PO_3^-, analogous to a carbonyl acylium ion) formed by a dissociative route was postulated as an intermediate in phosphorylation reactions by Todd (1959). It was assumed to add rapidly to the acceptor nucleophile to yield product. Although good evidence for the metaphosphate ion intermediate was not available, the existence of the iminometaphosphate intermediate (Scheme 7.15) was demonstrated by a dramatically increased rate of hydrolysis of phosphoramidate esters and halides over that for substrate where the intermediate could not form (where $RNH-$ is replaced by R_2N-; Traylor & Westheimer, 1965), by an intermediate with rate-limiting leaving group bond fission (Williams & Douglas, 1972), by the stereochemistry (Gerrard & Hamer, 1968) and by trapping results (Williams & Douglas, 1973). The geometrical changes are similar to those observed for the S_N1 pathway at saturated carbon, where a four-coordinate tetrahedral centre is converted to a three-coordinate one. Numerous experiments demonstrated that analogues (as above) could be made stable and even isolated (Scheme 7.16) and gas-phase work showed that the isolated metaphosphate ion is stable in its isolated state (Meyerson *et al.*, 1978; Westheimer, 1981; Henchman *et al.*, 1985). However, the case for the metaphosphate intermediate in solution is difficult to make because in aqueous solution the putative intermediate appears to be very reactive indeed (Herschlag & Jencks, 1989).

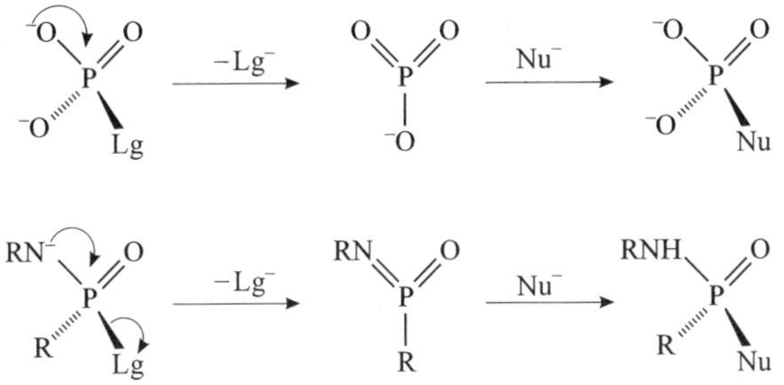

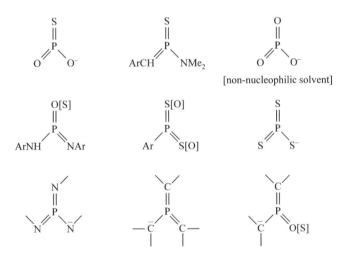

Scheme 7.15 Dissociative path for transfer of phosphoryl and aminophosphoryl groups.

Scheme 7.16 Some metaphosphate ion analogues.

Phosphoryl group transfer in aqueous solution is thought to usually involve a concerted mechanism. The stereochemistry of transfer of the — PO_3^{2-} group from $ArO—PO_3^{2-}$ to an alcohol involves inversion of configuration at the phosphorus centre (Buchwald *et al.*, 1984). Inversion at phosphorus also occurs in the transfer of the phosphoryl group to methanol in the Conant–Swan reaction (Eqn [7.20]) (Calvo, 1985). Inversion could arise from a dissociative path, provided that the intermediate reacted with the nucleophile prior to separation of the leaving group from the solvent cage (Eqn [7.21]).

$$\begin{array}{c} Br \\ \diagdown \\ CH-CHBr-PO_3^{2-} \\ \diagup \\ Ph \end{array} \xrightarrow{\quad CH_3OH \quad} PhCH{=}CHBr \; + \; CH_3OPO_3^{2-} \qquad [7.20]$$

$$Nu + Lg-PO_3^{2-} \rightleftharpoons [Lg-PO_3^{2-}.Nu] \rightleftharpoons [Lg.Nu.PO_3^{2-}]$$
$$\rightleftharpoons [Lg.Nu-PO_3^{2-}] \rightleftharpoons Nu-PO_3^{2-} + Lg \qquad [7.21]$$

Reaction of substituted pyridines with pyridine-*N*-phosphonates (analogously to Eqns [7.1] and [7.2]) (Bourne & Williams, 1984; Skoog & Jencks, 1984) gives *N*-phosphoryl-substituted pyridines. The rate constants obey a Brønsted dependence with no evidence for a break at the pK_a of the attacking pyridine corresponding to that of the leaving group (Fig. 7.3). The linearity of the plot over a range of pK_a values well above and below the calculated breakpoint $(\Delta pK = 0)$ is consistent with a concerted process.

Positional isotope exchange (PIX; Section 2.16) has been shown to occur in the reaction of ADP in acetonitrile or acetonitrile–*t*-butanol solvents (Scheme 7.17) (Cullis & Nicholls, 1987), where intervention of a metaphosphate ion intermediate would be favoured by the low nucleophilicity of the solvent.

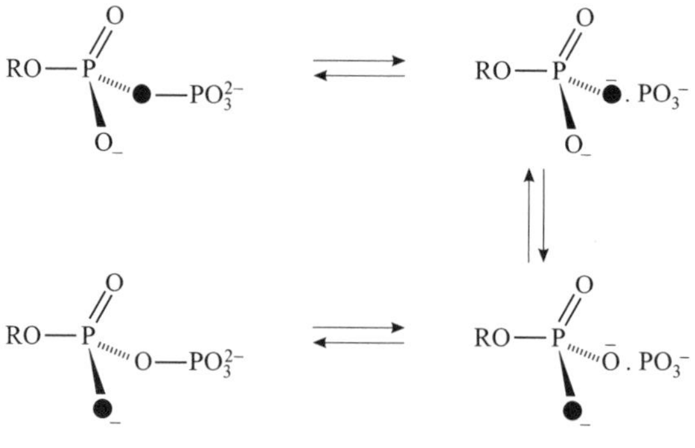

Scheme 7.17 Positional isotope exchange demonstrates stepwise phosphoryl group transfer in weakly nucleophilic solvent.

Associative concerted mechanism

Linearity of the Brønsted plot for the attack of aryloxyanions on the 4-nitrophenyl ester of the diphenylphosphate and diphenylphosphinate (Fig. 7.4) demonstrates concerted transfer of a neutral phosphyl group (R_2PO-) between weakly basic nucleophiles. It is reasonable to assume that phosphyl group transfer in general is associative and when the entering and leaving ligands are weak nucleophiles the mechanism becomes concerted. The transfer reactions of the monoanionic phosphoryl group is involved in many reactions in nucleotide chemistry and species with good leaving groups

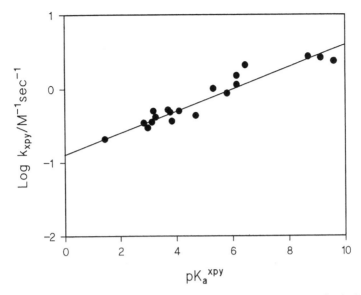

Fig. 7.3 Brønsted-type plot for the quasi-symmetrical attack of substituted pyridines on isoquinoline-*N*-phosphonate (data from Bourne & Williams, 1984).

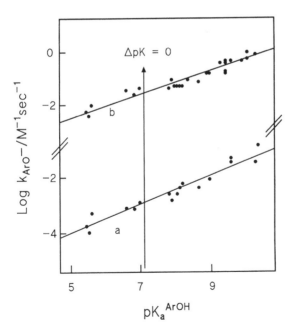

Fig. 7.4 Brønsted-type dependencies for the attack of phenoxide ions on the 4-nitrophenyl esters of diphenylphosphoric (a) and diphenylphosphinic (b) acids, indicating a concerted mechanism. The vertical scales are different for a and b.The plots are compiled from data in Bourne *et al.* (1988) and Ba-Saif *et al.* (1990).

probably involve concerted processes. Scheme 7.18 is an effective charge, ε, description for the base-catalysed cyclization of aryl nucleotide esters (Davis *et al.*, 1988a) and it is interesting to compare this transition state with that of the ribonuclease-catalysed reaction (Scheme 7.19; Davis *et al.*, 1988b).

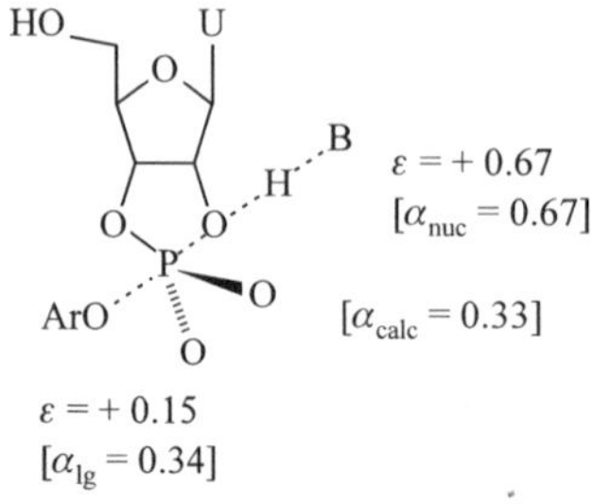

Scheme 7.18 Effective (ε)charge map for base-catalysed cyclization of aryl nucleotides; values in brackets are the Leffler α parameters for the bond indicated (U = uracil base).

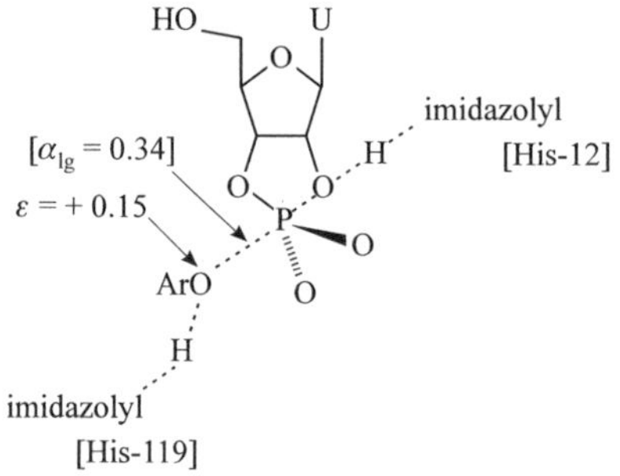

Scheme 7.19 Effective charge (ε) map for the bovine ribonuclease-A-catalysed cyclization of aryl nucleotides (U = uracil base).

Associative stepwise mechanism

The phosphyl group transfer reaction involving a pentacoordinate intermediate is preferred for strong donor and acceptor nucleophiles such as alkoxide ions. The existence of a pentacoordinate intermediate begs the question of pseudo-rotation (Westheimer, 1968; Hudson & Brown, 1972) whereby the ligands in the trigonal bipyramidal structure interchange apical and equatorial positions, as in Eqn [7.22]; interchange in the analogous tetrahedral intermediate is not of structural significance. The basis of the mechanistic interpretation is the assumption that incoming nucleophiles take up an apical position and that leaving groups depart from an apical position. Furthermore, it is usually assumed that small ring structures around a central phosphorus prefer an apical–equatorial orientation to the diequatorial one, which certainly will be strained.

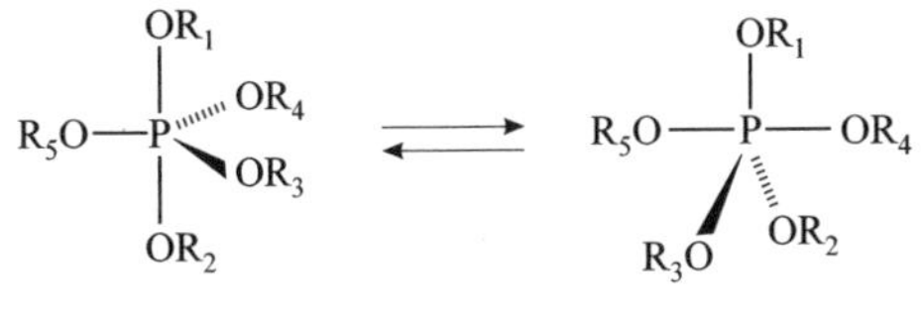

$$[7.22]$$

The rate of pseudo-rotation is crucial; if the effects of the isomerization are to be felt it must exceed that of the degradation of the intermediate. Acid-catalysed hydrolysis of the cyclic phosphate in Scheme 7.20 involves P—O fission some 10^6-fold faster than that of the acyclic analogue trimethylphosphate. Moreover, the reaction involves some 70% ring fission and the phosphoryl oxygen undergoes exchange with the solvent. The rate enhancements, ring fission and exchange are explained in Scheme 7.20, where the initially formed trigonal bipyramidal intermediate has less strain in its five-membered ring than has the reactant, because apical equatorial bridging at phosphorus by a five-membered ring is favourable. Pseudo-rotation is readily achieved because the tendency for the non-cyclic ligands to go into apical or equatorial positions in the trigonal bipyramidal structure is similar, and fission of the apical bond will lead to hydrolysis and exchange. Pseudo-rotation involving cyclic ligands will only be possible between isomers with favourable apical–equatorial bridges.

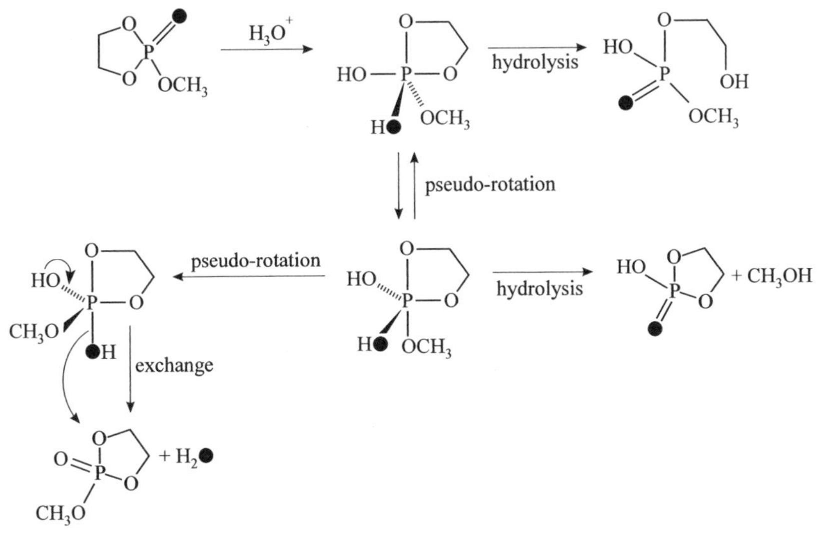

Scheme 7.20 Effect of pseudo-rotation in partitioning of cyclic pentacoordinate phosphyl species.

Sulphyl group transfer

Sulphuryl group transfer

Sulphuryl group transfer is closely analogous to that of the phosphoryl group; sulphur trioxide, a trigonal planar molecule, has been postulated as an intermediate (for example, Eqn [7.23]).

$$\text{ArO}-\text{SO}_3^{2-} \xrightarrow{-\text{ArO}^-} [\text{SO}_3] \xrightarrow{+\text{Nu}^-} \text{Nu}-\text{SO}_3^- \qquad [7.23]$$

Stereochemical analysis of the product of hydroxyl attack on phenyl sulphate, a molecule possessing a chiral sulphur atom by virtue of its oxygen isotopes (Eqn [7.24]), indicates that inversion of configuration occurs (Lowe, 1991a,b) consistent with either a concerted mechanism or a mechanism involving formation of sulphur trioxide in a cage which reacts with nucleophile faster than rotation can occur. The linear free-energy correlation for the quasi-symmetrical attack of substituted pyridines on isoquinoline-*N*-sulphonate (Fig. 7.1) indicates that transfer of the sulphuryl group is concerted. Analogues of sulphur trioxide have been demonstrated as intermediates and some (**4**, **5** and **6**) have even been isolated. It should be noted that although sulphur trioxide is readily obtained commercially, it is not available in its monomeric form; the monomer is so reactive that in the condensed phases sulphur trioxide exists only as a polymer. Commercially available sulphur trioxide liquid has stabilizer (usually dioxan) present which bonds with the sulphur. The monomer only exists in the dilute gas phase and it is not surprising that it has never been detected as a discrete intermediate in transfer reactions of the sulphuryl group in solution.

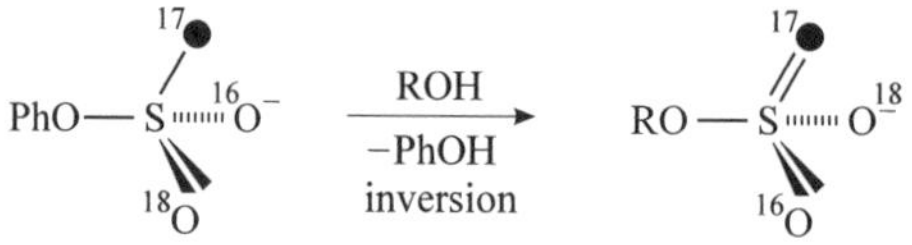

$$[7.24]$$

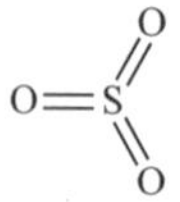

4 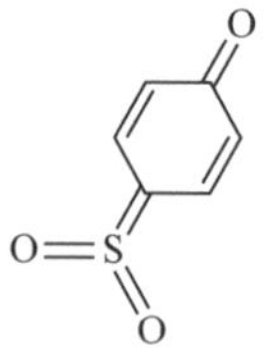   5 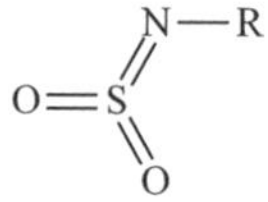   6

Sulphenyl group transfer

Reaction of nucleophiles with the sulphenyl group (Eqn [7.25]) resembles methyl group transfer between nucleophiles (Eqn [7.26]) and disulphide group reactions are of extreme importance in protein folding mechanisms.

$$CH_3SSAr \xrightarrow{+ \ Ar'S^-} \left| Ar'S \overset{\delta^-}{-}{-}{-} \overset{\overset{\displaystyle CH_3}{|}}{S} {-}{-}{-} \overset{\delta^-}{S}Ar \right|^{\ddagger} \xrightarrow{- \ ArS^-} Ar'SSCH_3 \qquad [7.25]$$

$$CH_3SAr \xrightarrow{+ \ Ar'S^-} \left| Ar'S \overset{\delta^-}{-}{-}{-} \overset{\overset{\displaystyle H}{\underset{\displaystyle H}{|}}}{C} {-}{-}{-} \overset{\delta^-}{S}Ar \right|^{\ddagger} \xrightarrow{- \ ArS^-} Ar'SCH_3 \qquad [7.26]$$

A linear Brønsted plot for the quasi-symmetrical attack of aryl thiolate ions on aryl methyl disulphides is consistent with a concerted mechanism (Wilson *et al.*, 1977). Chemical models have been prepared which are iso-electronic with transition states for the various types of sulphyl group transfer (structures **7** and **8**; Martin, 1983; Perkins *et al.*, 1985).

7   **8**

Further reading

Bender, M. L. (1951) Oxygen exchanges as evidence for the existence of an intermediate in ester hydrolysis, *J. Am. Chem. Soc.*, **73**, 1626.

Bender, M. L. (1960) Mechanisms of catalysis of nucleophilic reactions of carboxylic acid derivatives, *Chem. Rev.*, **60**, 53.

Bordwell, F. G. (1970) Are nucleophilic bimolecular concerted reactions involving four or more bonds a myth? *Acc. Chem. Res.*, **3**, 281.

Botting, N. P. (1994) Isotope effects in the elucidation of enzyme mechanisms, *Nat. Prod. Rep.*, 337.

Burgi, H. B. & Dunitz, J. D. (1983) From crystal statics to chemical-dynamics, *Acc. Chem. Res.*, **16**, 153.

Capon, B., Ghosh, A. K. & Grieves, D. McL. A. (1981) Direct observation of simple tetrahedral intermediates, *Acc. Chem. Res.*, **14**, 306.

Deslongchamps, P. (1983) *Stereoelectronic Effects in Organic Chemistry*, Pergamon, Oxford.

Dewar, M. J. S. (1948) *The Electronic Theory of Organic Chemistry*, Oxford University Press, Oxford, p. 117.

Giese, B. (1986) *Radicals in Organic Synthesis: Formation of Carbon Carbon Bonds*, Pergamon, Oxford.

Hine, J. (1977) The principle of least nuclear motion, *Adv. Phys. Org. Chem.*, **15**, 1.

Kice, J. C. (1980) Mechanisms and reactivity of organic oxyacids of sulphur and their anhyrides, *Adv. Phys. Org. Chem.*, **17**, 65.

Kirby, A. J. (1983) The anomeric effect and related stereoelectronic effects at oxygen, Springer, Berlin.

McClelland, R. A. & Santry, L. J. (1983) Reactivity of tetrahedral intermediates, *Acc. Chem. Res.*, **16**, 394.

Melander, L. & Saunders, W. H. (1980) *Reaction Rates of Isotopic Molecules*, John Wiley, New York, p. 266.

O'Leary, M. H. (1988) Transition state structures in enzyme catalysed decarboxylations, *Acc. Chem. Res.*, **21**, 450.

Saunders, W. H. (1975) Distinguishing between concerted and nonconcerted eliminations, *Acc. Chem. Res.*, **8**, 19.

Sinnott, M. L. (1988) The principle of least nuclear motion and the theory of stereoelectronic control, *Adv. Phys. Org. Chem.*, **24**, 113.

Sneen, R. A. (1973) Organic ion pairs as intermediates in nucleophilic substitution and elimination reactions, *Acc. Chem. Res.*, **6**, 46.

Thatcher, G. R. J. & Kluger, R. (1989) Mechanism and catalysis of nucleophilic substitution in phosphate esters, *Adv. Phys. Org. Chem.*, **25**, 99.

Westheimer, F. H. (1968) Pseudo-rotation in the hydrolysis of phosphate esters, *Acc. Chem. Res.*, **1**, 70.

Westheimer, F. H. (1981) Monomeric metaphosphate, *Chem. Rev.*, **81**, 313.

Westheimer, F. H. (1987) Why nature chose phosphates, *Science*, **235**, 1173.

Williams, A. (1989) Concerted mechanisms of acyl group transfer reactions in solution, *Acc. Chem. Res.*, **22**, 387.

Williams, A. & Douglas, K. T. (1975) Elimination–addition mechanisms of acyl transfer reactions, *Chem. Rev.*, **75**, 627.

8 Catalysis

8.1 Reactivity

Of the several hundred amino acid residues which may be possessed by an enzyme only a few are chemically involved in the bond-making and -breaking steps in catalysis. The main purpose of the other amino acid residues is to provide the three-dimensional framework required to maximize the binding energy between substrate and enzyme.

The majority of reactions occur by mechanisms involving a redistribution of electron density between ground state and transition state; catalytic groups stabilize these electron-density changes or provide an alternative pathway for the reaction. Stability may derive from charge complementarity, which is simply an extension of Pauling's electroneutrality principle. Electron-rich nucleophiles, bases and reducing agents tend to pair with electron-deficient electrophiles, acids and oxidizing agents, respectively. Stability may also arise from 'spreading' charge over several atoms, for example by delocalization.

Occasionally nature can reveal new chemistry but generally the mechanisms used by enzymes are of types that are already known to physical organic chemists. Sometimes the 'chemical' mechanism itself makes a large contribution to the rate enhancement brought about by the enzyme compared with the rate of reaction in the absence of similar catalytic groups used by the enzyme. This is especially true in reactions formally involving carbanions, where often the adduct-forming coenzymes provide routes to stabilized carbanions derived from the substrate. However, even in such cases it is important to remember that the 'non-reacting' parts of the protein and substrate contribute to the rate enhancement of the enzyme-catalysed reaction by utilizing the binding energy to compensate for energetically unfavourable processes.

Chemical catalysis derives its efficiency by avoiding the formation of unstable intermediates (Jencks, 1981), which often simply means bypassing the development of charge on a particular atom. Of course, it is imperative that intermediates are not too stabilized by the enzyme—otherwise, unwanted enzyme-bound derivatives would accumulate and inhibit catalysis. Catalysis is needed simply because, in its absence, the pathways available to reactions would have high activation energies and would not occur.

For example, substitution reactions at the acyl centre often involve expulsion of a leaving group Lg (structure **1**). This bond breaking may be very difficult if Lg is expelled as an unstable Lg⁻ anion but will be greatly facilitated if Lg is first protonated and then expelled as the more stable LgH. Similarly the nucleophile, NuH, often contains an ionizable hydrogen, and proton removal will render it more powerful. Although the practice is potentially misleading, sites of catalysis can often be predicted by examining the bonds to be made and broken in a reaction.

The hydrolysis of peptides (Eqn [8.1]) must involve the making of a bond between the carbonyl carbon and the water oxygen, the breaking of a bond between the carbonyl carbon and nitrogen and the transfer of several protons. If either of the heavy-atom transfer steps occurred without proton transfer, high-energy-charged intermediates, for example **2**, would be formed with a large change in the acidity and basicity of the constituent atoms. For example, the attacking water molecule changes from a weak acid into a very strong one and both the amide nitrogen and the carbonyl oxygen change from very weak bases to strong ones. Proton transfer between

$$RCONHR' + H_2O \rightleftharpoons RCO_2^- + R'NH_3^+ \qquad\qquad [8.1]$$

reactants and the solvent water changes from being thermodynamically unfavourable to favourable on formation of such unstable intermediates as **2**. Catalysis can increase the rate of reaction by 'trapping' such intermediates, by stabilizing an intermediate and the transition states leading to it, or by providing an alternative mechanism not requiring the formation of unstable intermediates. The mechanism of catalysis of many of these types of reactions is *enforced* by the lifetime and acid–base properties of the initially formed intermediate (Jencks, 1976; see also Section 7.1).

The acid-catalysed hydrolysis of acetals usually involves specific acid catalysis because protonation of the alcohol oxygen leaving group is required to facilitate bond fission (Scheme 8.1). The conjugate acid of the acetal is deprotonated at a rate which is faster than that of carbon–oxygen bond fission to generate the unstable oxocarbocation. Electron-withdrawing substituents in the leaving alcohol may make the conjugate acid so unstable that it is not formed and proton transfer from general acids is sufficient to cause bond fission (Scheme 8.1). Similarly, the conjugate acid may be 'bypassed' by general acid catalysis if structural changes make the oxocarbocation more stable (Fig. 8.1).

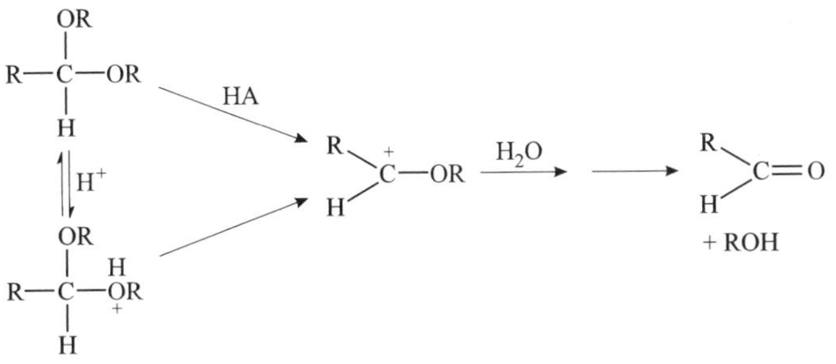

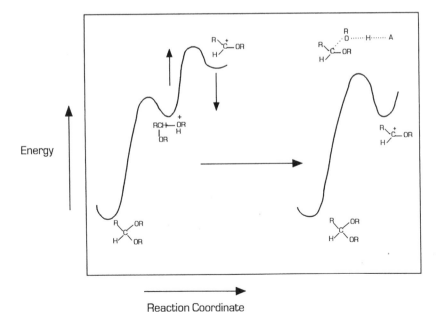

Scheme 8.1 Specific and general acid-catalysed formation of carbocation intermediates from acetals.

Covalent catalysis is effective because the transient chemical modification that occurs by forming a covalent bond to the substrate can give an intermediate which either facilitates the bond-making and -breaking processes or enables an active intermediate to be formed at a higher concentration than is possible in the absence of the catalyst. Most coenzymes function by one or both of these mechanisms. A requirement for efficient covalent catalysis is that the intermediate must react rapidly to prevent its accumulation and consequent return to reactants.

Energy

Reaction Coordinate

Fig. 8.1 Energy profiles for the change from specific to general acid-catalysed degradation of acetals, by raising the energy of the conjugate acid or by stabilizing the carbocation intermediate.

A simple illustration of the rate enhancement obtainable by chemical transformation is the conversion of carbonyl groups into imines (Schiff bases) (Eqn [8.2]).

$$\text{\Large$\diagdown$}\!\!\!\overset{\displaystyle}{\underset{\displaystyle}{C}}\!\!\!\text{\Large$\diagup$}=O \; + \; RNH_2 \; \rightleftharpoons \; \text{\Large$\diagdown$}\!\!\!\overset{\displaystyle}{\underset{\displaystyle}{C}}\!\!\!\text{\Large$\diagup$}=NR \; + \; H_2O \qquad\qquad [8.2]$$

Replacement of oxygen by nitrogen in a molecule makes it much more basic. For example, the pK_a of the conjugate acid of acetone is $ca -7$ compared with a value of $ca\ 7$ for the corresponding imine, and at pH 7 the fraction of the ketone protonated is $ca\ 10^{-14}$ (Eqn [8.3]) whereas half of the imine exists in the protonated form (Eqn [8.4]).

$$\begin{array}{c} CH_3 \\ \diagdown \\ C=O \; + \; H^+ \\ \diagup \\ CH_3 \end{array} \rightleftharpoons \begin{array}{c} CH_3 \\ \diagdown \\ C=OH^+ \\ \diagup \\ CH_3 \end{array} \qquad\qquad [8.3]$$

$$\begin{array}{c} CH_3 \\ \diagdown \\ C=NR \; + \; H^+ \\ \diagup \\ CH_3 \end{array} \rightleftharpoons \begin{array}{c} CH_3 \\ \diagdown \\ C=NHR^+ \\ \diagup \\ CH_3 \end{array} \qquad\qquad [8.4]$$

Unprotonated acetone is more susceptible to nucleophilic attack than the neutral imine group because of the greater electronegativity of oxygen. However, the protonated imine is many millions of times more reactive towards nucleophiles than the carbonyl group, and the predominant reaction at neutral pH will occur by attack on the imine conjugate acid. Therefore, a significant rate enhancement of a reaction involving nucleophilic attack on a carbonyl carbon can result from simply converting the carbonyl to an iminium group.

Enamines are the nitrogen analogues of enols but their formation from imines is thermodynamically more favourable than enol formation from ketones (Scheme 8.2); the equilibrium constant for enol formation is $ca\ 10^{-8}$ compared with a value of 10^{-5} for enamine formation. However, at pH 7 half of the imine exists as the iminium ion and the proportion of enamine present is 10^8-fold greater than the proportion of enolate anion (Page, 1984). In general this implies that loss of an electrophile from the α-carbon of a charged iminium ion to form a neutral enamine (Eqn [8.5]) is much easier than that from a neutral ketone to form a charged enolate ion (Eqn [8.6]).

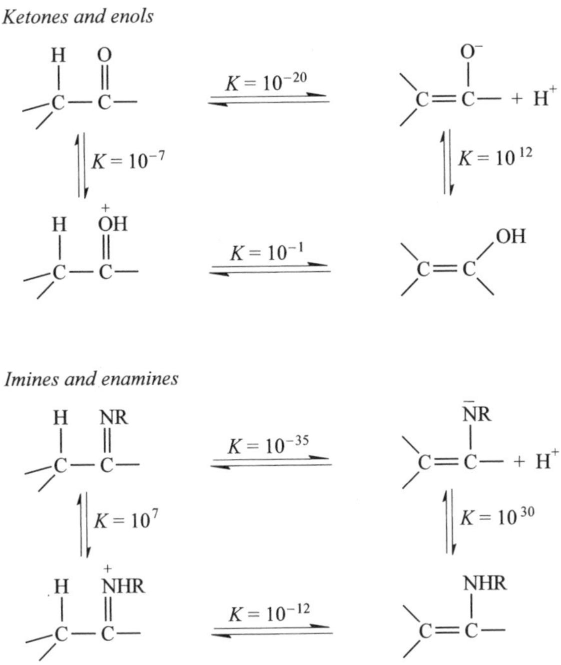

Ketones and enols

Imines and enamines

Scheme 8.2 Equilibrium constants for enol and enamine formation.

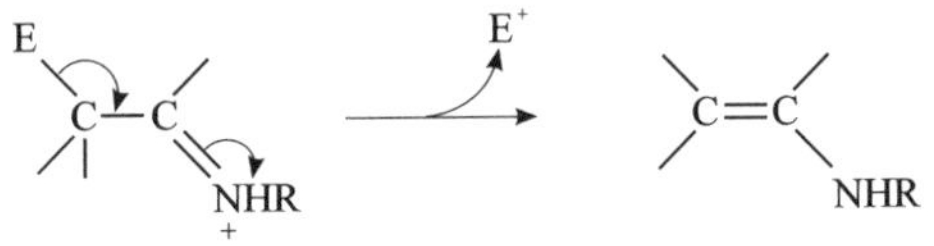

[8.5]

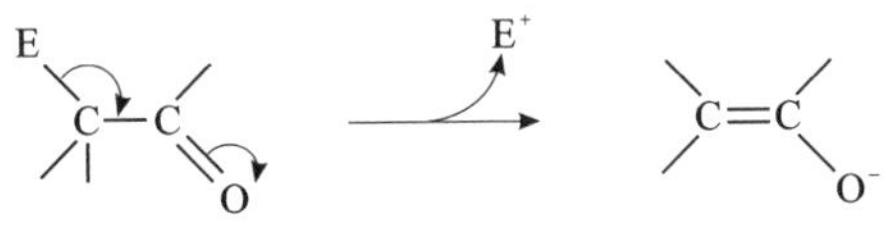

[8.6]

For example, the loss of a proton from the α-carbon (E = H, Eqns [8.5] and [8.6]) from the iminium ion is *ca* 10^9 times faster than that from the corresponding ketone. The neutral enamine is, of course, a delocalized structure **3**, which readily explains the nucleophilic character of the enamine α-carbon.

3

8.2   Coenzymes

Introduction

Most coenzymes are effective in catalysis because they stabilize normally reactive intermediates or prevent their formation. Coenzymes are not strictly catalysts but they often become involved in catalysis by acting as transfer agents. In the examples which follow, we delineate the reaction of co-enzymes with substrate and, where appropriate, leave consideration of the interaction with the enzyme to Chapter 10. The coenzyme is often bound tightly to the apo-enzyme to yield a holo-enzyme which catalyses the reaction between substrates.

Pyridoxal phosphate coenzyme

All living organisms use pyridoxal phosphate (PLP), a derivative of vitamin B_6, as a coenzyme to synthesize, degrade and interconvert amino acids (Akhtar *et al.*, 1984; Martell, 1989). The chemical logic of this system is that the pyridine ring can act as an electron sink and stabilize carbanions. Amino groups readily condense with the aldehyde of PLP to form a conjugated imine (**4**) (Scheme 8.3). The imines derived from amines may readily lose an electrophile, E (e.g. H^+, CO_2), to form a delocalized carbanion (**5**) which is especially stable if the pyridine nitrogen is protonated.

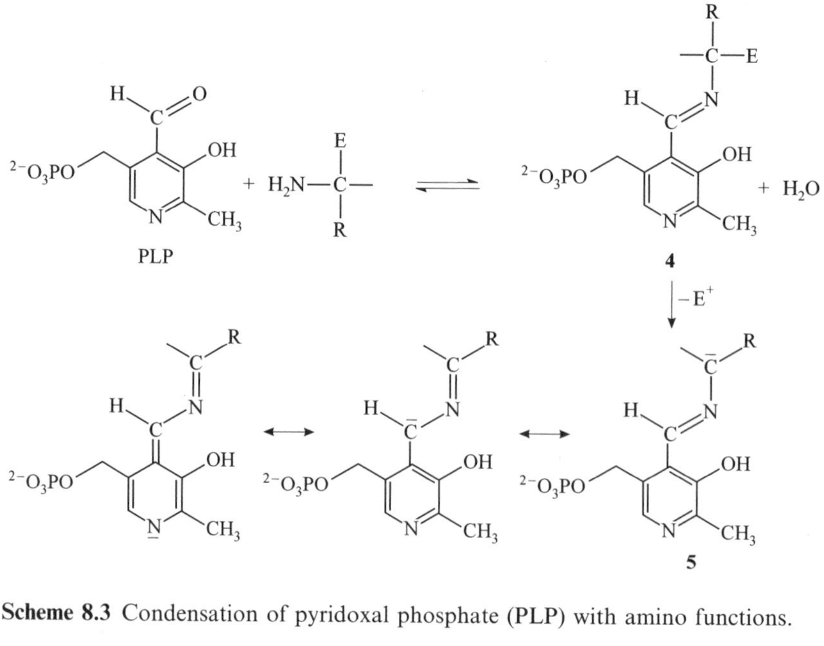

Scheme 8.3 Condensation of pyridoxal phosphate (PLP) with amino functions.

There has been much interest in the position of the hydroxyl proton in pyridoxal Schiff bases. Given that the imine nitrogen and phenolic hydroxyl are hydrogen-bonded, it makes little difference to mechanistic conclusions whether the proton is nearer nitrogen or oxygen. The pK_a of iminium ions is normally 6–7 and although pyridoxal Schiff bases are tribasic the important species near neutrality are the protonated pyridinium zwitterion (6) or its tautomer (7) and the corresponding structures with the pyridinyl nitrogen unprotonated.

Although pyridoxal phosphate is involved in a wide variety of reactions, the processes are all related mechanistically by the electron withdrawal towards the cationic or hydrogen-bonded imine nitrogen and into the electron sink of the pyridine/pyridinium ring.

Loss of the electrophile E from the imine 4 forms the carbanion 5, which may be attacked by another electrophile at two places. Addition to the α-carbon of the original amine generates a new imine, which, upon hydrolysis, gives a transformed amine derivative. Addition at the pyridoxal carbonyl carbon, usually by a proton, generates an isomeric imine (8) which, upon hydrolysis, gives pyridoxamine phosphate (PMP) and the α-ketone derivative of the original α-amine (Scheme 8.4). Because the steps are reversible, this system allows a mechanism for nitrogen transfer (transamination) from one substrate to another, with PMP temporarily storing the nitrogen.

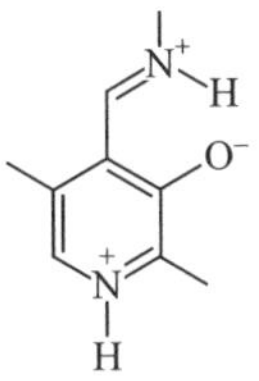

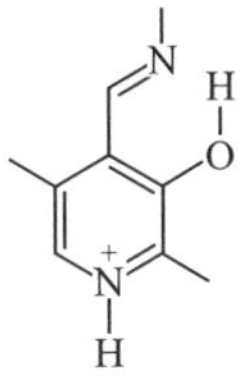

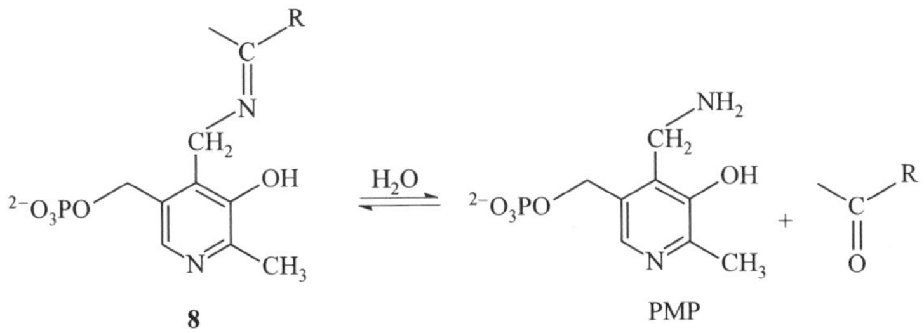

Scheme 8.4 Expulsion of an oxo derivative from the complex between PLP and an amine.

Three types of electrophile (E) (Scheme 8.3) have been identified:

(a) $E = H^+$, where deprotonation, reprotonation and hydrolysis could give the racemic amino acid or the amino acid with inverted configuration;

(b) $E = CO_2$, where decarboxylation at the α-position followed by protonation at the α-carbon and subsequent hydrolysis would yield an amine and PLP; protonation of the decarboxylated adduct at the pyridoxal carbonyl carbon and subsequent hydrolysis would yield an aldehyde and PMP;

(c) $E = R\!-\!CHO$, where removal of an aldehyde from the initially formed imine is followed by a retroaldol condensation, as exemplified by the conversion of serine to glycine by loss of an equivalent formaldehyde unit.

Which electrophile is lost from the amino acid residue is controlled by the enzyme (see Schemes 10.11 and 10.12), which may bind the PLP imine so that the electrophile is in close proximity to a suitable base to aid abstraction and also so that the orbital of the bond to be broken is periplanar with the π acceptor system, i.e. orthogonal to the plane of the pyridine ring (**9**). Maximal orbital overlap will lower the activation energy for the reaction. Aldol-type reactions can also occur with PLP; the key to making carbon–carbon bonds is the formation of a stabilized carbanion. Proton abstraction from the initially formed imine gives a 'masked' carbanion which can attack electrophilic centres. For example, attack on coenzyme A is reminiscent of a Claisen ester condensation (**10**).

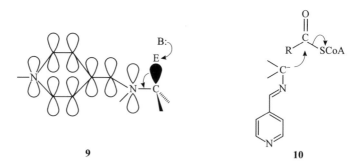

9   10

Another interesting feature of PLP catalysis is the stabilization of both electron-poor and electron-rich centres generated at the β-carbon of the original amino acid, illustrated in Scheme 8.5. Pathway a involves α,β-elimination by proton abstraction at C_α and expulsion of the leaving group at C_β. This generates an electrophilic electron-deficient β-carbon in **11** which can be attacked by nucleophiles. Conversely, pathway b gives the isomeric imine seen earlier, which can now lose X as an electrophile at C_β to generate an enamine with a nucleophilic electron-rich β-carbon in **12** which can be attacked by electrophiles.

Pyridoxal enzymes are discussed further in Chapter 10, Section 10.3.

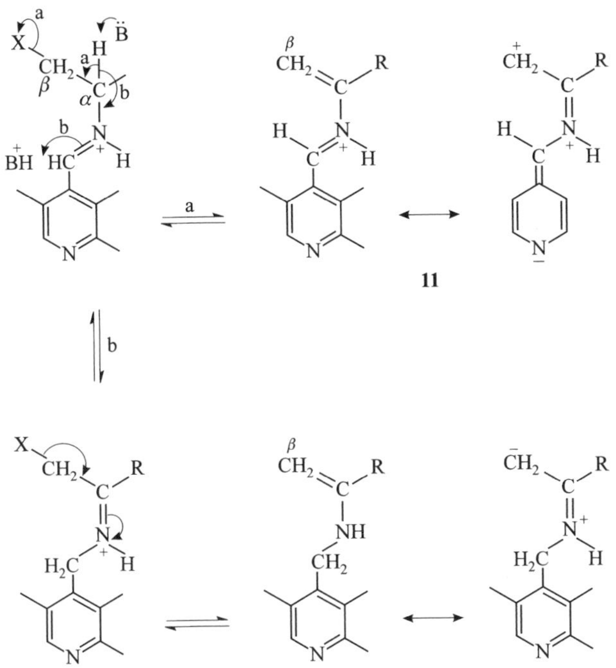

Scheme 8.5 Activation at the β-carbon mediated by the PLP complex with an amine.

Thiamine pyrophosphate coenzyme

The coenzyme thiamine pyrophosphate **13** contains an *N*-alkylated thiazole and a pyrimidine, and can stabilize electron density by spreading and neutralizing charge; it catalyses the decarboxylation of α-keto acids, the formation of α-ketols (acyloins) and transketolase reactions (Haake, 1987; Kluger, 1987).

Proton loss from C2 of the *N*-alkylthiazolium salt **14** gives an ylide-type carbanion (**15**) which is stabilized by the positive charge on nitrogen and by the adjacent sulphur—ionization gives a non-bonded pair of electrons in an sp^2 orbital which *cannot* therefore be stabilized by p delocalization. The acidity of C—H bonds is significantly enhanced by attached sulphur; although this is usually explained by d–p bonding between the carbanion lone pair and sulphur 3d orbitals, theoretical calculations indicate that this bonding is not significant and sulphur carbanions may owe their stability simply to the large polarizability of the relatively large sulphur atom.

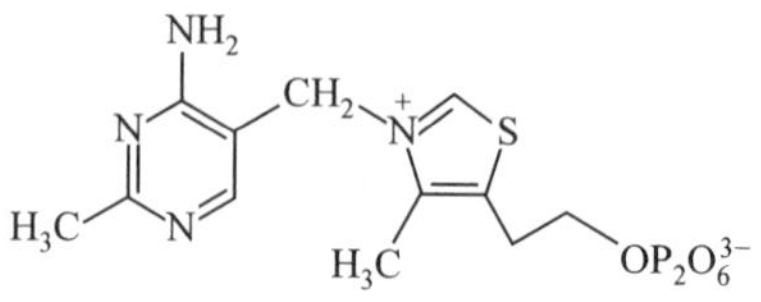

13

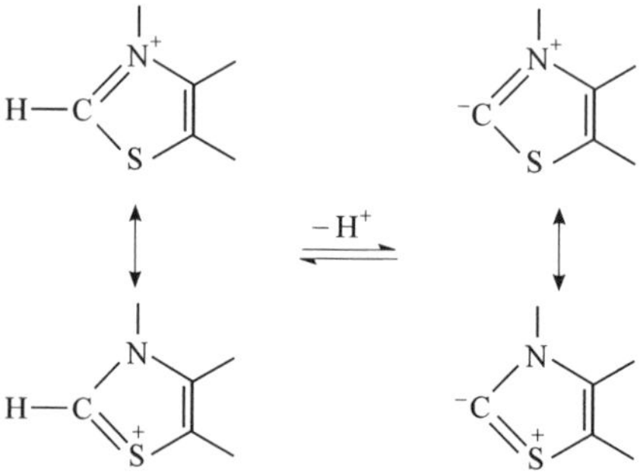

14   15

As with all reactive intermediates, it is important that they are not too stabilized to prevent facile further reaction. The thiazolium ylide is a potent carbon nucleophile but also a good leaving group. This is reminiscent of cyanide ion in the benzoin condensation and, in fact, the chemical logic of that reaction mechanism is similar to the thiamine-catalysed decarboxylations of α-keto acids (Scheme 8.6).

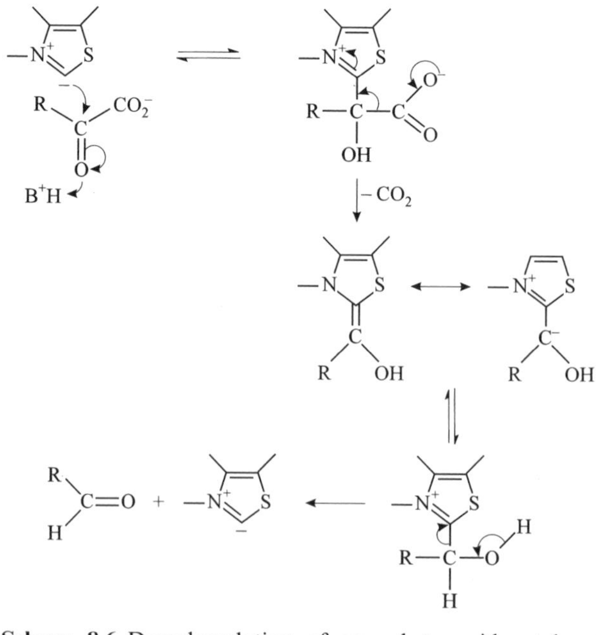

Scheme 8.6 Decarboxylation of an α-keto acid catalysed by thiamine pyrophosphate derivatives.

Carbon–carbon bond fission of an α-keto acid to give carbon dioxide generates an unstable carbanion. However, nucleophilic attack of the thiamine carbanion on the α-carbonyl group of the α-keto acid gives an intermediate from which carbon dioxide is readily lost because the carbanion now generated is a delocalized system resembling an enamine. This intermediate, a hydroxyalkylthiamine pyrophosphate, is in fact the form of much of the coenzyme *in vivo*. The process is completed by expulsion of the catalytic thiamine ylide.

The usefulness of thiamine pyrophosphate in carbon–carbon bond-forming reactions is exemplified by the condensation of aldehydes and ketones to give α-hydroxyketones (Scheme 8.7) and the mechanistically similar α-ketol transfers (Scheme 8.8).

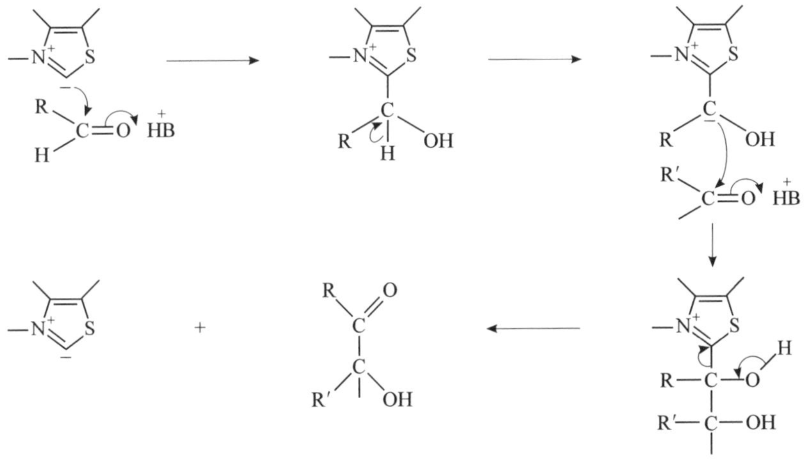

Scheme 8.7 Condensation of aldehydes and ketones to give α-hydroxy ketones.

In summary, the chemical usefulness of thiamine pyrophosphate depends on its ease of carbanion formation at C2; this carbanion is not only a good nucleophile but also a reasonably stable leaving group. In addition, the cationic imine stabilizes the formation of a carbanion on the *adjacent* carbon, bonded to C2.

Adenosine triphosphate (ATP)

Adenosine triphosphate (ATP; **16**) is a purine ribonucleoside triphosphate which, at neutral pH, exists mainly as the tetra-anion; it is a molecule which is important in almost all biological processes, acting mainly as an energy transfer agent. All reactions involving ATP usually require Mg^{2+} or other divalent cations which reduce the negative charge density on the

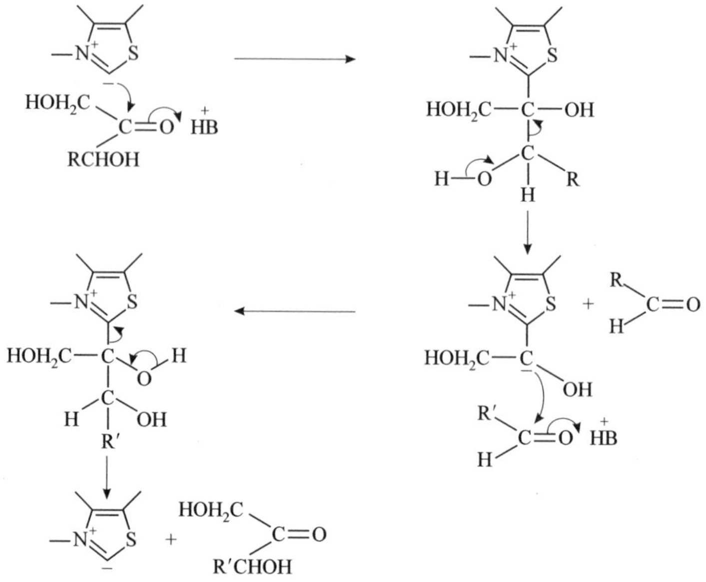

Scheme 8.8 Acyl group transfer catalysed by thiamine pyrophosphate derivatives.

molecule and control the site of nucleophilic attack. ATP allows many reactions to take place by phosphorylation of the substrate, which provides a good leaving group for carbon—carbon bond synthesis and elimination reactions. Although the rôle of ATP is often discussed in terms of its very exergonic hydrolysis to adenosine diphosphate and monophosphate, these reactions rarely occur directly and it is more useful to consider the mechanistic rôle (see also Section 7.2).

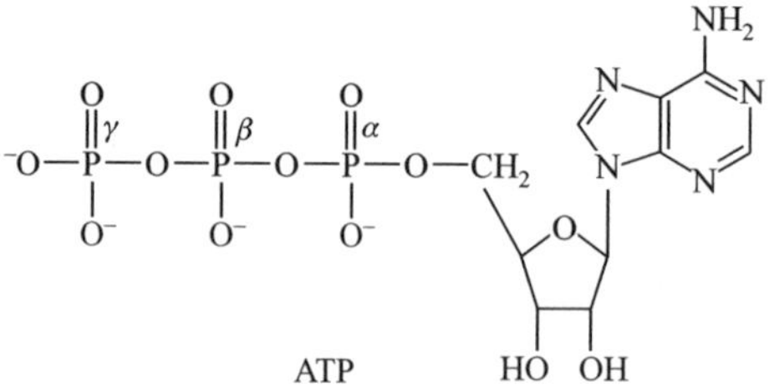

ATP

16

The two terminal linkages of the triphosphate are phosphoric acid anhydrides, i.e. they are activated phosphoric acid derivatives and nucleophilic attack at the electrophilic γ or β-phosphorus is thermodynamically very favourable. The advantage of phosphoric anhydride derivatives over acyl derivatives is that the former are kinetically stable in neutral aqueous solution unless a suitable enzyme is present to catalyse their reactions.

The direct expulsion of OH groups from saturated or unsaturated carbon is a difficult process because hydroxide ion is a strong base and a poor leaving group (Scheme 8.9); this provides a ready example of the chemical logic of using ATP as a phosphorylating agent to generate substrates with a good leaving group. If the hydroxyl group is first converted to a phosphate ester, expulsion will occur easily because the leaving group is now a resonance-stabilized phosphate anion, which is weakly basic (Eqn [8.7]).

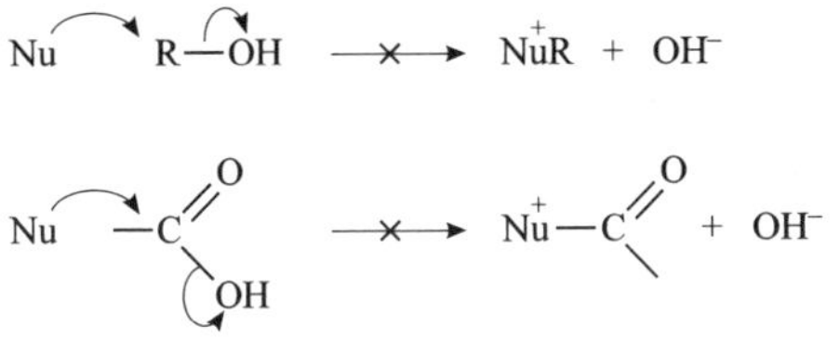

$$Nu \quad R-OH \xrightarrow{\quad\times\quad} Nu^{+}R + OH^{-}$$

$$Nu \quad -C \overset{\displaystyle O}{\underset{\displaystyle OH}{\big\langle}} \xrightarrow{\quad\times\quad} Nu^{+}-C \overset{\displaystyle O}{\big\langle} + OH^{-}$$

Scheme 8.9 The hydroxide ion requires activation to convert it into a good leaving group.

$$ATP^{4-} + ROH \xrightarrow[\quad]{ADP^{3-},\,H^{+}} ROPO_3^{2-} \xrightarrow[\quad]{NuH} RNu + HPO_4^{2-} \qquad [8.7]$$

A carboxylate anion is an unactivated, low-energy, resonance-stabilized system that is even less susceptible to nucleophilic attack, but it may be converted into an activated substrate, an acyl phosphate, by phosphoryl transfer from ATP (Eqn [8.8]) in which the real or hypothetical metaphosphate, PO_3^{-}, is transferred to the hydroxyl group (Section 7.2).

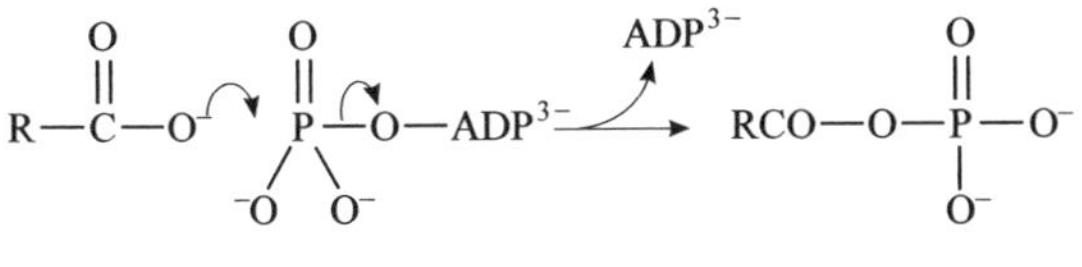

$$R-\overset{\displaystyle O}{\overset{\displaystyle \|}{C}}-O \quad \underset{\displaystyle ^{-}O\quad O^{-}}{\overset{\displaystyle O}{\overset{\displaystyle \|}{P}}}-O-ADP^{3-} \xrightarrow{ADP^{3-}} RCO-O-\overset{\displaystyle O}{\underset{\displaystyle O^{-}}{\overset{\displaystyle \|}{P}}}-O^{-} \qquad [8.8]$$

The product is a mixed acid anhydride and is readily attacked by nucleophiles at the carbonyl centre (Eqn [8.9]).

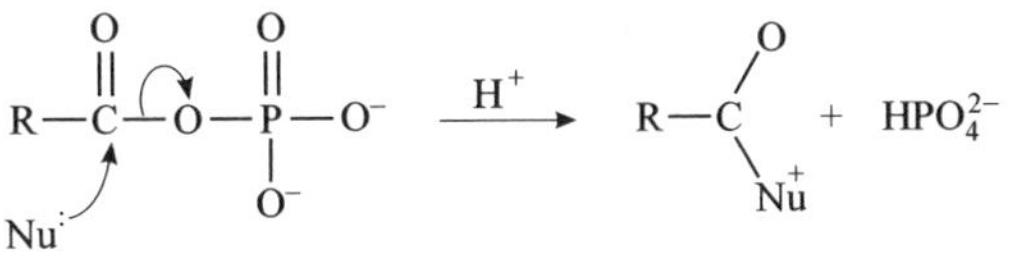

$$R-\overset{\displaystyle O}{\overset{\displaystyle \|}{C}}-O-\overset{\displaystyle O}{\underset{\displaystyle O^{-}}{\overset{\displaystyle \|}{P}}}-O^{-} \xrightarrow{H^{+}} R-C\overset{\displaystyle O}{\underset{\displaystyle Nu^{+}}{\big\langle}} + HPO_4^{2-} \qquad [8.9]$$

Acyl phosphates are too reactive to react generally as acyl transfer agents but they do occur, for example, in the biosynthesis of acetyl coenzyme A, the intermediate used for acetyl transfer (Eqn [8.10]). Here ATP is effectively acting as a dehydrating agent by activating the carboxylate oxygen anion for elimination by intermediate conversion to a phosphate ester from which phosphate can be displaced by the thiolate anion nucleophile of coenzyme A.

$$CH_3CO_2^- + ATP \xrightarrow{\text{acetate kinase}} CH_3CO-OPO_3^{2-} \xrightarrow[\substack{\text{phospho} \\ \text{transacetylase}}]{\text{CoASH}} CH_3CO-SCoA + HPO_4^{2-}$$

[8.10]

8.3   Proton transfer

Introduction

Proton transfer can catalyse reactions by stabilizing reactive intermediates either by neutralizing a strongly basic intermediate or by ionization of a strongly acidic intermediate. Many important reactions involve proton transfer and this step can itself often be catalysed. The varied mechanisms available for proton transfer are summarized as follows. (Jencks, 1976, 1980.)

Stepwise proton transfer (trapping)

Proton transfer between simple electronegative atoms of simple acids and bases occurs in a series of steps. Diffusion together is followed by proton transfer and then diffusion apart (Chapter 6, Section 6.4, Eqn [6.5]). If proton transfer is thermodynamically favourable, i.e. the pK_a of the proton donor HA (pK_a^{HA}) is less than that of the conjugate acid of the acceptor BH^+ (pK_a^{HB}), the rate-limiting step is diffusion together of the acid–base pair, k_1. If proton transfer is thermodynamically unfavourable, i.e. the pK_a of the proton donor HA is greater than that of the conjugate acid of the acceptor BH^+, the rate-limiting step is diffusion apart of the acid–base pair, k_2 (Eqn [6.5]).

Reactive intermediates which are generated as a result of covalent bond formation or fission are often proton donors or acceptors. For example, the nucleophilic addition of water to the carbonyl carbon of an amide (Eqn [8.11]) generates the reactive tetrahedral intermediate **2**, which is unstable with respect to proton transfer to or from water and with respect to breakdown back to reactants. Expulsion of water from **2** will occur much faster than expulsion of the unstable amine anion, $R'NH^-$. Removal of H^+ from the attacking water of **2** or proton addition to oxygen or nitrogen of **2** will generate relatively more stable intermediates, although proton addition to the oxyanion of **2** requires proton transfer from the oxonium ion to give products. A rate increase will be observed if these proton transfers occur to or from added general acids or bases rather than to solvent water.

$$
\underset{\textbf{RCNHR}'}{\overset{\text{O}}{\overset{\|}{}}}
\quad
\underset{k_{-1}}{\overset{k_1\,\text{H}_2\text{O}}{\rightleftarrows}}
\quad
\underset{\substack{|\\ {}^+\text{OH}_2 \\[2pt] \textbf{2}}}{\overset{\overset{\text{O}^-}{|}}{\text{R}-\text{C}-\text{NHR}'}}
\quad
\underset{k_{\text{A}}(\text{A}^-)}{\overset{k_{\text{HA}}(\text{HA})}{\longrightarrow}}
\quad
\underset{\substack{|\\ {}^+\text{OH}_2 \\[2pt] \textbf{17}}}{\overset{\overset{\text{O}^-}{|}}{\text{R}-\text{C}-\overset{+}{\text{N}}\text{H}_2\text{R}'}} + \text{A}^-
$$

$$[8.11]$$

The extent of catalysis depends critically upon the *stability* of the intermediate **2**. If the rate of expulsion of H_2O from **2**, k_{-1}, were slower than proton transfer to solvent water, the rate of formation of the intermediate, k_1, would be the rate-limiting step and no catalysis would be observed. The rate of protonation of the amine nitrogen of **2** by solvent water, k_{HA} (Eqn [8.11], HA = H_2O) depends on the basicity of the nitrogen and is given by $k_{\text{A}}K_{\text{w}}/K_{\text{a}}$ where k_{A} represents the diffusion-controlled abstraction of a proton by hydroxide ion with a value of approximately $10^{10}\,\text{M}^{-1}\,\text{s}^{-1}$ and K_{a} is the acid dissociation constant of the ammonium group of **17**. If the rate of expulsion of H_2O from **2** to regenerate reactants is faster than or similar to the rate at which **2** is 'trapped' by proton abstraction from water, the rate-limiting step occurs after the formation of intermediate **2** and the addition of general acids (i.e. other proton donors) may increase the observed rate.

If HA is a stronger acid than the ammonium function of **17** the rate of proton transfer to **2**, k_{HA}, would be a diffusion rate and the *observed rate constant would be independent of the acidity of HA*. If HA is a weaker acid than the ammonium function of **17**, the proton transfer from general acids, HA, to the nitrogen of **2** (Eqn [8.11]) is given by $k_{\text{A}}K_{\text{a}}^{\text{HA}}/K_{\text{a}}$ where K_{a}^{HA} is the acid dissociation constant of HA and k_{A} is the diffusion-controlled rate. *The observed rate would now be dependent upon the acidity of the catalyst HA and obey a Brønsted law with slope -1.*

In our hypothetical reaction the intermediate **2** could also be stabilized by proton transfer from oxygen to give **18** (Eqn [8.12]).

$$
\underset{\substack{|\\ {}^+\text{OH}_2 \\[2pt] \textbf{2}}}{\overset{\overset{\text{O}^-}{|}}{\text{R}-\text{C}-\text{NHR}'}}
\quad
\underset{k_{\text{BH}}[\text{BH}^+]}{\overset{k_{\text{B}}[\text{B}]}{\rightleftarrows}}
\quad
\underset{\substack{|\\ \text{OH} \\[2pt] \textbf{18}}}{\overset{\overset{\text{O}^-}{|}}{\text{R}-\text{C}-\text{NHR}'}} + \text{BH}^+
\qquad [8.12]
$$

The proton acceptor B could be solvent water or a general base catalyst. In this case also, the reaction will only be catalysed if the rate of breakdown of **2** to regenerate reactants is faster than the rate of proton transfer. In this case such catalysis would be independent of the basic strength of the catalyst B, as proton transfer would invariably be thermodynamically favourable and hence occur at the maximum diffusion-controlled rate. As proton transfer to solvent is thermodynamically favourable, proton donation to 55 M water is likely to be faster than to, say, 1 M added base. Therefore, any observed catalysis by base must be due to another mechanism such as transition-state stabilization by a hydrogen-bonding or a concerted process, as described later.

It is of interest to examine the general case when a nucleophile attached to hydrogen attacks a carbonyl group, eventually to displace the group Lg, but becomes acidic when the intermediate is formed (Eqn [8.13]).

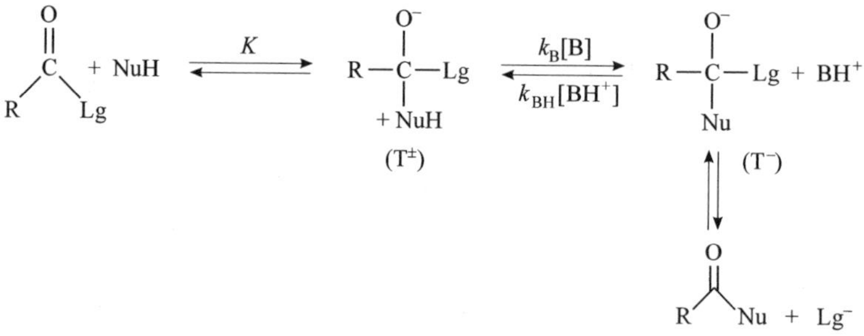

[8.13]

Catalysis by the general base B will be observed when the intermediate $T^\pm$ breaks down to reactants *faster* than $T^\pm$ transfers a proton to water. The rate of formation of T^-, which may or may not represent the overall rate of reaction, is given by $k_B[B]K$, where K is the equilibrium constant for formation of the intermediate $T^\pm$ and k_B is the rate of proton transfer from $T^\pm$ to the catalyst B.

An example will illustrate the typical rate enhancement brought about by general acid and base catalysts. The reaction of amines with penicillins requires the removal of a proton from the attacking amine after the initial formation of the tetrahedral intermediate (Scheme 8.10), (Morris & Page, 1980a).

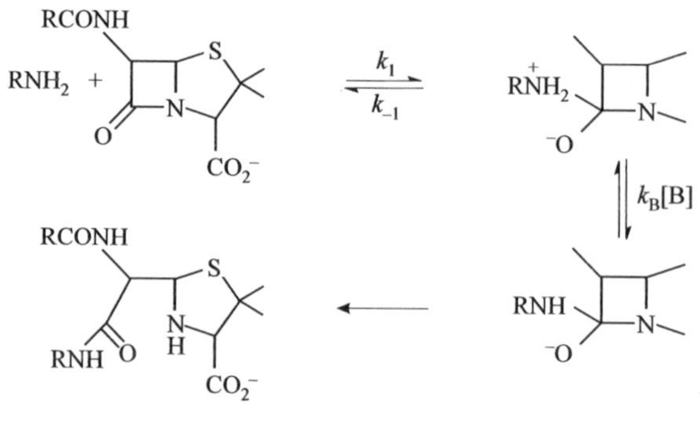

Scheme 8.10 Aminolysis of penicillins catalysed by general bases.

The rate of expulsion of RNH_2 to regenerate the reactants, k_{-1}, is between 10^6 and $10^9 s^{-1}$ and much faster than that of proton transfer to water ($B = H_2O$), between 10^{-1} and $10^2 s^{-1}$. When the general base catalyst is a second molecule of amine ($B = RNH_2$) proton transfer is much more efficient and gives a rate enhancement of about 1000-fold. The greater basicity of the amino group compared with water is offset by a solvent water concentration of 55 M; amine catalyst at 1 M is then only about 20 times more efficient than water. In general, rate enhancements of 10–100 are brought about by general acids and bases up to a concentration of about 1 M. It is unlikely that a greater rate enhancement is obtained when the general acid or base catalyst of an enzyme acts within the enzyme–substrate complex.

Preassociation

When the rate-limiting step of a reaction such as that of Eqn [8.13] is the diffusion-controlled encounter of two reagents, an enzyme may increase the rate simply by having the nucleophile and catalyst preassociated so that they do not have to diffuse through bulk solution before reaction can occur. A similar situation occurs in simple intermolecular reactions when the rate of breakdown of the intermediate to regenerates reactants (k_{-1}; see Eqn [8.14] and Fig. 8.2) is faster than the rate of separation of the intermediate and catalyst (between 10^{10} and $10^{11} s^{-1}$) to give $T^{\pm}$ and B. The formation of $T^{\pm}.B$ is then forced to take place by a preliminary association of the reactants and catalyst (Eqn [8.14] and Fig. 8.2).

$$B + NuH + \overset{\diagdown}{\underset{\diagup}{}}C{=}O \underset{k_{-1}}{\overset{k_1}{\rightleftarrows}} B.HNu. \overset{\diagdown}{\underset{\diagup}{}}C{=}O \underset{k_{-2}}{\overset{k_2}{\rightleftarrows}} B.\overset{+}{H}Nu{-}\overset{\mid}{\underset{\mid}{C}}{-}O^- \qquad [8.14]$$

$$(Int) \qquad\qquad (B. T^{\pm})$$

The preassociation mechanism is more efficient than the trapping mechanism (Section 8.3) because it generates an intermediate (Int) which is immediately trapped by an ultrafast proton transfer and thus avoids the diffusion-controlled step bringing catalyst and intermediate together. This mechanism is sometimes called a 'spectator' mechanism because although the catalyst is *present* in the transition state it is not undergoing any transformation (Jencks, 1976).

The rate enhancement for preassociation compared with 'trapping' is given by the ratio of the rate of breakdown of the intermediate $T^{\pm}.B$ to generate reactants to the rate of its dissociation to $T^{\pm}$ and B (Fig. 8.2). The *maximum* lowering of the free energy of activation obtainable is the activation energy for diffusion apart from the catalyst and intermediate, i.e. *ca* 15 kJ mol^{-1}, a rate enhancement of *ca* 400. There can be no rate advantage from a preassociation mechanism when the proton transfer to $T^{\pm}$ from B is thermodynamically unfavourable.

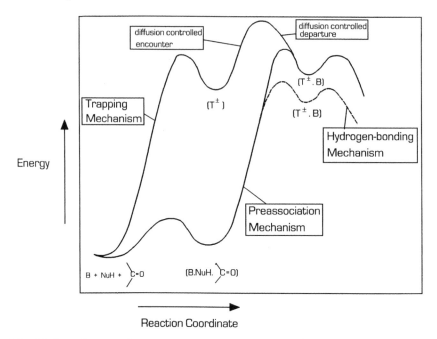

Fig. 8.2 Reaction coordinate diagram showing how a reaction will proceed through a preassociation complex (B.NuH ⌒C = 0) when the associated intermediate and catalyst $(T^{\pm}.B)$ breaks down to reactants faster than it dissociates to separated intermediate $(T^{\pm})$ and catalyst (B) (see Section 8.3, Eqn [8.14]). Additional stabilization of $(T^{\pm}.B)$ may occur through hydrogen bonding (below).

Hydrogen bonding

As an intermediate becomes progressively more *unstable*, catalysis by hydrogen bonding becomes progressively more important. In the preassociation mechanism for intermolecular reactions the catalyst is in close proximity to the reactants because they are in the same solvent cage. If the catalyst is correctly located for subsequent proton transfer there is the possibility of stabilization of the transition state by hydrogen bonding, which must be more favourable than that of the substrate to water for there to be a rate enhancement. The dependence of the stability of hydrogen bonds upon acidity is small, with a typical Brønsted β-value of 0.2. Changing a proton-donor, hydrogen-bonding catalyst from water to one of pK_a 5 is therefore calculated to increase the rate, compared with the preassociation mechanism, by *ca* 150-fold. The importance of catalysis by hydrogen bonding increases sharply as the intermediate becomes less stable. The *maximum* rate advantage of a preassociation mechanism with hydrogen bonding compared with the trapping mechanism (see p.177) in the case of the proton donor with pK_a 5 is $400 \times 150 = 6 \times 10^4$-fold.

Concerted proton transfer

When the 'intermediate' is so unstable that it cannot exist, it would have a lifetime less than that of a bond vibration ($\simeq 10^{-13}$ s) and the reaction *must* proceed by a concerted mechanism (see Section 7.1, p. 134). The proton transfer steps and other covalent bond-forming and -breaking processes occur simultaneously but with a varying degree of coupling between their motions. However, it is still not clear whether a concerted mechanism can occur when the intermediates, which would be formed in a stepwise mechanism, have a significant lifetime. This is an important question for reactions catalysed by enzymes, because the nature of the intermediate itself will control whether the enzyme- and non-enzyme-catalysed mechanisms are forced to be similar if the sole criterion for a concerted mechanism is the stability of the intermediate.

If there is little coupling between the motions of bond making and breaking because of unfavourable geometry and orbital overlap, a concerted mechanism can only occur when it is enforced, i.e. when there is no activation barrier for decomposition of the hypothetical intermediate.

If there is an energetic advantage from coupling two steps of a reaction mechanism into one, the mechanism may become concerted. This is unlikely to occur if the energy barriers to bond-making or -breaking steps are large. The barriers for proton transfer between electronegative atoms are usually smaller than those for carbon, so the advantage for coupling is more likely to arise in the former type of reaction.

8.4   Metal ions

Introduction

The coordination of electron-rich donors to electron-deficient metal ions is normally an exergonic process. One role for metals in metalloenzymes is as an electrophilic catalyst which acts by stabilizing the increased electron density or negative charge that is often developed during reactions. Metal ions have been frequently referred to as 'superacids' because the metal ion may be multiply charged and could appear to be a better catalyst than a proton. However, there is little evidence to support this claim for either the binding or the activation of substrates. A proton binds more strongly to monodentate and even bidentate ligands than do most mono-, di- or even tri-positively charged metal ions. The equilibrium of Eqn [8.15] invariably lies more to the right than that of Eqn [8.16]; this is often true even if L is a bidentate ligand and M is a metal.

$$L + H_3O^+ \rightleftharpoons HL^+ + H_2O \qquad\qquad [8.15]$$

$$L + M(H_2O)_x^{n+} \rightleftharpoons ML(H_2O)_{x-1}^{n+} + H_2O \qquad\qquad [8.16]$$

Comparison of ionization potentials indicates that a proton 'binds' to an electron more strongly than any metal ion. This is not surprising in view of the electron density surrounding the nucleus of a metal ion compared with that of the 'bare' proton, which could be regarded as the supreme 'super-acid'—perhaps as a 'hyperacid'.

The strong binding of a proton to basic sites increases the reactivity of adjacent bonds more so than does coordination of a metal ion. For example, coordination of a proton to a water molecule changes the acidity from a pK_a of 15.7 to -1.7, whereas coordination of a divalent metal ion changes it to about 8 and even a tripositively charged ion changes it only to about 3. There is very little evidence that the binding of metal ions to substrates causes a larger rate enhancement than that of protons. For example, although zinc(II) ions greatly increase the rate of proton abstraction from 2-acetylpyridine (**19**), they are no more efficient than the coordination of a proton to the carbonyl group, as in **20**.

Metal ions are effective electrophilic catalysts for a wide variety of reactions but they generally owe this efficiency to either (a) the model substrates invariably having a second coordination site available that is much more basic than the reactive site, or (b) the limited concentration of hydronium ions in neutral aqueous solution. If the two sites in case (a) are suitably situated, the substrate can act as a bidentate ligand. When reaction occurs, this is usually accompanied by an increase in basicity of the reactive site coordinated to the metal ion which leads to more favourable binding and consequently a lowering of the activation energy. The metal ion binds more strongly to the transition state than it does to the ground-state structure of the substrate. In case (b), for example, the metal-ion concentration may be orders of magnitude higher than the hydronium ion concentration at pH 7.

Extrapolation of observations obtained from model systems to enzymes must be treated with caution. Metalloenzymes do not operate with high concentrations of metal ions, and for those cases that have been studied the substrate usually acts as a monodentate ligand when it is directly coordinated to a metal ion—as for example in carboxypeptidase, where the carbonyl oxygen of the amide link to be cleaved coordinates to a zinc atom. The protein itself is probably responsible for a large fraction of the binding energy resulting from the interaction of the substrate with the metalloenzyme. In fact, the major function of the metal in metalloenzymes of known structure appears to be structural (see Chapter 2, Scheme 2.2; and Schemes 10.19 and 10.20 and structure **3** in Chapter 10).

When the metal ion of the metalloenzyme acts as an electrophilic catalyst it serves an important function of stabilizing the negative charges developed in the substrate. However, there is little evidence to suggest that it is markedly more efficient at this task than proton donors in the protein. For example, an indication of the stabilization of a negative charge on oxygen brought about by a metal ion can be estimated from the binding energies of this species resulting from coordination to the metal ion. The equilibrium constant for zinc(II) ion binding hydroxide ion (Eqn [8.17]) is *ca* 10^5 estimated from K_a/K_w where K_a is the dissociation constant for the ionization of zinc-bound water and K_w is the dissociation constant of water.

$$\text{Zn(H}_2\text{O)}_n^{2+} + \text{OH}^- \rightleftharpoons \text{Zn(OH)(H}_2\text{O)}_{n-1}^+ + \text{H}_2\text{O} \tag{8.17}$$

$$\text{H}-\text{A} + \text{OH}^- \rightleftharpoons \text{H}_2\text{O} + \text{A}^- \tag{8.18}$$

The 'stabilization' of hydroxide ion by a proton donor HA may be estimated from Eqn [8.18] for which the equilibrium constant is given by K_a^{HA}/K_w, where K_a^{HA} is the dissociation constant of HA. For a pK_a^{HA} of 7, the equilibrium constant for Eqn [8.18] is about 10^7. Although the stabilization of the negatively charged oxygen by the zinc(II) ion is considerable, it is not better than a weak general acid catalyst.

The carbonyl oxygen of the amide substrate is probably coordinated to the zinc(II) in carboxypeptidase, which will stabilize the tetrahedral intermediate by binding to the alkoxide anion (Eqn [8.19]). Such stabilization on oxygen would only definitely lead to a rate enhancement if formation of the intermediate were rate-limiting. If breakdown of the intermediate is the rate-limiting step, there is little advantage in coordination. Although coordination would increase the concentration of the tetrahedral intermediate, it would decrease its rate of breakdown because the electron density on oxygen required to execute carbon—nitrogen bond fission (21) is decreased by coordination. The net effect would depend on the relative charge density on oxygen in the transition state (Prince and Wooley, 1972).

$$\begin{array}{ccc} & & R \\ & & | \\ \text{R}-\text{C}\!=\!\text{O} \cdots \text{Zn}^{2+} & \xrightarrow{\text{Nu}} & \overset{+}{\text{Nu}}-\text{C}-\text{O}^- \cdots \text{Zn}^{2+} \\ | & & | \\ \text{NHR} & & \text{NHR} \end{array} \tag{8.19}$$

Knowledge of the binding of the carbonyl oxygen of amides to zinc(II) of carboxypeptidase based on X-ray studies of enzyme–inhibitor complexes (see Section 10.2, p. 223 and Fig. 10.2) is ambiguous because the inhibitors may bind to the enzyme differently from the substrate. An alternative mechanism, which makes chemical sense, involves metal-ion coordination to the amide nitrogen; this would not only stabilize the tetrahedral intermediate (22) but also facilitate carbon—nitrogen bond cleavage; such a mode of catalysis has been observed in the metal-ion-catalysed hydrolysis of penicillins. Another possibility is general acid-catalysed breakdown of the tetrahedral intermediate by zinc(II)-bound water (**23**).

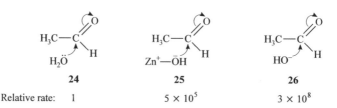

A nucleophile with an ionizable proton shows an increase in acidity upon coordination to a metal ion (Eqn [8.20]). Metal ions could therefore provide a high concentration of the deprotonated nucleophile at neutral pH.

$$M^{n+}(NuH) \rightleftharpoons M^{n+}(Nu^-) + H^+ \qquad [8.20]$$

The coordinated, deprotonated nucleophile will show a reactivity intermediate between that of the free ionized and un-ionized species. At a given pH, if the increase in *concentration* of deprotonated nucleophile which results from coordination to a metal ion more than compensates for its decreased reactivity, a rate enhancement will be obtained for a reaction of the coordinated nucleophile compared with the uncoordinated species. For example, in water at pH 7 there is 10^{-7} M hydroxide ion, but as water bound to zinc(II) has a pK_a of *ca* 9 the amount of hydroxide ion bound to zinc is 10^{-2} M. Coordination of hydroxide ion to a metal ion does not remove as much electron density from oxygen as does coordination to a proton, and consequently the nucleophilicity of metal-coordinated hydroxide ion will be only slightly less than that of the solvated species. For example, the relative rates of water (**24**), zinc(II)-bound hydroxide ion (**25**) and hydroxide ion (**26**) attack on acetaldehyde are 1, 5×10^5 and 3×10^8, respectively (Scheme 8.11).

Scheme 8.11 Metal bound hydroxide ion is generally less nucleophilic than free hydroxide ion in water solvent.

Electron transfer with metals

The ability of a metal to act as an electron donor or acceptor is very dependent upon the type and arrangement of the ligands surrounding the metal. This is true both thermodynamically and kinetically. The electron density at the metal is ligand-dependent, as is the overall stability of the complex. The ease of the electron transfer process itself is dependent not only upon the reduction potential but also upon the geometry of the complex, because of Franck–Condon restrictions.

Thermodynamic stability of metal complexes

Metals are conventionally assigned an oxidation state but, of course, the charge density of the metal itself in a complex is dependent upon the nature of the surrounding ligands and upon the solvent. A consequence of this convention is that it is common to discuss the change in redox potential of a given oxidation state of a metal *brought about by the ligands*. It is important to realize that this, too, then becomes a convention and in reality one should discuss the thermodynamic stability of the whole complex. One could equally well bias a discussion of changes in redox potentials towards the stabilizing influence of the metal *on* the ligands.

There is no single generalization which predicts the values of formation constants of complexes, but a number of useful correlations exist. For a given ionic size, increase in charge almost invariably results in a substantial increase in complex stability. Complexes with the metal in the higher oxidation states are normally more stable than those for lower oxidation states. There are exceptions to this rule; for example, complexes of 1,10-phenanthroline, 2,2-bipyridyl, carbon monoxide and cyanide are more stable with the metal in the lower oxidation state. These ligands have relatively low-energy antibonding orbitals available to accept electrons 'back' from the metal.

'Hard' acids or class 'a' cations bind more strongly to 'hard' bases, and 'soft' acids or class 'b' cations bind more strongly to 'soft' bases—the principle of hard and soft acids and bases. This principle, however, lacks a satisfactory quantitative basis.

Kinetic effects

If the metal complex is inert with respect to ligand exchange, electron transfer usually takes place by a tunnelling or an outer-sphere mechanism. The *outer-sphere mechanism* involves electron transfer from reductant to oxidant, with the coordination shells (or spheres) of each staying intact. The act of electron transfer takes place much faster (10^{-15} s) than the rate of change of nuclear configuration ($\simeq 10^{-13}$ s). The nuclei effectively remain static during electron transfer so that there are restrictions on changes in spin angular momentum and in geometry (Franck–Condon principle). An *inner-sphere mechanism* is one in which the reductant and oxidant *share* a ligand in their inner coordination sphere, the electron being transferred across the bridging group.

The rates of outer-sphere reactions between metal complexes vary widely. For example, the rate of electron transfer between two octahedrally co-ordinated low-spin complexes which differ from one another only by the presence of an extra electron in the t_{2g} orbitals of the complex, with the metal in the lower oxidation state, tend to be very fast ($10^3 \, M^{-1} \, s^{-1}$). Transfer of an electron from low-spin Fe(II) t_{2g}^6 to low-spin Fe(II) t_{2g}^5 involves the transfer of an electron from one t_{2g} orbital to another. Low-spin metal complexes of this kind have very similar metal—ligand bond lengths in their two oxidation states. For example, Fe—C in $[Fe(CN)_6]^{4-}$ is 1.92 Å and in $[Fe(CN)_6]^{3-}$ it is 1.95 Å.

The rate of electron transfer can also be retarded if the metal—ligand bond distances change considerably with oxidation state. For example the Fe—O distance in $[Fe(H_2O)_6]^{2+}$ is 2.21 Å but that in $[Fe(H_2O)_6]^{3+}$ is 2.05 Å; therefore, if electron transfer took place from $[Fe(H_2O)_6]^{2+}$ to $[Fe(H_2O)_6]^{3+}$ with both complexes in their ground states, the product would be a compressed Fe(II) ion and a stretched Fe(III) ion. It is therefore necessary that the molecules become vibrationally excited *before* electron transfer takes place, i.e. the Fe(III) and Fe(II) geometries must be expanded and compressed, respectively.

Outer-sphere reactions between complexes of different metals are often very fast, and are, of course, assisted by the decrease in free energy accompanying chemical reaction.

8.5   Intramolecular reactions

Introduction

The two dramatic effects of enzymic catalysis are molecular recognition and rate acceleration. The increase in the rate of reactions brought about by enzymes can be enormous, but the quantitative evaluation of their rate enhancement compared with a non-enzyme catalysed reaction is not simple. The mechanism by which the enzyme-catalysed reaction occurs is often different from that of the non-enzyme-catalysed reaction. It is probably this simple fact which has been responsible for the speculation that it is the mechanism or mode of chemical catalysis which is responsible for the large *rate enhancement* brought about by enzymes and it is the shape and electrical complementarity of the binding surface of the enzyme which is responsible for *specificity or recognition of the substrate*. There is little advantage to be gained by treating these two aspects of enzymic catalysis separately (Page, 1984b, 1987).

Chemical catalysis alone can rarely explain the rate enhancement brought about by enzymes. Groups on both the substrate and the enzyme which are *not* involved in the chemical mechanism of bond making and breaking, i.e. those not involved in the 'curved arrows', can make an enormously important contribution to catalysis. This is illustrated by examining a system where the mechanism is the same for both the enzyme-catalysed and the non-enzyme-catalysed reactions. The mechanism of action of succinyl CoA acetoacetate transferase involves nucleophilic attack of the enzyme's glutamate carboxylate on the thioester, succinyl CoA (**27**) to give an anhydride intermediate (White & Jencks, 1976). The second-order rate constant for this reaction is 3×10^{13}-fold greater than the analogous reaction of acetate with the same ester (**28**). Although the nucleophilicities of the carboxylates may be slightly different because of solvation effects and the enzyme may provide some other forms of catalysis, these contributions will not be large. The *non-reacting* part of the enzyme therefore lowers the free energy of activation by up to $78 \, kJ \, mol^{-1}$. Similarly, changing those parts of the substrate structure away from the atoms involved in bond making and breaking may also significantly affect catalytic efficiency. The enzyme also forms anhydride intermediates from 'non-specific' substrates (**29**). Even though the chemical reactivities of the two thioester substrates (**27** and **29**) are similar, for example towards alkaline hydrolysis, the enzyme reaction proceeds up to 3×12^{12}-fold faster with the so-called specific substrate (**27**). The presence of the *non-reacting* part of the substrate—the CoA residue—lowers the activation energy by $72 \, kJ \, mol^{-1}$ compared with **29**.

It is apparent that enzymes may bring about enormous increases in the rates of reactions, even when the mechanism of the enzyme-catalysed and the non-enzyme-catalysed reactions are similar.

$$27 \qquad\qquad 28 \qquad\qquad 29$$

Quantifying the effect of the spatial approximation of reactants

The first step in enzyme-catalysed reactions is the bringing together of the substrate and the enzyme. This binding process brings the reacting groups on the enzyme and substrate into close proximity (Scheme 8.12). Attempts to understand the contribution that this *spatial approximation* of reactants makes towards catalysis have centred on an analogy with intramolecular reactions in which the reactants are *covalently* linked together (Scheme 8.12). Intramolecular reactions often proceed at a very much faster rate than the analogous intermolecular reaction (Scheme 8.12) and they provide an approach to estimating the maximum effect of binding the reactants together at the active site of an enzyme.

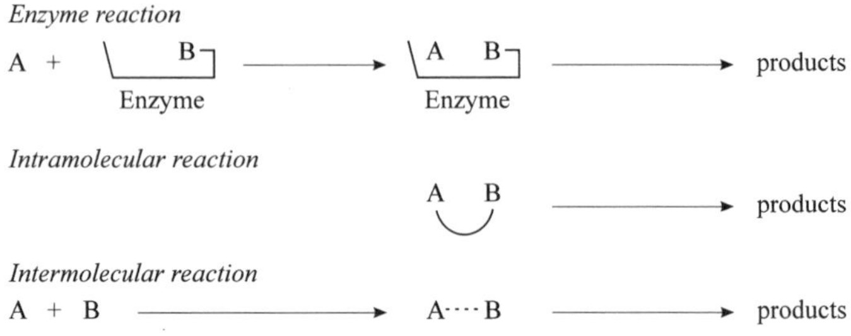

Enzyme reaction

A + ⎣___ B⎤ ⟶ ⎣A B⎤ ⟶ products
    Enzyme   Enzyme

Intramolecular reaction

                   A B
                    ⌣ ⟶ products

Intermolecular reaction

A + B ⟶ A····B ⟶ products

Scheme 8.12 Relationship between enzyme catalysis, intermolecular reactions and intramolecular reactions.

The interaction of two functional sites within a molecule can affect ground-state and transition-state structures, the nature and relative stability of products and the rates of reactions. The interaction in the transition state (Eqn [8.21]) may only be transient and involve only weak forces of attraction so that the sites are held together in a loose geometrical configuration. At the other extreme, a covalent bond may be formed between the two sites so that they are held in a specific, relatively rigid geometry. *Neighbouring-group participation* can thus involve a type of ring closure and include intramolecular catalysis and reactions giving cyclic products (Page, 1973; Capon & McManus, 1976; Kirby, 1980; Illuminati & Mandolini, 1981; Mandolini, 1986).

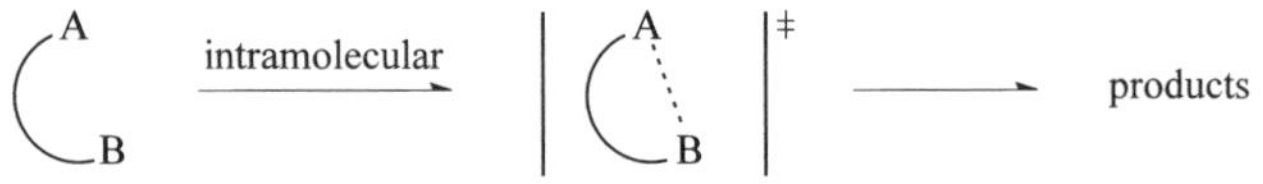

$$\left(\begin{smallmatrix}A\\B\end{smallmatrix}\right. \xrightarrow{\text{intramolecular}} \left|\left(\begin{smallmatrix}A\\B\end{smallmatrix}\right.\right|^{\ddagger} \longrightarrow \text{products} \qquad [8.21]$$

It is not only that the *rates* of intramolecular reactions (Eqn [8.21]) are faster than those of analogous intermolecular ones; *the equilibrium constants* for a cyclization reaction are usually more favourable than those for the corresponding intermolecular process. Typical rate enhancements and favourable equilibria of intramolecular reactions involving the formation of five-membered rings are shown in Scheme 8.13. Cyclization occurs either in the product or transiently in the transition state and despite their involving the formation of the same ring size, a comparison of the equilibrium or rate constants for these reactions with those of analogous intermolecular reactions gives ratios varying from $0.5\,\text{M}$ to $2 \times 10^{12}\,\text{M}$ (Page, 1973). The rate enhancements and favourable equilibrium ratios have units of concentration because a unimolecular reaction is being compared with a bimolecular one. For this reason the rate enhancement is sometimes called the 'effective concentration' or 'effective molarity', which is the hypothetical concentration of one of the reactants in the intermolecular reaction required to make that reaction proceed at the same rate or to the same extent as the intramolecular one.

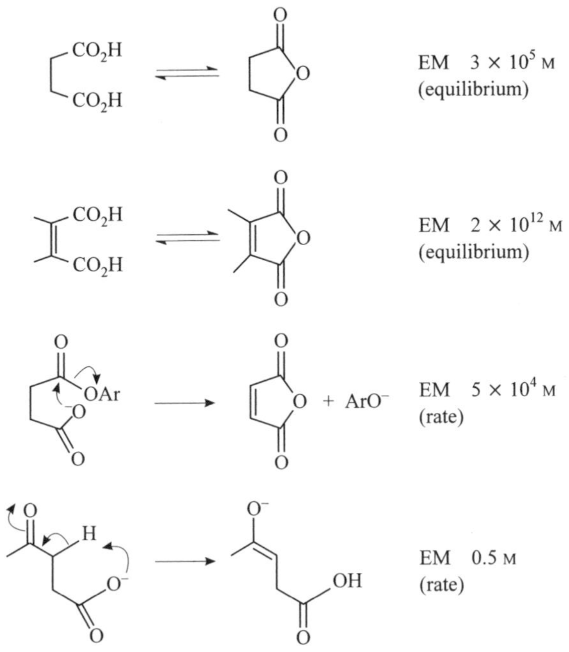

Scheme 8.13 Effective molarities (EM) for some intramolecular reactions involving five-membered rings.

It was widely believed in the 1960s that completely surrounding a molecule of one reactant by many molecules of the other was the maximum advantage to be gained by approximation. Thus the number 55 M, the molarity of pure water, was often used to standardize the increase in rate 'expected' in intramolecular reactions. However, completely surrounding one reactant molecule by another does not give rise to a special proximity effect. This simple approach to quantify the advantage of intramolecularity does not take account of the enormous difference in entropy changes between bringing two reactant molecules loosely together and separated by their van der Waals radii and tightly linking them together by covalent bonds as in an intramolecular reaction.

Another problem with some of these special explanations of intramolecular reactions is that they assume that a *rate* phenomenon is involved. However, the variation in the effectiveness of intramolecular reactions is shown by both equilibrium and rate constants (Scheme 8.13); furthermore there is often a linear relationship between these constants. Attempts to explain the fast rates of intramolecular reactions in terms of critical distances, critical angles, reaction windows, enhanced methods of energy transfer and other phenomena usually concerned in the rates of reactions are therefore not addressing the full problem.

Changes in equilibrium constants are reflected by variations in the free-energy difference between reactants and products. This thermodynamic approach is enhanced by a knowledge of the geometrical structures of reactants and products so that enthalpy and entropy differences can be examined (Page, 1991). A similar approach may be used for the rates of reactions if the transition-state theory of reaction rates is accepted. Although the geometrical structure of the transition-state is not known directly, transition-state theory allows the application of thermodynamics to an understanding of the relative rates of reactions. It is clear that the large effective concentrations are normal and expected, and that it is often the low values which require a 'special' explanation (Page & Jencks, 1971; Page, 1973).

It is quite easy to set an upper limit to the rate enhancement, in the absence of strain, that may be brought about by covalently binding the reactants together as in an intramolecular reaction, or by binding them to the active site of an enzyme. This may be done by considering the different entropy changes that occur in bimolecular and unimolecular reactions.

There are different degrees of freedom lost in intramolecular and analogous intermolecular reactions which gives rise to large differences in the entropy change between the two systems. For a non-linear molecule containing n atoms, its freedom to move along the axes in space is registered by three degrees of translational freedom; it also possesses three degrees of rotational freedom representing the freedom of the whole molecule to rotate about its centre of gravity and $3n - 6$ degrees of freedom associated with internal vibrational and rotational motions. There are losses of three degrees of translational freedom and three degrees of rotational freedom on forming the product of a bimolecular association reaction, and there is a gain of six new vibrational modes in the product (Table 8.1). However, there is no net change in the number of degrees of freedom of translation, rotation and vibration upon forming the product in a unimolecular and intramolecular reaction (Table 8.2). The entropy associated with these motions may easily be calculated from the partition functions, which are simply an indication of the number of quantum states that the molecule may occupy for each type of movement.

Some typical values of entropy terms are shown in Table 8.3. The translational entropy is normally about $120–150 \, \text{J K}^{-1} \, \text{mol}^{-1}$ for a standard state of $1 \, \text{M}$ and is not very dependent upon the molecular weight. The entropy of

Table 8.1 Distribution of degrees of freedom in a bimolecular association reaction.

	A	+	B	⇌	A—B
Translation	3		3		3
Rotation	3		3		3
Vibration	$3n - 6$		$3n' - 6$		$3n + 3n' - 6$

Table 8.2 Distribution of degrees of freedom in an intramolecular 'association' reaction.

	A B $\rightleftharpoons$ A—B
Translation	
3	
3	
Rotation	

rotation is about $85–115\,\mathrm{J\,K^{-1}\,mol^{-1}}$ for the majority of medium-sized

Table 8.3 Typical entropy contributions from translational, rotational and vibrational motions at 298 K.[a]

Motion	$S^{\circ}\,(\mathrm{J\,K^{-1}\,mol^{-1}})$
Three degrees of translation freedom	
Molecular weights 20–200; standard state, $1\,\mathrm{mol\,l^{-1}}$	122–148
Three degrees of rotational freedom	
Water	44
n-Propane	90
endo-Dicyclopentadiene, $C_{10}H_{12}$	114
Internal rotation	13–21
Vibrations	
$\omega\ =1000\,\mathrm{cm^{-1}}$	0.4
$800\,\mathrm{cm^{-1}}$	0.8
$400\,\mathrm{cm^{-1}}$	4.2
$200\,\mathrm{cm^{-1}}$	9.2
$100\,\mathrm{cm^{-1}}$	14.2

[a] From Page & Jencks (1971).

molecules and is not normally very dependent upon the size and structure of the molecule. The exceptionally small rotational entropy for water, $44\,\mathrm{J\,K^{-1}\,mol^{-1}}$, is due to nearly all the mass being concentrated at the centre of gravity: hence the moment of inertia, and thus the entropy, are small. The entropy of internal rotation makes a comparatively small contribution to the total entropy of the molecule, and the vibrational contribution is even smaller except for very low vibrational frequencies.

The total loss of translational and rotational entropy for a standard state of 1 M and at 298 K for a gas-phase bimolecular association reaction is about $-220\,\mathrm{J\,K^{-1}\,mol^{-1}}$ and this value has only a small dependence upon the masses, sizes and structures of the molecules involved. However, this loss is often compensated to varying extents by low-frequency vibrations in the product or transition state. For several reactions having 'tight' transition states or covalently bonded products, the change in internal entropy is about $+50\,\mathrm{J\,K^{-1}\,mol^{-1}}$ so the total entropy change is predicted to be about $-170\,\mathrm{J\,K^{-1}\,mol^{-1}}$, which is the value observed experimentally for many bimolecular gas phase reactions. In solution, the entropy change for a bimolecular reaction is estimated to be only about $20\,J\,K^{-1}\,mol^{-1}$ less than that in the gas phase, at the same standard state of 1 M (Page, 1973).

Since a large loss of entropy for a bimolecular reaction is avoided if the reactants are bound to an enzyme active site (Scheme 8.12) or converted to an intramolecular reaction (Eqn [8.21]), the maximum *entropic advantage* from approximation of reactants may now be estimated (Table 8.4). The approximate maximum effective concentration or rate enhancement for these reactions from entropic factors alone is about 10^8 M. In the comparatively rare situation of a bimolecular reaction having a very 'loose' transition state or product, then association will be even less entropically unfavourable since the entropy of the low-frequency vibrations in the 'loose' complex will counterbalance the large loss of translational and rotational entropy. The rate enhancement for the analogous intramolecular reaction will, therefore, be smaller, perhaps in the range of 100 M or less.

In summary, bringing two molecules together in a reaction is accompanied by a negative change in entropy because of the reduced volume of space available to the reactants. Mechanically, the increase in order in this bimolecular process is expressed mainly as a loss of translational and rotational entropy. The more severely the geometrical relationship between the reactant molecules is defined, the greater is the loss of entropy. For most molecules of average shape and size this entropy change makes bimolecular reactions, in respect to both rates and equilibria, unfavourable by factors of up to 10^8 M at a standard state of 1 M and at 25 °C (Table 8.4). These changes do not occur in unimolecular reactions and 10^8 M is therefore the maximum difference expected between an intramolecular reaction and an intermolecular one based *only* on the entropy difference between them and in the absence of strain and solvation effects. The entropy changes are theoretical and are not necessarily be reflected in the *observed* entropies of activation and reaction because of solvent effects in water and polar solvents.

Effective concentrations greater than 10^8 M are the result of strain energy differences between the two systems. Either the intermolecular reaction shows an unfavourable change in strain energy or the intramolecular one exhibits a release of strain upon ring closure. It has been shown that there is a good relationship between rates and equilibrium constants for intramolecular reactions and the strain energy changes accompanying ring closure. There is therefore nothing special about these very high effective concentrations. Their relevance to enzymic catalysis is limited because indu-

Table 8.4 Estimates of the maximum entropy advantage from spatial approximation of reactants.

A + B $\rightleftharpoons$	A---B	$\rightleftharpoons$	A—B
Separate reactants	'loose' transition state or product		'tight' transition state or product
$\Delta S\,(J\,K^{-1}\,mol^{-1})$	-40		-150
Effective molarity (M)	10^2		10^8

cing geometrical strain into substrates probably does not make a major contribution to the efficiency of most enzymes.

The reasons for the variation and the low values sometimes observed in the effective concentrations of intramolecular reactions are:

(a) an unfavourable potential energy change accompanying the intramolecular reaction where strain energy is introduced upon ring closure;
(b) unfavourable negative entropy changes in the intramolecular reaction resulting from the loss of internal rotation and a small loss of overall rotational entropy upon ring closure; the loss of internal rotation corresponds to a factor of only about 5–10 in rate per loss of internal rotation;
(c) favourable positive entropy changes in the intermolecular reaction resulting from a loose transition state where there are weak forces of interaction between the reactants giving a flexible geometry. These entropy changes compensate for the large negative loss of entropy associated with translation and rotation to give a smaller unfavourable entropy change for bimolecular reactions (Table 8.4).

The major reason for the variation in effective molarities for different reactions is that the intramolecular process often involves the introduction or removal of strain energy upon ring closure. If the cyclic transition state or product is more strained than the reactants, an effective molarity of lower than 10^8 M will be observed. Conversely, if strain is removed upon cyclization then the effective molarity can be greater than 10^8 M. It has been shown that there is a correlation between effective molarities of intramolecular reactions and the corresponding change in strain energies of cyclization. For example, the effective molarities of lactonization of hydroxy acids may be correlated with changes in their strain energy (Scheme 8.14).

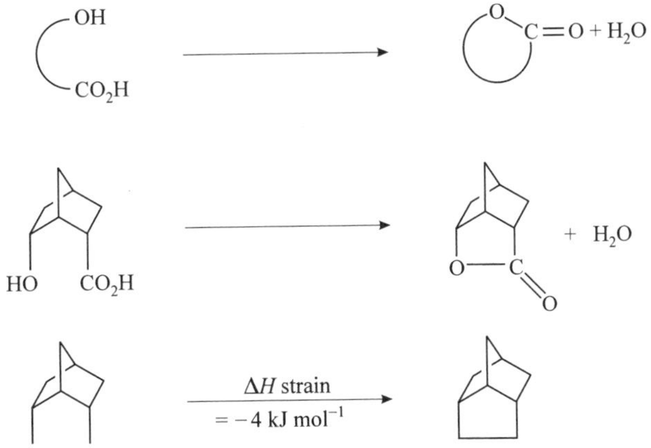

Scheme 8.14 Correlation of effective molarities with changes in strain energy (see text).

Upon cyclization the degree of freedom associated with internal rotation in the reactant becomes a vibrational motion in the product or transition state corresponding to an unfavourable entropy change of about $-18 \, K^{-1} \, mol^{-1}$, a rate factor of about 10 at 298 K. The loss of entropy upon freezing an internal rotation is partially compensated by a favourable enthalpy function change of about 2 kJ mol^{-1} per internal rotation, so that the loss of internal rotational freedom may reduce the rate by a factor of about 5. Several cyclization reactions occur where rate increases greater than fivefold result from 'freezing' of some internal rotations. Almost invariably these are due to differences in strain energy upon cyclization.

The limit of a 'loose' transition state may be considered as the encounter-controlled collision of two species which are independently freely rotating within the collision complex. The formation of the transition state involves no change in rotational freedom of the reactants as they are converted to the collision 'complex'. However, three degrees of translational freedom are converted into rotational modes representing the rotation of the whole complex around its new centre of gravity. For average-size molecules these changes correspond to an entropy loss of only about $-30 \, J \, K^{-1} \, mol^{-1}$ (standard state 1 M) and a collision frequency of about $10^{11} \, M^{-1} \, s^{-1}$. The analogous intramolecular reaction would perhaps involve a rate-limiting conformational change so that the major entropy change would be due to loss of internal rotation. The entropic advantage of intramolecularity in such cases is minimal, giving effective molarities of less than 10 M.

A less extreme 'loose' transition state could involve some interaction between the reactants so that they are not freely rotating in the transition state of the bimolecular reaction. In this case, the intermolecular reaction would involve the conversion of translational and rotational degrees of freedom into internal rotation and low-frequency vibrational motions which, for average-sized molecules, would correspond to a total entropy change of -40 to $-100 \, J \, K^{-1} \, mol^{-1}$. The positive entropy contribution to this change comes from the entropy of internal rotation and low-frequency vibrations which have no analogy in the intramolecular case. Consequently, the effective molarity could be less (10^2–10^5 M) than the normal maximum of 10^8 M, even in an intramolecular system involving no increase in strain energy upon ring closure. In general it appears that the absolute and relative reactivities of intramolecular reactions can be explained by differences in entropy and strain energy.

Reactions involving proton transfer between electronegative atoms invariably have 'loose' transition-state structures and intramolecular general acid or general base catalysis would exhibit low effective concentrations. This appears to be true for proton transfer to or from both electronegative atoms and carbon atoms, presumably because such reactions either have a transition state which is a very loose hydrogen-bonded complex or have a rate-limiting step which is diffusion-controlled. Thus there does not seem to be much of a rate advantage upon changing an intermolecular general acid- or general base-catalysed reaction to an intramolecular or enzyme process. This appears to be true for mechanisms which occur by a stepwise or a coupled concerted process. An example of the former is offered by the stepwise trapping mechanism in the general base-catalysed aminolysis of benzylpenicillin. Amines react with penicillin to form an unstable tetra-hedral intermediate which may be trapped by a diffusion-controlled encoun-ter with a strong base as shown in Scheme 8.10. Diamines also undergo this reaction, but at a much faster rate than monoamines of the same basicity; this is attributed to intramolecular general base catalysis by the second amino group acting as a proton acceptor (**30**). However, the effect of intra-molecularity itself is small and the fact that the catalyst is held in close proximity to the reaction site makes it only as effective as a concentration of 1 M intermolecular catalyst, of the same basicity, freely diffusing through the solution.

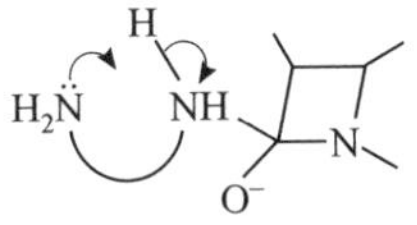

30

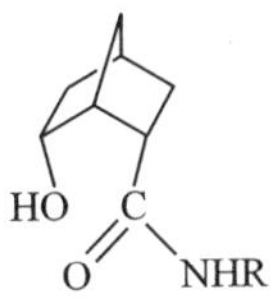

31

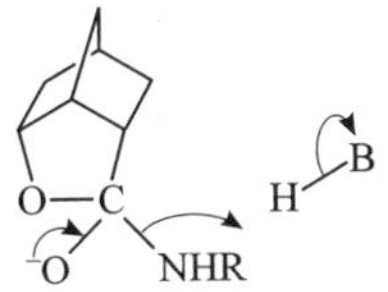

32

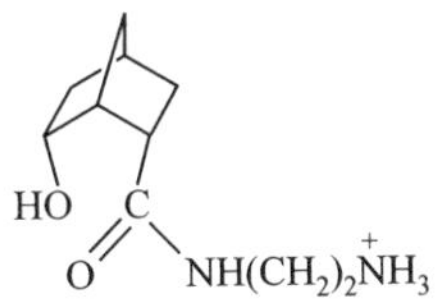

33

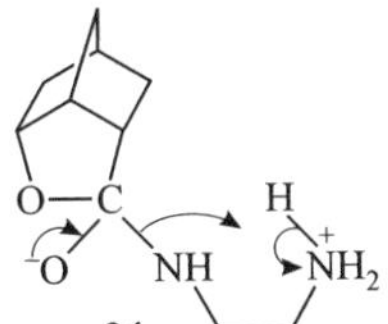

34

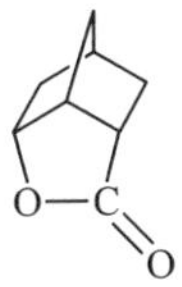

35

The weak advantage upon changing intermolecular to intramolecular general acid–base catalysis in a concerted process is exemplified by the hydrolysis of the hydroxy amide (**31**). The mechanism has been shown unequivocally to proceed by concerted general acid-catalysed breakdown of the tetrahedral intermediate in which there is coupling between proton transfer and carbon–nitrogen bond fission (**32**) (Morris & Page, 1980d). The presence of an intramolecular general acid catalyst such as the ammonium group of **33** causes a rate enhancement attributable to intramolecular catalysis. This rate enhancement is due almost entirely to the difference in acidities of the proton donors: water in the intermolecular reaction (32) (B = OH) and the ammonium ion in the intramolecular process (34). However, the contribution of intramolecularity itself is small and the intramolecular general acid catalyst, although in close proximity to the reaction site, is only as efficient as an intermolecular catalyst of the same acid strength at a concentration of 1 M.

Most intramolecular general acid–base-catalysed reactions for which effective concentrations have been reported involve a stepwise mechanism in which the proton itself is in a potential energy well in the transition state of the rate-limiting step. The stepwise mechanism for acid–base catalysis represents the limit of a 'loose' transition state in which the reactant molecules are diffusing together and it is not surprising, therefore, that the entropy change associated with such bimolecular steps is small and that there is little rate advantage to be gained by covalently linking the reactant molecules together in an intramolecular reaction. It appears that the same is also true even if the proton is 'in flight' in the transition state of the intermolecular reaction; such a mechanism is one of concerted proton transfer. This is expected in view of the relatively small entropy changes associated with hydrogen-bonding equilibria.

There is an interesting contrast between the large contribution (a factor of *ca* 10^8) to the rate enhancement of intramolecular and enzyme-catalysed reactions by nucleophilic catalysis, and the much smaller contribution (*ca* 1–10) of general acid–base catalysis. An example of the pure entropic advantage of covalent catalysis is given by the hydrolysis of the hydroxy amide (31), which is a good model for the serine proteases (Morris & Page, 1980c). The intramolecular nucleophilic hydroxy group causes a rate enhancement of *ca* 10^8 M compared with the equivalent intermolecular reaction; this is attributed to neighbouring group participation to form a lactone intermediate, (35) (equivalent to the acyl-enzyme in the serine proteases). The rate enhancement is due entirely to the entropy effect—the intramolecular reaction involves little change in strain energy on forming the lactone and solvation effects are minimal, so that the nucleophilicity of the nucleophiles in the intra- and inter-molecular reactions are very similar.

Further reading

Capon, B. & McManus, S. P. (1976) *Neighbouring Group Participation*, Plenum Press, New York.

Gerlt, J. A. & Gassman, P. G. (1993) An explanation for rapid enzyme-catalysed proton abstraction from carbon acids: importance of late transition states in concerted mechanisms, *J. Am. Chem. Soc.*, **115**, 11552.

Illuminati, G. & Mandolini, L. (1981) Ring closure reactions of bifunctional molecules, *Acc. Chem. Res.*, **14**, 95.

Jencks, W. P. (1969) *Catalysis in Chemistry and Enzymology*, McGraw-Hill, New York.

Jencks, W. P. (1976) Enforced general acid–base catalysis of complex reactions and its limitations, *Acc. Chem. Res.*, **9**, 425.

Kirby, A. J. (1980) Effective molarities for intramolecular reactions, *Adv. Phys. Org. Chem.*, **17**, 183.

Menger, F. M. (1993) Enzyme reactivity from an organic perspective, *Acc. Chem., Res.*, **26**, 206.

Page, M. I. (1977) Entropy, binding energy and enzymatic catalysis, *Angew. Chem., Int. Ed. Engl.*, **16**, 449.

Page, M. I. (1984) The mechanisms of chemical catalysis used in enzymes, in *The Chemistry of Enzyme Action*, Page, M. I. (ed.) Elsevier, Amsterdam, p. 229.

Page, M. I. (1991) The energetics of intramolecular reactions and enzyme catalysis, *Philos. Trans. R. Soc. London, Ser. B.*, **332**, 149.

Page, M. I. & Jencks, W. P. (1987) In defence of entropy and strain as explanations for the rate enhancement shown in intermolecular reactions, *Gazz. Chin. Ital.*, **117**, 455.

9 Complexation catalysis

9.1 General considerations

Catalysis requires an interaction of catalyst with substrate sufficient to facilitate the reaction with subsequent expulsion of the products. For example, acid catalysis of acetal hydrolysis involves bonding to a proton at one of the ether oxygens to induce sufficient positive charge to facilitate carbon–oxygen bond fission, and the catalytic proton is released after several subsequent steps from another site (see Scheme 8.1). This form of 'simple' catalysis results from a perturbation of the electron density of the reactant molecule, by making sites richer or more deficient in electrons and generally stabilizing entities which would be formed in the absence of catalysis. A large body of work has been carried out in recent years specifically aimed at modelling the action of enzymes which complex with the substrate prior to the catalytic step, thus bringing reagent groups into close proximity with the substrate. Intramolecular reactions (Chapter 8) have often been studied in order to model the reaction in a complex, whereby the complexation step is bypassed by prior synthetic work. Such studies do not provide working *catalysts*, but are nevertheless of the highest importance to our understanding of catalysis because they allow a separation of the complexation process from the direct catalytic components.

Simply recreating known enzymes by chemical synthesis does not assist our understanding of enzyme action, but it can indicate that our knowledge of a particular structure is correct. The heroic work on the synthesis of ribonuclease-A (Gutte & Merrifield, 1969, 1971; Hirschman *et al.*, 1969) exemplifies this approach. Schneider & Kent (1988) synthesized sufficient quantities of the important protease from human immunodeficiency virus type-1 to determine its three-dimensional structure (Jaskolski *et al.*, 1991); the sequence on which the structure is based was derived from the gene structure and only minute quantities of the natural protease were available. An important use of chemical synthesis will be the incorporation of arrangements of novel amino acid residues which can alter specificity and reactivity and can test theories of enzyme action (Ulmer, 1983; Fersht *et al.*, 1984).

The study of model systems can provide very useful knowledge about catalysis in general; it can also assist in the understanding of details of mechanism and the geometrical requirements between reacting groups, because such systems do not possess the complexity of enzymes.

The association step in complexation catalysis can involve the formation of a covalent bond between reagent and substrate; or it may involve a *non-covalent* linkage, utilizing a combination of relatively weak binding forces between reagent and substrate comprising ionic (electrostatic), hydrogen-bonding and apolar π-bonding and hydrophobic interactions. Non-covalent complexes can include molecular aggregates—emulsions, liposomes, micelles, membranes and the like; they also include inclusion complexes and complexes involving such species as reactive linear polymers, reactive microgels, reactive dendrimers and polysoaps (Page & Crombie, 1984).

Variation of the environment of molecules in ground and transition states by complexation will affect their energies and hence effect a change in the rate of the reaction. Association of reactant (A) with a host (B) to give the complex (A.B) is governed by the general equation [9.1].

$$
\begin{array}{ccc}
\text{A} + \text{B} & \xrightleftharpoons{K} & \text{A.B} \\
\Big\downarrow k_1 & & \Big\downarrow k_2 \qquad\qquad [9.1] \\
\text{product} & & \text{product}
\end{array}
$$

The mechanism of Eqn [9.1] gives rise to Eqns [9.2] and [9.3] when [B] $\gg$ [A]. A system that commonly occurs is where k_1 and k_2 contain the concentration term of a further reactant such as the hydroxide ion.

$$
\frac{1}{k_1 - k_{\text{obs}}} = \frac{1}{(k_1 - k_2)} + \frac{K}{(k_1 - k_2)[\text{B}]} \qquad [9.2]
$$

$$
k_{\text{obs}} = \frac{k_1 \cdot K + k_2 \cdot [\text{B}]}{K + [B]} \qquad [9.3]
$$

The above equations predict experimental profiles as illustrated in Fig. 9.1.

The host (B) in Eqn [9.1] need not involve itself chemically in the reaction but could merely facilitate (or inhibit) reaction of A as a 'spectator', providing a micro-environment for A which is different from that in the uncomplexed state.

Effective catalysis by complexation requires relatively rapid binding of reactants and product release. A major difficulty with many of the synthetic systems is that the product binds tightly and hence the 'catalyst' does not regenerate (such a system is said to involve no 'turnover').

The 'complex' (A.B) may not necessarily have a single structure because the forces holding it together can be such that significant relative motion can occur within it (Kintzinger *et al.*, 1981; Bergeron & Burton, 1982); relative motion even occurs in a complex which is in the crystalline state (Fyfe, 1973). Only one of the conformations comprising A.B may be active and the Curtin–Hammett principle (Chapter 2) is therefore an important consideration in studies of such systems. Complexation may occur with more than one molecule of the reactant A. The form of the equation is

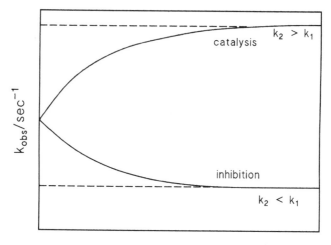

Concentration of host

Fig. 9.1 Effect of host concentration on guest reactivity; see Eqn [9.1].

the same as that for the 1:1 complexation, provided the concentration of reactant A is less than that of the 'host', so that no interaction occurs between individual molecules of A in the complex. Another important consideration is that host molecules often 'stack' in solution (Tso *et al.*, 1963), particularly if they have a flat shape; as the host concentration is increased progressively more than one host could bind to the reactant (Eqn [9.4]).

$$A + B \rightleftharpoons A.B \overset{+B}{\rightleftharpoons} A.B_2 \overset{+B}{\rightleftharpoons} A.B_3 \overset{+B \text{ etc}}{\rightleftharpoons} \dots \rightleftharpoons A.B_n \quad [9.4]$$

The experimental difference between this system and that of 1:1 stoichiometry is very small and it is difficult to distinguish between multiple and simple complexation. The aggregation phenomena discussed later in this chapter represent an extreme form of this stacking process.

When the host molecule has a group which reacts with the guest, rather than the host acting as a 'spectator', the existence of a 'saturation' curve of the type in Fig. 9.1 is consistent with complexation. However, the equation would have the same form if k_2 (Eqn [9.1]) were not important and reaction only occurred through the bimolecular collision pathway (where B reacts directly with A via k_1). Non-catalytic complexation would lead to inhibition by the host, whereas complexation of one stereochemical form of the guest would give stereoselective catalysis or inhibition.

9.2   Covalent catalysis

Covalent bond formation between a potential reactant site and the complexing catalyst often modifies the properties of a substrate, by enhancing its electrophilic or nucleophilic character or even by completely reversing its normal reactivity. The principal feature of covalent catalysis is that complexation involving covalent bond formation must be rapid and reversible, otherwise the system would not be catalytic (Section 8.1). The mechanism of many enzymes involves covalent catalysis (Spector, 1982) and this aspect is described in Chapter 10. The hydrolysis of methyl 2-formylbenzoate catalysed by morpholine provides a non-enzymic example of covalent catalysis (Scheme 9.1) (Bender *et al.*, 1965) where covalent bond formation is followed by reaction to give an overall enhancement of hydrolysis over the uncatalysed reaction. Morpholine rapidly reacts with aldehyde to form a labile carbinolamine complex, with an oxyanion in close proximity to the ester, which expels the methanol to give overall hydrolysis rather than aminolysis.

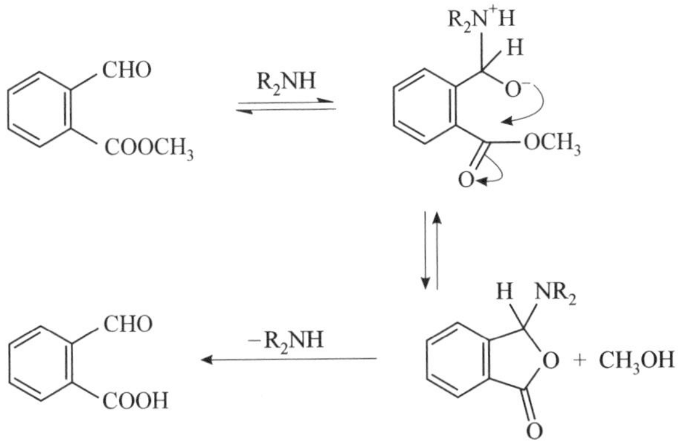

Scheme 9.1 Morpholine-catalysed hydrolysis of methyl 2-formylbenzoate.

9.3   Inclusion complexation

A cycloamylose molecule (cyclodextrin) is a naturally occurring cyclic oligosaccharide which offers a binding cavity (Tee, 1994; Breslow, 1995). Cyclohepta-amylose (**1**; Fig. 9.2) (Van Etten *et al.*, 1967) binds the aromatic part of phenyl esters, and the secondary alcohols, forming the rim of the cavity, react with the ester in their ionic form to release phenoxide ion and give acylated cycloamylose. This process has been compared with the acylation step of catalysis by proteolytic enzymes; although all phenyl esters studied complex with the cycloamylose, only the *meta*-substituted phenyl acetates undergo rapid reaction. Inspection of models indicates that the *meta*-substituted phenyl ring of the ester could bind in the cycloamylose torus with the substituent at the bottom of the cavity (**2**). This arrangement places the ester in the *meta* position adjacent to one of the secondary hydroxyl groups on the rim of the host; *para*-substituted esters do not allow the hydroxyl and ester to come in close proximity to each other (**3**).

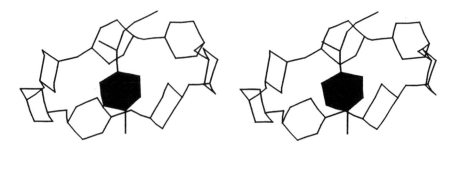

1

Fig. 9.2 A stereo-diagram of a framework model of cyclohepta-amylose (**1**) incorporating a guest molecule. See the note below the Contents list (p. vii) for viewing instructions.

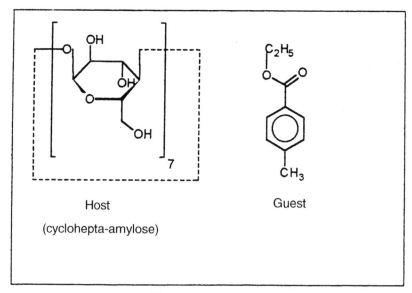

Key to stereo-diagram

The cycloamylose torus can be used to protect part of an aromatic nucleus towards electrophilic substitution (**4**) (Breslow & Campbell, 1969, 1971). The attachment of groups to the rim of the cycloamyloses can enhance the reaction rate by reducing the loss of entropy required for catalysis (**5**) (Breslow *et al.*, 1978).

The observation of complexation with cycloamyloses stimulated investigations of other cyclic systems and such species as crown ethers, cyclophanes, calixarenes and the like have all been demonstrated to act as vehicles for catalytic function through inclusion of the substrate within the 'cavity' formed by the host molecule (Page & Crombie, 1984). A classical example is due to Chao & Cram (1976), where an ammonio ester guest binds to the polyether cavity of a host and the attack of a pendant thiol group is enhanced. Eqn [9.6] illustrates a recent example (Lehn & Sirlin, 1978); since the cavity is chiralthere is significant enantioselectivity on reaction between D- and

L-forms of an amino acid ester. A more recent example is the methylation of quinoline by methyl iodide (Eqn [9.7]) catalysed by complexation with a cyclophane (**7**) (Stauffer *et al.*, 1990) where inclusion provides a microscopic medium for the transition state which stabilizes it more efficiently than the bulk medium. Application of a methodology similar to that for transition-state acidities (see Chapter 3 and Scheme 9.3) indicates that the cyclophane host complexes the transition state of the reaction better than it does the ground state. This indicates that the developing positive charge is more stabilized by the donor properties of the cyclophane—in other words, the microscopic environment is more conducive to the reaction than is water. The catalysis is inhibited by quaternary ammonium salts, in agreement with this hypothesis; and since the product is a quaternary ammonium salt, product inhibition is observed. Product inhibition is a common problem with this type of catalysis.

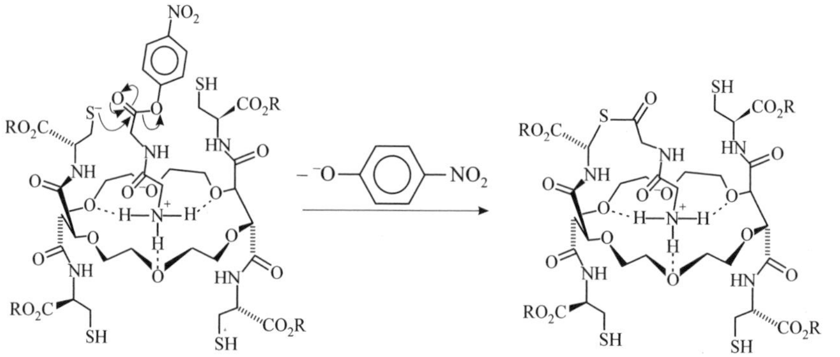

[9.5]

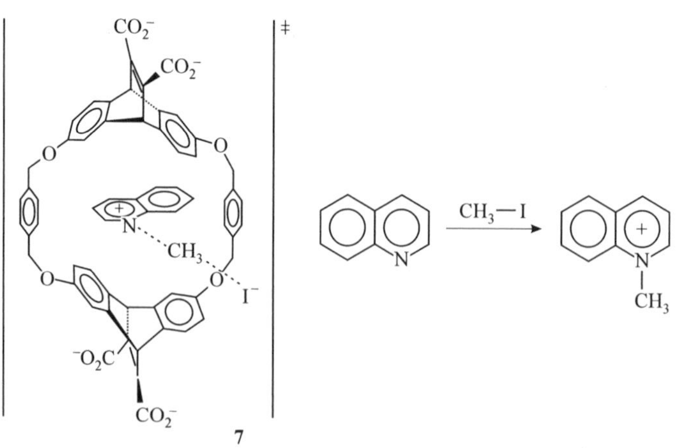

[9.6]

Detailed analyses of interactions leading to association have attempted to identify the relative importance of solvation, hydrogen bonding, etc. (Diederich *et al.*, 1992). The equilibrium constant for complexation is, of course, the result of differences between the two states. The energetic difference between a hydrogen bond to the substrate from water and that from the host is often very small.

The molecular inclusion agents discussed above are relatively complicated molecules; to achieve their preparation all except the naturally occurring cycloamyloses require experts in synthesis with plenty of time. Although nature achieves the preparation of such high value-added species as enzymes with consummate ease, it is likely that enzyme mimics synthesized by regular chemical methods will eventually compete with these natural enzymes as relatively cheap syntheses are invented. An enzyme mimic should have the advantage that it can act in conditions not suitable for natural enzymes, which have limited industrial use due to their relative instability and solvent incompatibility. Some of the molecules prepared as hosts possess beautiful structures, exemplified by cyclophane (**8a,b**) (Diederich *et al.*, 1988) and by cucerbituril (**9**; Fig. 9.3) (Mock & Shih, 1989) and they possess undoubted molecular recognition properties (Rebek, 1990; Wintner *et al.*, 1994).

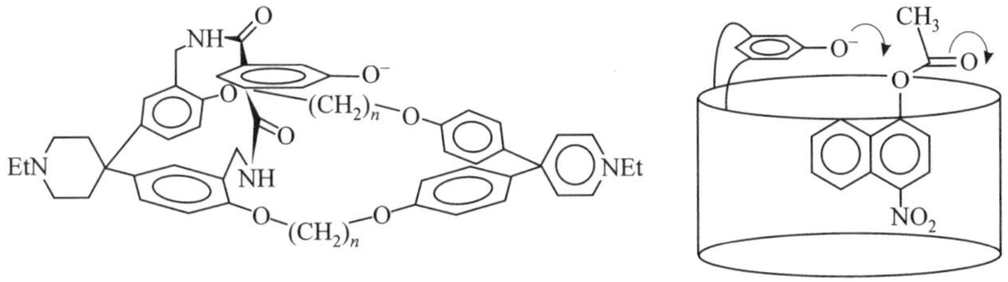

8a

8b

Modification of the substrates/reactants can be carried out to make them compatible with a particular host. For example, cyclic porphyrin trimers can bind dienes and dienophiles having pyridine residues which act as ligands for metal ions. The entropic advantage of bringing the reactant molecules together for the Diels–Alder reaction contributes towards a rate acceleration of 10^3 and the system is also stereoselective, giving only the *exo* adduct (**10**; Walter *et al.*, 1993).

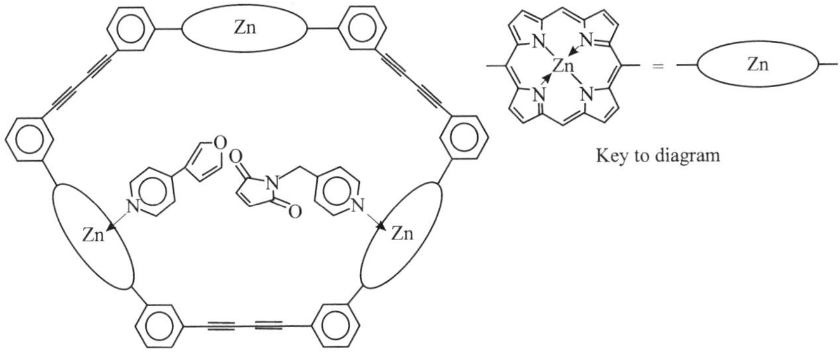

Key to diagram

10

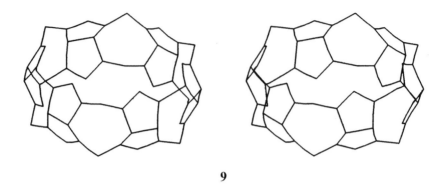

9

Fig. 9.3 A stereo-view of the framework structure of cucerbituril (**9**). See the note below the Contents list (p. vii) for viewing instructions.

Cucerbituril

Key to stereo-diagram

A receptor 'cryptand' for ATP not only binds the triphosphate strongly by a combination of electrostatic and hydrogen-bonding effects ($K = 1.5 \times 10^{-5}$ M) but also cleaves the ATP by a phosphorylation process yielding aminophosphate and ADP (**11**; Hosseini *et al.*, 1983, 1989).

11

9.4   Catalysis by organized aggregates and phases

Catalysis by micelles

Detergent molecules, which consist in the main of long-chain aliphatic groups with polar heads, often form micellar aggregates in aqueous media. The nature and especially the shape of the aggregates are still in doubt (Menger, 1979) but light-scattering experiments indicate an average diameter, if the aggregates were spherical, of about 5 nm. The classical spherical model is probably not strictly true; nevertheless the surface must exhibit a polar-type region (Fig. 9.4) in order to be compatible with the water solvent.

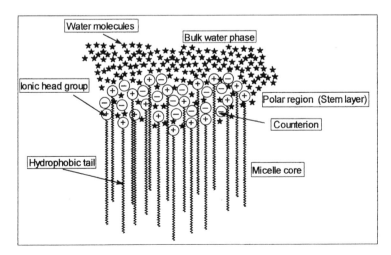

Fig. 9.4 Representation of a micellar surface for a cationic detergent in water; the surface does not have a sharp demarcation between water, Stern layer and micelle interior.

The micelles rapidly exchange detergent molecules with any bulk medium as well as with other micellar aggregates. Solutes will be attracted to the micellar region, especially if they are counterions to the head charge or if they have structures compatible with the long chains. A cationic micelle will complex with anionic counterions and also with esters of long-chain fatty acids. The bringing together of these two reagents by adsorption to the surface of the micelle causes a rate acceleration (Bunton, 1984), by reducing the entropy loss and increasing their effective concentrations. For example, micelles of cetyl trimethylammonium bromide (CTAB) increase the hydroxide-ion-catalysed hydrolysis of penicillins by up to 50-fold (Gensmantel & Page, 1982). The 'loose' interactions between the detergent molecules within the micelle and the reactant substrate are in sharp contrast with the 'tight' interactions available between the atoms within one molecule of enzyme and its substrate (see Chapter 10); such contrasting properties mean that micelles do not compete with enzymes as catalysts.

Attachment of reagent groups to the detergent molecule effectively enhances the coming together of reagent and reactant by hydrophobic interactions. For example, a mixed micelle between lauryl acetohydroxamic acid and cetyltrimethylammonium bromide reacts with 4-nitrophenyl acetate to yield an acetylated hydroxamic acid (**12**) and there is no 'turnover' of the catalyst. When an imidazolyl group is incorporated in the surfactant molecule (as in the mixed micelle **13**) deacetylation occurs (Fig. 9.5) and the hydroxamic acid can then act as a true catalyst (Kunitake *et al.*, 1976).

Since micellar systems are not strictly homogeneous phases, the term 'pseudo phase' has been applied to the micellar aggregate, and the site of reactivity is probably at the interface between bulk solvent and micelle. The interface should not be considered as a definitive demarcation nor as a time-stable region and it is probably best illustrated as in Fig. 9.4 (Menger *et al.*, 1978), where there is a merging of bulk solvent molecules with the polar

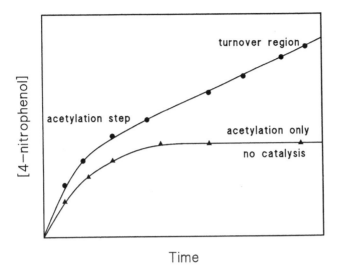

Fig. 9.5 Turnover induced in micelle-catalysed fission of 4-nitrophenyl acetate by detergents possessing imidazolyl and hydroxamate ion groups (**13**). The figure is compiled from the data of Kunitake *et al.* (1976).

heads of the detergent molecules and the counterions (when present). The interface at an inverse micelle involves merging of non-polar solvent molecules with the non-polar detergent tails while the polar heads reside in the bulk of the micelle, merging with polar or aqueous solute.

Aggregates with structures more complex than a micelle arise when the detergent molecule is not a simple chain terminated by a polar group (as for example in phospholipids); these aggregates can comprise vesicles and membranes, and, when the constituent molecules are armed with suitable groups, they can also act as complexation catalysts.

Pseudo-phase model of micellar catalysis

The explanations of the catalytic activity of micelles rest largely on conjecture based on rate enhancements, which must be significant if reasonable hypotheses are to be proposed. The enhanced reactivity of bimolecular reactions is due in the main to the increased concentration caused by adsorption of reactants in the micellar Stern layer (Bunton, 1984; Bunton & Savelli, 1987). The molar volume of a Stern layer can be estimated and thus a second-order rate constant within the Stern layer of the catalysed reaction (k_{cat}) may be calculated and compared with that of the non-catalysed rate constant (k_{uncat}). Bunton showed that the experimental values are often close to each other in magnitude and that the reactants are therefore located in a water-rich region of the micellar pseudo-phase. Apolar reactants should be located more deeply in the micelle; so it is not possible to define an effective micellar volume appropriate to all micelle catalysed systems.

Ion-exchange type processes complicate the kinetics if the ionic ratios and identities are not held constant. Thus, increasing CTAB concentration enhances the rate of hydroxide-ion-catalysed hydrolysis of penicillins, but the effect comes to a maximum and then decreases due to the increasing bromide ion concentration of the CTAB competing with hydroxide ion in the Stern layer (Gensmantel & Page, 1982). A regular saturation plot is obtained (Fig. 9.6) if the bromide ion concentration is maintained constant throughout the concentration range of the detergent.

Catalysis by polymers

The study of complexation and catalysis by synthetic polymers possessing pendant reactive groups is of special interest because enzymes themselves are polymers. Polymer catalysts of many types have been studied, such as linear homo- and hetero-polymers, cross-linked colloidal polymer aggregates (microgels), macropolymers (as, for example, various ion-exchange resins), starburst dendrimers and linear irregularly branched polymers such as polyethyleneimine. Most of the above types of polymer can act as polysoaps

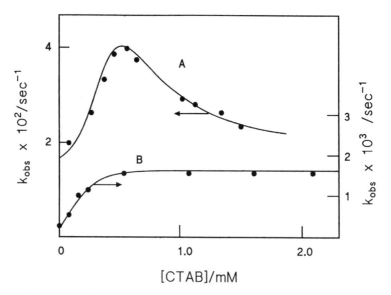

Fig. 9.6 Ion-exchange effect during catalysed hydrolysis of 4-nitrophenyl laurate in the presence of increasing concentrations of cetyltrimethylammonium bromide. A, without added bromide ion; B, total bromide ion kept constant by addition of KBr. The figure is drawn from the data of Al-Awadi & Williams (1990).

when suitable ionic groups are attached to the backbone; under these circumstances the action of the polymers is similar to that of micelles and vesicles. The systems, unlike those of the aggregates discussed above, do not involve exchange between the components of the aggregates because of the covalent linkage within the polymer.

Linear polymers present opportunities for binding and catalysis but, unlike enzymes, are not structurally discrete. Enzymes are constituted from linear polymers but in their case the internal 'cross-linking' caused by various hydrogen-bond, non-polar bonding and disulphide links effectively produces a discrete structure. Undoubtedly the structure at the active site of an enzyme could, in principle, be emulated by non-peptide copolymers. There have been attempts to make polymeric catalysts by incorporating a transition-state analogue in the polymerization process so that its shape and possible binding parameters are imprinted in the polymer. Difficulties with this technique are the removal of the transition-state analogue, accessibility of the reactants and removal of the product (Page & Crombie, 1982, 1984). Some linear polymeric catalysts exhibit relatively large rate enhancements and under certain conditions specificity can be attained. A substantial catalytic effect was obtained in hydrolysis of an ester catalysed by a polymer possessing pendant hydroxamate anions and imidazolyl groups (to provide turnover) (Scheme 9.2) (Kunitake & Okahata, 1976).

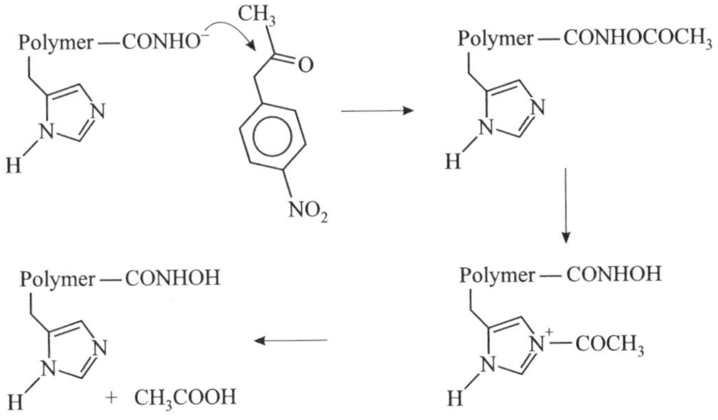

Scheme 9.2 Polymer catalyst for ester fission incorporating 'turnover' capacity.

Microgels are bead-like polymer aggregates prepared by emulsion polymerization of a monomer feed containing a cross-linking monomer. The diameters of the beads can range from about 40 to 300 nm and in a given preparation the population of diameters is quite uniform. The aggregates

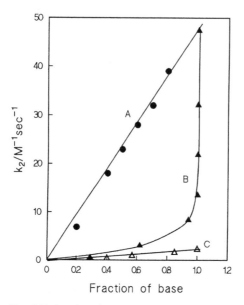

Fig. 9.7 Acceleration of 4-nitrophenyl ester fission by hydroxamate groups attached to microgel aggregates. A, ester with acetohydroxamic acid; B, ester with polymer; C, adipate monoester with polymer. Data are drawn from Weatherhead *et al.* (1980); the rate constants have been corrected for background hydrolysis. The acceleration at high fraction of base corresponds to high pH, and care was taken to correct for background hydrolysis of the ester under these conditions.

form colloidal solutions and can be prepared with suitable pendant groups to effect reaction and also to render the beads 'soluble' in the desired solvents.

Judged from the slope of the line in Fig. 9.7 the reaction of hydroxamate ion pendant on microgels can be enhanced by some 10^6-fold (Fig. 9.7) (Weatherhead *et al.*, 1980) in attack on esters compared with that of acetohydroxamate ion. Microgels complex the ester and at a high fraction of free base the reaction will occur in the interior of the microgel, where the microscopic medium could be favourable compared with the conditions in water or at the surface of the aggregate. The microgel aggregate offers some advantage in that there is little of the diffusional constraint possessed by macrogels (for example ion-exchange resins); microgels have time-stable structures, in contrast with the fluxional nature of linear and branched polymers or of micelles.

The branched polymer polyethyleneimine has been extensively investigated (Klotz, 1986). In a typical example, polyethyleneimine 15% *N*-alkylated with imidazolylmethyl and 10% with dodecyl groups is remarkably active as a catalyst for hydrolysis of 2-hydroxy-5-nitrophenyl sulphate compared with simple imidazole.

Non-covalent complexing with small molecules

There are several examples of organic reactions in water between poorly soluble reactants. The rate constants for these homogeneous reactions often exceed those in polar solvents. For example, rate constants of Diels–Alder reactions are up to three orders of magnitude larger in water than in a hydrocarbon solvent. The rate acceleration is not due to any form of association or aggregation of the diene and dienophile in water. It appears that the main driving force is enforced pairwise hydrophobic interactions and a reduction of the hydrophobic surface area of the reactants in the transition state (Blokzijl *et al.*, 1993).

Complexes readily form in aqueous solution between hydrophobic molecules; the behaviour of caffeine in aqueous solution is a classical example of the self-association observed with purines and pyrimidines which is akin to the 'stacks' of heterocyclic bases constituting the structure of RNA and DNA helices.

Complexation may *inhibit* reactions. This phenomenon has been exploited in pharmaceutical studies to stabilize drugs (Connors, 1987). Addition of caffeine to solutions of the ester benzocaine is well known to depress its rate of hydrolysis. Similar rate inhibitions, for example that of 3,5-dinitrobenzoate ion on the hydrolysis of indolylacryloylimidazole, are also very difficult to explain on the basis of 1:1 complexation (Menger & Bender, 1966).

The type of inhibition recorded above can be considered to result from a change in the microscopic environment excluding the nucleophile, hydroxide ion, and disfavouring the transition state for hydrolysis. In contrast, the solvolysis of the tosyl (Ts) ester **14** in acetic acid is *enhanced* by the addition

14

of aromatic donor molecules (Colter *et al.*, 1964), and the function of the catalyst is possibly to stabilize the incipient carbenium ion by provision of a donating micro-solvent. Application of a technique analogous to that for determining transition-state acidity (Section 3.6) reveals that the transition state is bound in the donor complex more tightly than is the ground state (Scheme 9.3) (Eq [9.7]) (Tee, 1994).

$$K^{\ddagger} = k_2.K/k_1$$
$$(K^{\ddagger} \approx 55\,\text{M}^{-1} \text{ and } K \approx 3\,\text{M}^{-1})$$

[9.7]

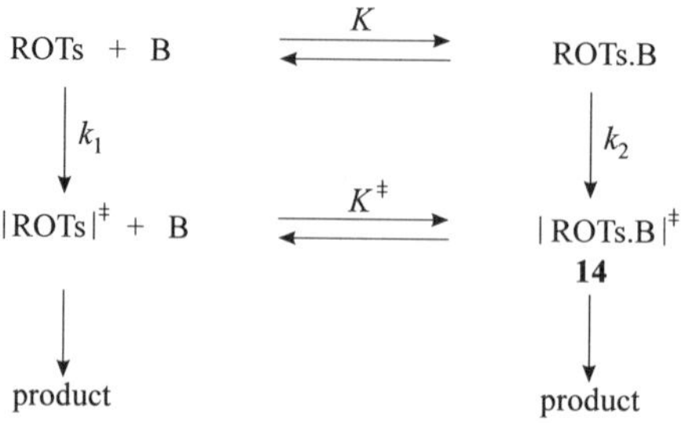

$$\text{ROTs} + \text{B} \xrightleftharpoons{K} \text{ROTs.B}$$

$$\downarrow k_1 \qquad\qquad\qquad \downarrow k_2$$

$$|\text{ROTs}|^{\ddagger} + \text{B} \xrightleftharpoons{K^{\ddagger}} |\text{ROTs.B}|^{\ddagger}$$

14

$$\downarrow \qquad\qquad\qquad \downarrow$$

product   product

Scheme 9.3 Complexation of ground and transition states by the donor molecule.

The action of many molecular complexing agents is often interpreted by reference to a single structure as if there were a single complex with a defined geometry. This is certainly not the case, despite the fact that X-ray crystallography will produce a single structure for the complex; in order to use X-ray crystallography, it is required to obtain crystals which naturally derive from single structures. Such observed structures may not be the ones that lead to reaction. This problem (see the Curtin–Hammett principle—Section 2.6) is very important for the simple model systems where scope for movement of substrate relative to donor is far greater than it is in enzyme–substrate complexes. Since relatively simple molecules aggregate, more work is needed to study the influence of the aggregation properties of the potentially simple complexation catalysts.

9.5 Future directions

Knowledge of mechanism derived from enzymic catalysis is being increasingly applied to incorporate structures into synthetic model compounds, to attempt to recreate the catalytic prowess of the naturally occurring species. In the same way that the technique of unambiguous synthesis is employed to confirm a structure, studies of synthetic model enzymes are being used to confirm mechanism. This approach has the additional goal of producing cheap selective catalysts which can survive the conditions which might be needed, for example in an industrial chemical reactor.

Few model systems approach the catalytic activity of their enzyme analogue. At the time of writing we are witnessing a change in focus of supramolecular chemistry from synthesis and structure to dynamics and function. The past decade or so has been noted for an enormous investment of time into the synthetic routes to novel molecules possessing cavities, together with related structural studies. The future holds promise of many remarkable investigations of this type of compound in regard to its recognition, complexation, transport, sensing, control and transformational aspects. Such studies require the ability of the chemist to synthesize the cavity molecules with precisely defined structures: this is made possible by the superlative achievements in supramolecular chemistry. We confidently expect this type of cavity molecule to figure extensively as a residue in many catalytic systems in the future, although we believe that successful systems will probably arise by chance or designed chance rather than by design.

Further reading

Attwood, J. L. (ed.) (1990) *Inclusion Phenomena and Molecular Recognition*, Plenum, New York.

Bender, M. L. & Komiyama, M. (1978) *Cyclodextrin Chemistry*, Springer, Berlin.

Breslow, R. (1972) Biomimetic chemistry, *Chem. Soc. Rev.*, **1**, 553.

Cram, D. J. & Cram, J. M. (1974) Host–guest chemistry. Complexes between organic compounds simulate the substrate selectivity of enzymes, *Science*, **183**, 803.

Cram, D. J. & Cram, J. M. (1978) Design of complexes between synthetic hosts and organic guests, *Acc. Chem. Res.*, **11**, 8.

Diederich, F. (1988) Complexation of neutral molecules by cyclophane hosts, *Angew. Chem., Int. Ed. Engl.*, **27**, 362.

Diederich, F. (1991) *Cyclophanes*, Royal Society of Chemistry, Cambridge.

Fendler, J. H. & Fendler, E. J. (1975) *Catalysis in Micellar and Macromolecular Systems*, Academic Press, New York.

Gokel, G. (1991) *Crown Ethers and Cryptands*, Royal Society of Chemistry, Cambridge.

Gutsche, C. D. (1989) *Calixarenes*, Royal Society of Chemistry, Cambridge.

Keehn, P. M. & Rosenfeld, S. M. (1983) *Cyclophanes*, Academic Press, New York.

Kirby, A. J. (1966) Enzyme mechanisms, models and mimics, *Angew. Chem. Int. Ed. Engl.*, **35**, 307.

Knowles, J. R. (1987) Tinkering with enzymes: what are we learning? *Science*, **236**, 1252.

Lehn, J. M. (1978) Cryptates: the chemistry of macropolycyclic inclusion complexes, *Acc. Chem. Res.*, **11**, 49.

Lehn, J. M. (1988) Supramolecular chemistry—scope and perspectives, molecules, supermolecules and molecular devices, *Angew. Chem., Int. Ed. Engl.*, **27**, 89.

Murakarri, Y., Kikuchi, J., Hisaeda, Y. & Hagashida, O. (1996) Artificial enzymes, *Chem. Rev.*, **96**, 721.

Roberts, S. M. (ed.) (1989) *Molecular Recognition: Chemical and Biochemical Problems*, Special Publn 78, Royal Society of Chemistry, Cambridge.

Stoddart, J. F. (1988) Molecular Lego, *Chem. Br.*, **24**, 1203.

Szejtli, J. (1982) *Cyclodextrins and Their Inclusion Complexes*, Akademiai Kioda, Budapest.

Tabushi, I. (1982) Cyclodextrin catalysis as a model for enzyme action, *Acc. Chem. Res.*, **15**, 66.

Vögtle, F., Lohr, H.-G., Franke, J. & Worsch, D. (1985) Host/guest chemistry of organic onium compounds—clathrates, crystalline complexes and molecular inclusion compounds in aqueous solution, *Angew. Chem., Int. Ed. Engl.*, **24**, 72.

10 Some enzymes

10.1 Introduction

In this chapter we focus on the overall mechanisms of some enzymes selected to illustrate underlying mechanistic processes. Our choice of examples is based on the level of knowledge of the mechanism rather than on the biochemical significance of the system. Structural knowledge is an invaluable prerequisite for understanding mechanism but, alone, does not define the region of the protein supporting the active site; structural studies on the enzyme complexed with substrate or inhibitor are required. Even this does not often provide rigorous proof of the location of the active-site region in an enzyme structure, and our knowledge of the identities of the active site constituents rests heavily on model building and a large measure of subjectivity derived from chemical intuition based on knowledge of simple chemical reactions. Even when the location of the active site is established in an enzyme structure the mechanism of the catalysis is often not obvious. It is difficult to exclude mechanisms simply on the grounds that distances and angles measured from static X-ray crystallographic structures do not appear to allow contacts to be made between substrate and atoms on the enzyme putatively involved in catalysis. The necessary movements of atoms in an enzymic process involve binding of the substrate, chemical reaction and product release, which are all dynamic processes; an increasing number of examples are known where conformational changes occur during the process leading to significant structural differences in the enzyme. Arguments concerning relatively small distances (such as 'an atom could not be involved in bonding because the structure indicates a distance 1 or $2\,\text{Å}$ greater than the optimal value') are often used to exclude mechanisms; considering the accuracy with which atomic coordinates may be measured in proteins and the assumptions made in 'docking' substrate into a putative active site, conclusions from such arguments must be treated with some caution. X-ray crystallographic studies of frozen enzyme systems (Douzou & Petsko, 1984; Petsko & Ringe, 1984) have enabled conformational changes to be measured in enzyme states along some reaction pathways; for example in α-chymotrypsin catalysis a shift of between 1 and $2\,\text{Å}$ in the imidazolyl group of histidine-57 towards the hydroxyl group of serine-195 occurs between ground and reactive states. In ribonuclease there is a movement of the phosphorus atom of about $2\,\text{Å}$ during the catalytic sequence.

The kinetics of many enzyme-catalysed reactions show a saturation phenomenon with respect to substrate concentration which is explained by the Michaelis–Menten scheme (Eqn [10.1a]) and the rate law (Eqn [10.1b]). $[E_0]$ and $[S_0]$ are the initial concentration of enzyme and substrate, respectively, and V_{max} is the maximal initial rate at saturation ($V_{max} = k_{cat}[E_0]$), and is independent of $[S_0]$. The Michaelis–Menten constant, K_m, is defined as the concentration of S_0 which gives half the maximal rate and is not necessarily numerically equal to the dissociation constant of the enzyme–substrate complex, ES.

$$E + S \underset{K_m}{\overset{}{\rightleftarrows}} ES \xrightarrow{k_{cat}} products + E \qquad [10.1a]$$

$$\text{Initial rate} = \frac{V_{max}[S_0]}{(K_m + [S_0])} = \frac{k_{cat}[S_0][E_0]}{(K_m + [S_0])} \qquad [10.1b]$$

The ratio $[S_0]/(K_m + [S_0])$ represents the fraction of enzyme bound to the substrate. When $[S_0] \ll K_m$, the initial rate is first-order in $[S_0]$ and is given by the expression $(k_{cat}/K_m)[E_0]$ when most of the enzyme is 'free' and unbound; when $[S_0] \gg K_m$, the initial rate is independent of $[S_0]$ and is equal to V_{max}, all of the enzyme is bound up with substrates and/or intermediates.

10.2   Enzymes catalysing acyl group transfer

Introduction

Many enzyme-catalysed acyl group transfer reactions (Eqn [10.2a]) proceed by the formation of an intermediate acyl-enzyme in which the acyl group has been initially covalently transferred to a nucleophilic group (X) on the enzyme (Eqn [10.2b]). This acyl-enzyme intermediate is subsequently transferred to the receiving nucleophile giving rise to, at least, a two-step process—acylation and deacylation (Eqn [10.2b]).

$$[10.2a]$$
$$RCO-Lg + Nu^- \longrightarrow RCO-Nu + Lg^-$$

$$RCO-Lg + Enz-XH \underset{k_{-1}}{\overset{k_1}{\rightleftarrows}} ES \xrightarrow{k_2} RCO-X-Enz + LgH$$
$$\Big\downarrow k_3, \text{Nu, deacylation}$$
$$RCO-Nu + Enz-XH$$
$$[10.2b]$$

For those processes occurring through the intermediate formation of an acyl-enzyme the various microscopic kinetic constants are given by Eqn [10.3]. At saturation ($[S_0] \gg K_m$), k_{cat} can reflect either deacylation

$$k_{cat} = \frac{k_2 k_3}{(k_2 + k_3)}; \qquad \frac{k_{cat}}{K_m} = \frac{k_1 k_2}{(k_{-1} + k_2)} \qquad\qquad [10.3]$$

(k_3) if $k_2 \gg k_3$, or acylation (k_2) if $k_3 \gg k_2$, where the enzyme can predominate as either the acyl-enzyme or the enzyme–substrate complex respectively. Below saturation conditions, the second-order rate constant, k_{cat}/K_m, always reflects the rate of *acylation* irrespective of the relative values of k_2 and k_3, so long as formation of the acyl-enzyme is effectively irreversible ($k_{-2} = 0$). The effect on k_{cat}/K_m of modifying the reaction conditions can only be used to identify parameters concerned with acylation. The advantage of employing k_{cat}/K_m is that, at least, the ground-state structures of the substrate and reactant are known. When effects on k_{cat}, alone, are employed, the detailed structure of the reactant (complexed or bonded to the enzyme) is often unknown, particularly in terms of effective charge distribution.

Evidence for acyl-enzyme intermediates

An acyl-enzyme intermediate was first demonstrated by 'burst' kinetics in the α-chymotrypsin-catalysed hydrolysis of 4-nitrophenyl ethyl carbonate indicating a build-up of an intermediate (Hartley & Kilby, 1954). This arises because $k_2 > k_3$ and the rapid release of the leaving group during the build-up of intermediate concentration is followed by a slower 'turnover' in the steady state controlled by k_3 (Eqn [10.2b]). The use of unreactive acyl groups gives intermediates which are relatively stable to hydrolysis and may even be isolated by recrystallization under slightly acidic conditions. The location of the acyl group at a serine hydroxyl group in chymotrypsin can be obtained from UV-vis studies (Bender *et al.*, 1964) and, more recently by electron spray mass spectrometry of cinnamoyl derivatives (Scheme 10.1). Acyl-enzyme intermediates in thiol protease-catalysed reactions were shortly afterwards demonstrated by use of thion ester substrates which produce a diagnostic dithioester chromophoric group (see, for example, Lowe, 1976; Brocklehurst, 1987); this experiment indicates that an acyl-enzyme intermediate is formed and also that the acyl group is located at a sulphur. Less direct methods involve examination of the kinetics for evidence of a common step. For example, papain-catalysed hydrolyses have identical k_{cat} values for aryl and alkyl *N*-benzoylglycinate esters, consistent with slow deacylation in a mechanism involving build-up of an acyl-enzyme intermediate; hindered esters and amides have k_{cat} values less than these due to acylation being rate-limiting (Lowe, 1976).

CT-OH + PhCH=CHCO-Lg $\longrightarrow$ CT-OCOCH=CHPh $\xrightarrow{\text{H}_2\text{O}}$ CT-OH + PhCH=CHCO$_2$H

α-chymotrypsin   cinnamoyl-α-chymotrypsin

PAP-SH + PhCONHCH$_2$CS-Lg $\rightarrow$ PAP-SCSCH$_2$NHCOPh $\xrightarrow{\text{H}_2\text{O}}$ PAP-SH + PhCONHCH$_2$COSH

papain   benzamidothionacetylpapain

Scheme 10.1 Kinetic demonstration of acyl-enzyme intermediates.

Acyl-enzyme intermediates have been trapped in the α-chymotrypsin system (Epand & Wilson, 1963) making use of benzoylglycine esters as substrates and hydroxylamine as the trapping agent (Section 2.11). The build-up of acyl-enzyme is usually not observed explicitly unless the concentration of enzyme is at a relatively high level consistent with the measuring technique.

Site of the catalytic function

A problem inherent in most enzyme studies is that a group involved in covalent inhibition is not necessarily involved in the catalytic mechanism; it could simply be close enough to the active site to prevent access of the substrate when modified by addition of the inhibitor. This ambiguity was solved in an ingenious way for α-chymotrypsin by reducing the size of serine-195 (a putative catalytic function) as well as altering it chemically. Weiner *et al.* (1966) treated the serine hydroxyl of α-chymotrypsin with phenylmethane sulphonyl fluoride and dehydrated the product with hydroxide ion to yield an anhydrochymotrypsin where the $>CHCH_2OH$ of serine-195 is replaced by $>C=CH_2$. The enzyme still retains its overall structure with an accessible active site, as evidenced by the results of binding studies with a chromophoric inhibitor, and the hydroxide ion concentrations employed do not destroy the enzyme.

Locating the active site in an enzyme structure is best effected by studying the structure of complexes with inhibitors and checking that substrate *protects* the enzyme against the action of these inhibitors. This procedure only gives an approximate location for substrate in the active site; precise information can only be obtained at present by model-building studies using energy-minimizing programs. Our earlier remarks about the significance of arguments based on precise measurements from structures are particularly relevant to the interpretation of such models. X-ray crystallographic studies of enzymes complexed with poor substrates or inhibitors may or may not indicate the mode of substrate binding.

Linear free-energy relationships

Application of linear free-energy relationships to enzyme mechanisms (Kirsch, 1972) has naturally been attempted, but the influence of the substituent may not simply be through the transmission path of the regular Hammett or Brønsted standard; this is because in addition to normal resonance and inductive effects, there may be direct interactions of the substituent with the protein which can affect the activation energy. Such problems are seen in the extreme in a non-enzymic example, namely the acylation of cyclodextrin by substituted phenyl acetates (Van Etten *et al.*, 1967), where a combination of Hantsch constants, Hammett σ and molar volumes are unable to correlate the rate constant data for a system manifestly less complex than an enzymic one.

In the case of chymotrypsin, the acylation constants k_2 and k_{cat}/K_m for substituted phenyl acetates give a tolerably good Hammett correlation but since the plot is relatively scattered there is probably substantial binding of the phenyl moiety within the tosyl pocket and it is doubtful if the ρ-value can be easily interpreted. The situation is much better in the case of substituted phenyl esters of acylamido acids where the side-chain binding of the substrate within the tosyl pocket will effectively force the phenoxide-ion leaving group to point into the bulk solution, where the interactions are with solvent alone and are thus similar to those of the standard ionization reaction used to define the Hammett σ-values. The low value of the Hammett ρ-value for these substrates points to substantial electrophilic assistance in the transition state for the acylation step. The obvious interpretation of this is that the electrophilic assistance comes from some type of interaction at the carbonyl of the ester (Williams, 1984; Thea & Williams, 1986) because phenoxide-ion leaving groups are unlikely to require assistance to enable them to depart from a tetrahedral intermediate and therefore the interaction can be identified as deriving from hydrogen bonding from peptide NH groups to the ester carbonyl oxygen.

In the above example the negative charge developed on the carbonyl oxygen in the transition state is substantially less than that developed on the leaving oxygen in the product. It is not often possible to disentangle the effect of the binding step on Hammett or Brønsted exponents from that of the catalytic steps, unlike the situation in normal solution reactions. The second-order rate constant k_{cat}/K_m reflects the free-energy difference between the transition state for acylation and the 'free' enzyme and substrate. Hence, the ground-state 'structure' of the substrate in terms of effective charge distribution is known. When k_{cat} is used as the diagnostic tool (and also, if it can be measured, k_2) it reflects the free-energy difference between the unknown transition-state structure and the often-unknown substrate structure in its enzyme-bound state.

Application of the Leffler parameter α (Chapter 3) is perfectly general and does not need to refer to polar effects; the generalized method has been applied to the binding steps of tyrosyl-tRNA synthetase, the enzyme responsible for catalysing formation of tyrosyl-tRNA and representative of the group of enzymes which assists in editing the sequence during protein biosynthesis. The logarithm of the rate constant for the reaction is linearly dependent on the logarithm of the equilibrium constant (Eqn [10.4]) for a range of directed point mutations of amino acid residues close to the proposed active site (Fersht *et al.*, 1986, 1987).

$$\log k = \alpha \log K + C \quad (\alpha = 0.79) \tag{10.4}$$

The Leffler α-value indicates that the change in binding is some 79% complete in the transition state compared with that in the final bound system. Moreover the Leffler α-value reaches very large values when the point-mutated residues are involved in binding in the transition state but not in the reactant and product states; such deviations bear a striking resemblance to those for identity reactions (Chapter 6, Fig. 6.9).

The constraints which should be applied in the use of this type of approach are the same as those discussed in Chapter 3, namely that the effect of the changes in the structure should be relatively small; in addition they should not destroy the integrity of the enzyme structure if the results of such changes are to be easily interpreted.

Structure

Charge relay systems

A structural feature discovered by Blow (1976) for α-chymotrypsin, whereby the amino acid residues aspartate-102 and histidine-57 are linked by hydrogen bonding, has since been observed in many other proteases acting via an active serine group, and similar arrangements may also be seen in the structures of the thiol proteases (cysteine replaces serine as a nucleophile) and also in some non-protease enzymes. It was originally thought that the aspartate enhanced the nucleophilicity because it would assist removal of a proton from the imidazolyl group, which in turn removes a proton from the serine. This proposal (Scheme 10.2) has long been doubted by organic chemists since the basicity of the imidazolyl group does not seem to be any different from normal and moreover an intrinsically reactive nucleophile does not fit in with the requirement of specificity in enzymic catalysis.

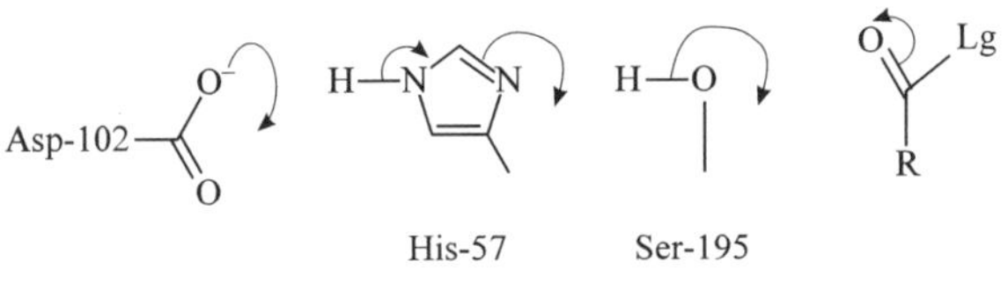

Asp-102

His-57   Ser-195

Scheme 10.2 Putative charge-relay mechanism for chymotrypsin and serine proteases.

Proton inventory work (see Section 4.4) indicates that only one proton is transferred in the transition state of the rate-determining step for non-specific substrates (Stein *et al.*, 1983); a charge-relay system would require two such protons 'in flight' involving the aspartate–imidazolyl and imidazolyl–serine pairs. The observation of more than one proton 'in flight' for some specific amide substrates is attributed to strengthening of hydrogen bonds as the tetrahedral intermediate is formed (Fink, 1987). The structural features of the triad of residues (histidine, aspartic acid and serine) are conserved through the serine proteases and are therefore important but it is not certain that the process, known as the 'charge relay', has much effect on intrinsic reactivity, because models with this network alone do not show enhanced reactivity (Rogers & Bruice, 1974); nor does chymotrypsinogen, the precursor of α-chymotrypsin, have any activity although it possesses an intact triad (Birktoft *et al.*, 1976). Recent studies have indicated that the distance between the serine hydroxyl and imidazolyl is too large for hydrogen bonding in many of the serine protease systems and a movement of between 1 and 2 Å is required before the imidazolyl group can abstract the proton from the hydroxyl group. The aspartate-102 of α-chymotrypsin has an abnormally low pK_a, indicating that during the catalytic process the proton probably stays on the imidazolyl, which therefore becomes cationic. The role of the imidazolyl–aspartate system is now thought to be simply to confine the location of the imidazolyl group (Fersht & Sperling, 1973) in the systems exhibiting these features.

Electrophilic assistance

The structure of the serine protease, subtilisin BPN$'$, possesses an array of peptide NH groups which could interact with the oxyanion produced as a result of nucleophilic attack of the serine on the amide or ester substrate (Robertus *et al.*, 1972). This array was termed the oxyanion hole or pocket. A similar three-dimensional array exists in the structures of tosyl chymotrypsin (Matthews *et al.*, 1967) and in indolylacryloyl chymotrypsin (Steitz *et al.*, 1969) and occurs in most other serine and thiol protease structures (Brocklehurst, 1987; Fink, 1987). Tosyl chymotrypsin is an excellent structural model of the tetrahedral adduct and also of the transition state for acylation (see Scheme 10.3) whereby the sulphonate group models the orthoester intermediate in acylation of the enzyme. The covalent adducts between *Streptomyces griseus* protease A and its naturally occurring inhibitor, chymostatin (Delbaere & Brayer, 1980) and the inhibitor Ac-Pro-Ala-Pro-NHCH(CH$_2$Ph)CHO (Brayer *et al.*, 1979) and that of benzeneboronic acid with subtilisin BPN$'$ (Matthews *et al.*, 1975) are also excellent models of the transition state for acylation and the tetrahedral adduct. The orthoester and boronate structures (Scheme 10.3) are only able to decompose to reactants and are stable enough to be studied by X-ray crystallography. In all the examples an oxygen of the adduct is close to the NH groups of the peptide backbone adjacent to the reactive serine. This is illustrated with a three-dimensional diagram of part of the tosyl chymotrypsin structure (Fig. 10.1).

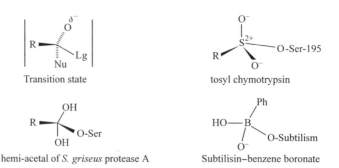

Scheme 10.3 Analogues of tetrahedral transition states in proteases.

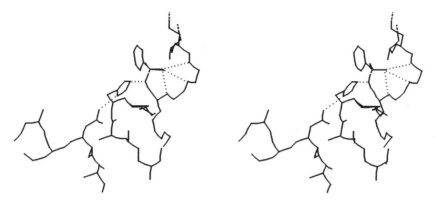

Fig. 10.1 Three-dimensional diagram of the structure of tosyl α-chymotrypsin drawn from the atomic coordinates deposited at the Brookhaven Protein Data Base (Bernstein *et al.*, 1977). For assistance with viewing see the note below the Contents list (p. vii).

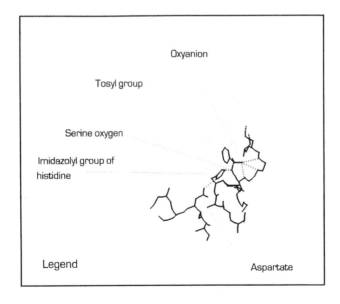

Key for stereo-diagram

The observation of only little change in effective charge in specific substituted aryl ester substrates of chymotrypsin, subtilisin and papain (Brocklehurst, 1987; Fink, 1987) is evidence that the oxyanion pocket assists the catalytic function by electrophilic interaction. Thion ester substrates of chymotrypsin show reduced activity consistent with the $C{=}S$ and the optimal $S{\cdots}HN$ bond lengths being larger than those for the oxygen analogues (Asboth & Polgar, 1983; Campbell *et al.*, 1983). The observation that chymotrypsinogen is inactive is also consistent with the participation of the oxyanion pocket because in this precursor it is not properly formed and the NH of Gly-193 is forced to point in the wrong direction for interaction with the putative oxyanion (Birktoft *et al.*, 1976).

Efficient catalysis requires that the tetrahedral intermediate is not too stabilized, otherwise it could accumulate and lead to a reduced rate of turnover. Probably the best example of such stabilization occurs in the interaction of trypsin inhibitor peptide with trypsin. Although electrophilic catalysis is needed for the formation of the tetrahedral intermediate, its breakdown requires the transfer of electron density from the oxygen 'anion' back to form the incipient carbonyl oxygen to expel the leaving group. Overstabilization of this negatively charged oxygen would thus raise the energy barrier for the next step in the reaction (Gensmantel *et al.*, 1980; Morris & Page, 1980b).

The oxyanion-pocket hypothesis is a case where little background knowledge is available from simple chemical models. A recent study of the pK_a values of phenolic groups intramolecularly hydrogen-bonded to carbonyl functions (Mock & Chua, 1995) concludes that regular hydrogen bonding of the oxyanion accounts for only a small amount of the acceleration observed in acylation; interaction between backbone dipoles may be the answer in this case.

Side-chain specificity

The structure of the active site of chymotrypsin was mapped (Bender & Kezdy, 1965) by correlating the structure of the substrate with the value of k_{cat}/K_m, essentially the second-order rate constant for formation of the transition state for acylation from the free enzyme and free substrate. Nonproductive binding modes are excluded from the ratio k_{cat}/K_m due to cancellation. Hydrophobic side chains of the substrate increase the reactivity and this is ascribed to the complexation with a hydrophobic site at the active centre denoted as the ρ_2 site; the ρ_1, ρ_3 and ρ_4 sites correspond to binding of the acylamido, the acyl function and the α-hydrogen respectively. The first structures of tosyl chymotrypsin confirmed these conclusions; there is a cavity close to serine-195 denoted the 'tosyl pocket' because it accepts the tosyl function of the inhibited enzyme and this is identified as the ρ_2 site. Comparing the high resolution X-ray crystallographic structure of α-chymotrypsin with that of its inactive precursor, chymotrypsinogen, indicates that the ρ_2 site or tosyl pocket is not properly formed in the zymogen, thus contributing to its inactivity (Birktoft *et al.*, 1976).

Some mechanisms for proteases

The enzymes described above form two classes of proteases—the serine proteases and thiol (or cysteine) proteases.

Two other classes of proteases, namely the zinc and aspartate proteases, together with the serine and thiol proteases, constitute most of the naturally occurring proteolytic enzymes. The differences between the classes seem to be due to the different ways in which the carbonyl group of the scissile amido function is activated in the substrate. Electrophilic activation is probably effected by complexation with zinc(II) in the case of the zinc proteases (Fig. 10.2), by an oxyanion pocket (Scheme 10.4) in the serine proteases, or by hydrogen bonding with a carboxylic acid system in the aspartate enzymes (Scheme 10.5).

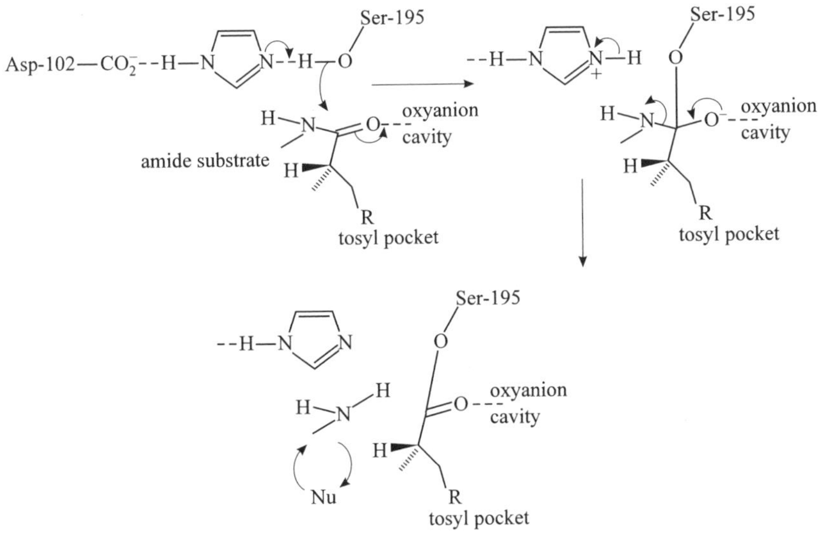

Scheme 10.4 A mechanism for α-chymotrypsin-catalysed transfer showing the 'half-reaction' formation of acyl enzyme; reaction of acyl enzyme is simply the reverse with nucleophile substituted for amine (Fink, 1987).

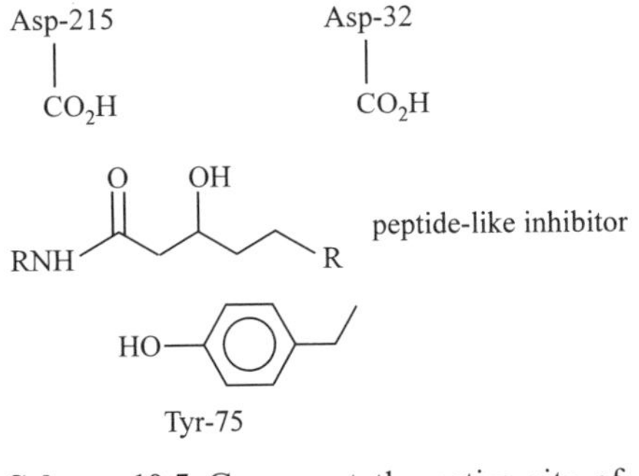

Scheme 10.5 Groups at the active site of endothiapepsin (Fischer, 1987; Foundling *et al.*, 1987; Chatfield & Brooks, 1995).

The carboxypeptidase substrate Tyr-Gly is shown in a non-productive binding mode (Fig. 10.2) and it is inactive probably because the terminal ammonium ion electrostatically attracts the glutamic acid anion (Glu-270), which is probably required in the catalytic function.

An analogue of a substrate with the carbonyl group replaced by a hydroxyl is a potent inhibitor of aspartate proteases; the structure exhibits hydrogen bonding between the aspartate residues (Asp-215 and Asp-32) and the oxygen of the hydroxyl group (Scheme 10.5). Both zinc and aspartate proteases have not been shown to involve covalently linked enzyme intermediates although both enzyme types have carboxylic acid anions at their active sites which could be expected to give transitory anhydride intermediates. An anhydride intermediate has been demonstrated for reaction of an aspartate protease, pepsin (penicillopepsin) with an unnatural substrate (a sulphite) (Nakagawa *et al.*, 1976). An intermediate where a carboxyl group of the aspartic acid protease is linked covalently with the amine leaving group of the peptide is excluded by the absence of ^{18}O-label in the enzyme when the reaction is carried out in labelled water (Antonov *et al.*, 1981). It is currently assumed that the enzymes catalyse the simple addition of water to the acyl carbonyl group. The currently accepted mechanism for a serine protease is illustrated in Scheme 10.4 and active-site constituents for aspartate and zinc proteases are illustrated in Scheme 10.5 and Fig. 10.2 respectively.

There are several examples where a knowledge of the mechanism of the enzyme-catalysed reaction has resulted in the design of transition-state analogues or mechanism-based inhibitors which may lead to the development of useful therapeutic agents (Page, 1990; Kreutter *et al.*, 1994; Ohba *et al.*, 1996).

10.3   Catalysis of proton transfer reactions

Introduction

Chapter 8 indicates the importance of proton transfer in catalysis. Whereas most enzymes involve proton transfer even when it may only be a subsidiary process, the enzymes described in the following sections catalyse reactions where proton transfer is central. The role of proton transfer in facilitating enzyme and organic reactions is to dissipate potentially unfavourable charge build-up during a reaction, effectively bypassing highly unstable acidic or basic intermediates. A proton transfer agent is required in proteases to assist departure of the leaving amine and also to assist attack of the nucleophilic group on the peptide bond.

The triose phosphate isomerase system (1,2-hydrogen transfer)

Triose phosphate isomerase catalyses the migration of a proton between C2 and C3 positions in the interconversion of G3P (glyceraldehyde-3-

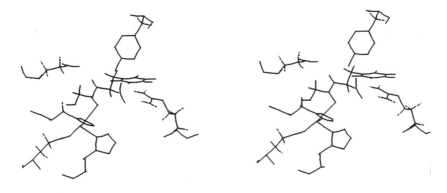

Fig. 10.2 Groups at the active site of carboxypeptidase-A (Auld, 1987); the three-dimensional diagram shows the binding of Tyr-Gly and is drawn from atomic coordinates deposited at the Brookhaven Protein Database (Bernstein *et al.*, 1977). For assistance with viewing see the note below the Contents list (p. vii).

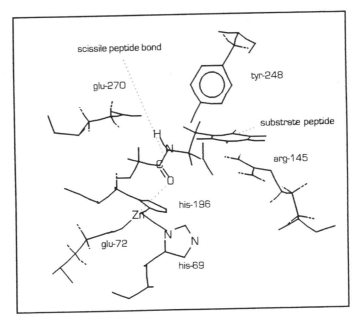

Key for stereo-diagram

phosphate) and DHAP (dihydroxyacetone phosphate) in an equilibrium which favours formation of DHAP. Since proton transfer is the key step, isotopes of hydrogen have proved very important in determining the overall mechanism and labelling reveals the stereoselectivity of incorporation of the 1-pro-(R) proton (Rieder & Rose, 1959). The 1[(R)-^{3}H]-labelled DHAP retains about 6% of the label in the C2 position of G3P (Herlihy *et al.*, 1976), which strongly suggests that a single base shuttles a proton to and from a single face of the enediol intermediate. Scheme 10.6 illustrates the overall mechanism of the catalytic reaction for the enzyme. Measurement of the initial rate of deuterium and tritium exchange and of the isotopic content of reactants and products in labelled water enable the individual rate and equilibrium constants to be determined (Albery & Knowles, 1976a,b). The three-dimensional structure of triose phosphate isomerase includes glutamic acid-165 close to the active site, consistent with the results of studies (Waley *et al.*, 1970; de la Mare *et al.*, 1972) involving the active-site-directed inhibitors **1** and **2**.

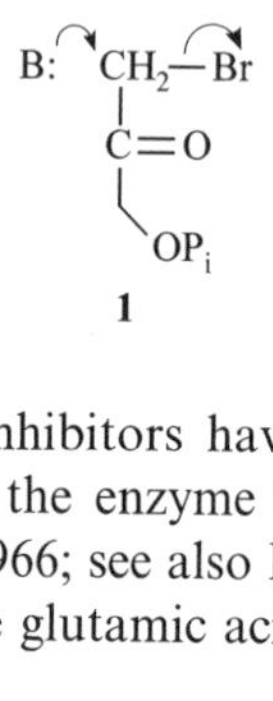

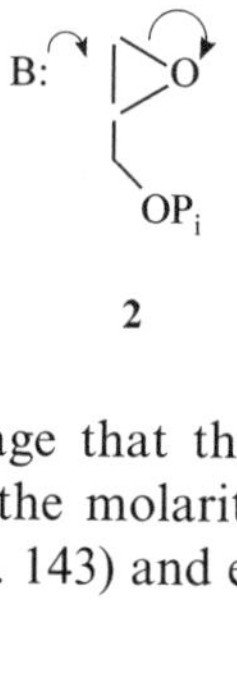

These inhibitors have the advantage that they may also be employed to 'titrate' the enzyme to determine the molarity of the active sites (Bender *et al.*, 1966; see also Fersht 1984, p. 143) and essentially react to form esters with the glutamic acid residue.

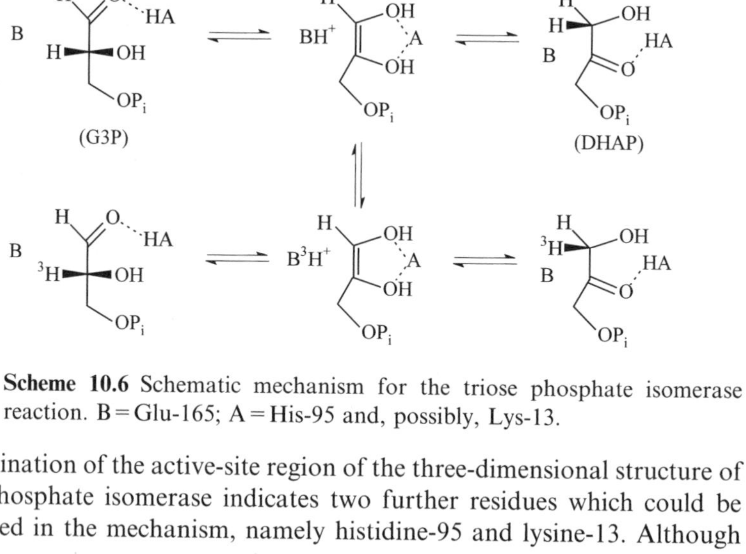

Scheme 10.6 Schematic mechanism for the triose phosphate isomerase reaction. B = Glu-165; A = His-95 and, possibly, Lys-13.

Examination of the active-site region of the three-dimensional structure of triose phosphate isomerase indicates two further residues which could be implicated in the mechanism, namely histidine-95 and lysine-13. Although the pH-dependence of the enzyme indicates a base with a pK_a of 3.9, it is not evident if this is involved in proton transfer by the glutamic acid residue. Point-directed mutation of the enzyme by replacement of the histidine-95 residue by a glutamine residue gives a mutant 400-fold less active than the native species (Nickborg *et al.*, 1988) consistent with the imidazolium ion acting as an important electrophilic species. One of the problems associated with assignment in proton transfer reactions is kinetic ambiguity (Section 2.13): the imidazolyl group could therefore act as the proton acceptor. Infrared studies on the carbonyl group of the triose phosphate–substrate complex indicate that it is polarized (Belasco & Knowles, 1980) but the polarization is lost on point mutation of glutamine for histidine-95 (Komives *et al.*, 1991), which is consistent with electrophilic interaction. Figure 10.3 illustrates the structure obtained from X-ray crystallography for the complex between triose phosphate isomerase and phosphoglycolylhydroxamic acid, which is an analogue of the enediol intermediate.

An interesting point concerning the abnormally low pK_a of histidine-95 (at least 2 units lower than normal) is that it is thought to occur by interaction of the imidazolyl group with the dipole from a length of α-helix (residues 95 to 102) (Knowles, 1991; Lodi & Knowles, 1993).

Pyridoxal enzymes (1,3-hydrogen transfer)

It is very difficult to achieve a Brønsted correlation utilizing structure variation of the enzyme itself because of the limited number of bases available as side chains in amino acid residues. Moreover, site-directed mutagenesis would also be required, making the system extremely laborious as each point in the Brønsted correlation would require preparation of a separate enzyme and its kinetic study. If the proton transfer agent (acid or base) were exogenous to the protein structure, it would be possible to obtain Brønsted correlations simply by variation of the transfer agent and therefore to obtain some very significant mechanistic data (see Chapter 3). Toney & Kirsch (1989) carried out a site mutation where the reactive lysine-258 residue is replaced by an alanine residue in aspartate aminotransferase. The mechanism of this pyridoxal-dependent aminotransferase (Scheme 10.7) involves 1,3-hydrogen transfer which is thought to be catalysed by lysine-258. The mutant enzyme is only weakly active as a catalyst of the transfer reaction, but activity is partially restored by exogenous bases which effectively replace the lysine in an intermolecular mechanism. The rate constants for various amines (B in Scheme 10.7) obey a linear free-energy relationship (Eqn [10.5]) which requires the inclusion of a steric parameter (namely the molecular volume). The relatively large effect of the molar volume is consistent with the requirement that the amine must be accommodated in a restricted cavity vacated when the lysine is exchanged for the alanine residue. The β-value indicates that some 40% of the full charge is developed on the nitrogen base, in agreement with that found by Auld & Bruice (1967) in studies of a model system.

$$\log k_{\mathrm{B}} = 0.39\, pK_a - 0.055 \times \text{molecular volume} - 0.7 \qquad [10.5]$$

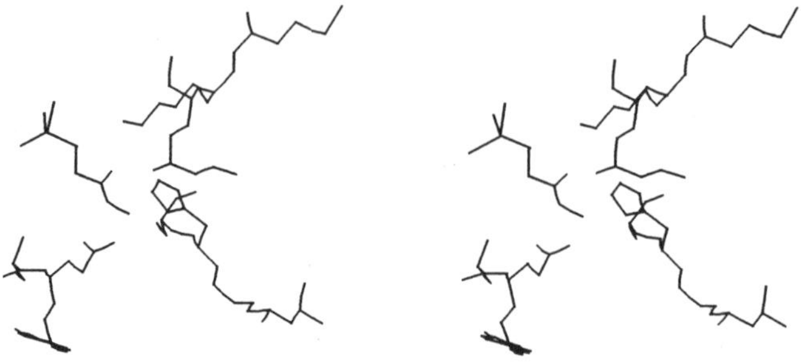

Fig. 10.3 Three-dimensional structure of the complex between triose phosphate isomerase and phosphoglycolylhydroxamic acid, which is analogous to the enediol of Scheme 10.5; the diagram is drawn from atomic coordinates deposited at the Brookhaven Protein Database (Bernstein *et al.*, 1977). For assistance with viewing see the note below the Contents list (p. vii).

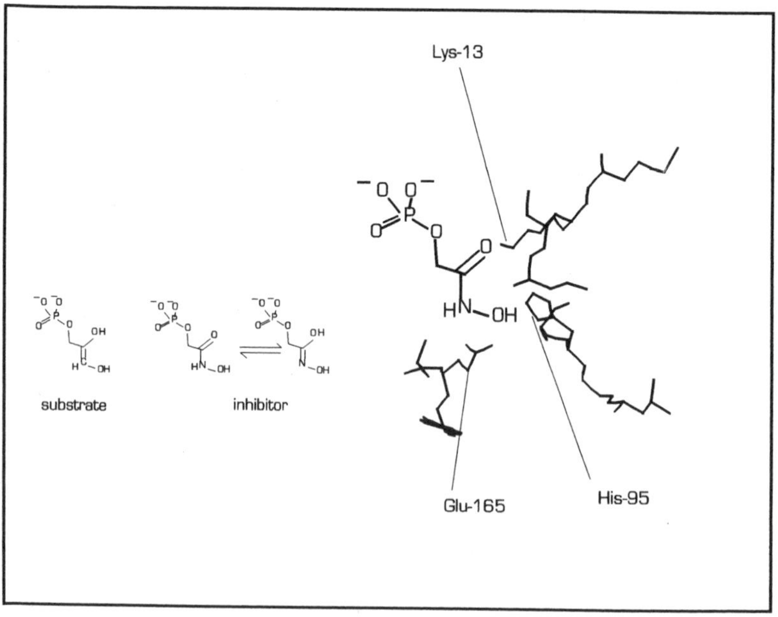

Key for stereo-diagram

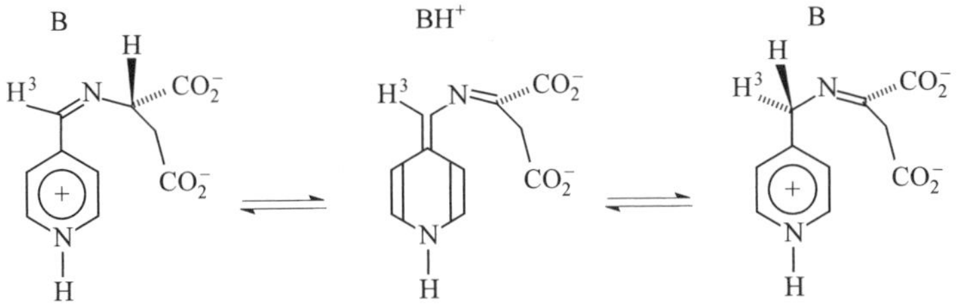

Scheme 10.7 The 1,3-hydrogen transfer reaction of the ^{3}H-labelled adduct of PLP and aspartic acid catalysed by lys-258 (B) of aspartate aminotransferase.

The stereochemistry of the proton transfer in aminotransferase catalysis is very useful as it enables the interpretation of the three-dimensional structure of the enzyme at the active site. Partial retention of label has been demonstrated for the 1,3-hydrogen transfer, which must therefore involve a *syn* elimination process consistent with a single base catalytic group (Dunathan, 1971). The rate of the 1,3-shift is less than that of proton exchange with the medium, indicating that the catalytic group is exposed to the bulk solvent in the ternary complex between pyridoxal phosphate coenzyme, substrate and enzyme. Julin & Kirsch (1989) carried out a double isotope fractionation experiment (Chapter 4) and showed that the primary kinetic isotope effect for the hydrogen on the C_α position is small and moreover essentially independent of the isotopic composition of the solvent. This result is consistent with concerted transfer of the hydrogen between the 1 and 3 positions; it is proposed that this occurs by the amine of lysine-258 both accepting a proton and donating its proton as in Scheme 10.8; the quinonoid species illustrated in Scheme 10.7 is now thought to be a blind-alley *intermediate*.

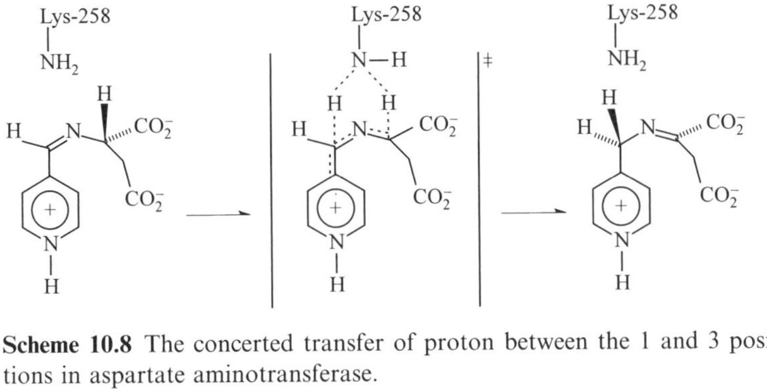

Scheme 10.8 The concerted transfer of proton between the 1 and 3 positions in aspartate aminotransferase.

Addition of $NaBH_4$ to the binary complex between aspartate aminotransferase and its coenzyme destroys catalytic activity and the stereochemistry of the resultant nucleotide product is derived from attack on the *Re* face of the imino group (Scheme 10.9). The stereochemical result requires that the binary complex with the pyridoxal covalently linked by imino group to Lys-258 has its *Re* face exposed to the solvent.

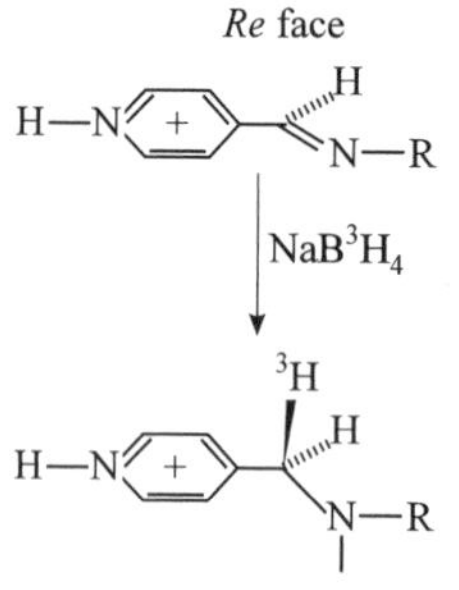

Scheme 10.9 Reduction of the binary complex between coenzyme and aspartate aminotransferase. The substituents on the pyridine residue have been omitted for clarity.

Stereochemistry of the transfer of the hydrogen from C_α of the aspartic acid group to the $C4'$ position on the pyridoxal in the ternary complex may be established by quenching the mixture of aspartate and enzyme with NaB^3H_4. Hydrogen transfer occurs on to the *Si* face, in contrast with the reaction in the binary complex (Scheme 10.10). The change of face at the $C4'$ position available for reduction by the NaB^3H_4 is important in interpreting the three-dimensional structure of the enzyme complex with substrate analogues. The structure of the binary complex of the enzyme with pyridoxal phosphate has a cavity above the *Re* face of $C4'$ of the coenzyme which is attached to Lys-258 and entry to the active site is not hindered,

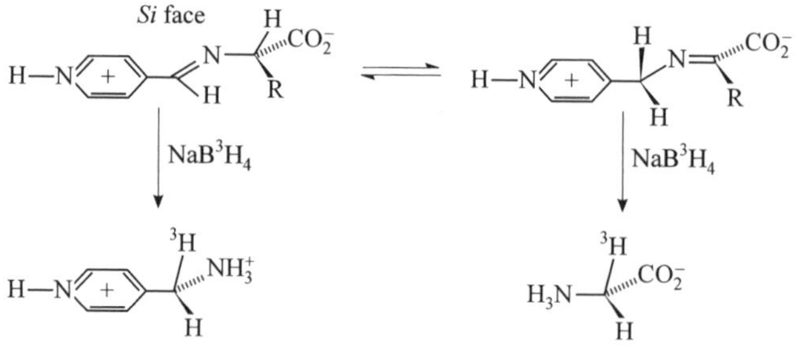

Scheme 10.10 Stereochemistry of the reduction of the ternary complex of aspartate aminotransferase by sodium borohydride.

facilitating ready diffusion of substrates to give the ternary complex. Transimination between Lys-258 and the α-amino group of the substrate is accompanied by a rotation of the C5–C5$'$ bond of about 30° and a change in protein conformation to give a site with an obstructed *Re* face of C4$'$ in the ternary complex.

Amino acid residues implicated in the mechanism by virtue of their coordinates in the crystal structure include Arg-386 and Arg-292 (Kirsch *et al.*, 1984), which are proposed to bind to proximal and distal carboxyl groups respectively of the aspartic acid substrate. Lysine-258 links covalently with the pyridoxal aldehyde and the imino group is transferred to the substrate prior to the 1,3-hydrogen shift. Once the imine is formed between pyridoxal and aspartic acid, the released lysine-258 is proposed to act as the acid–base catalyst in the 1,3-hydrogen transfer. The NH of the pyridoxal interacts with aspartate-222 and tyrosines-70 and -225 assist in binding the phosphate ion and the phenolic O⁻. The phosphate group also interacts with tryptophan-140 and serine-255 and the peptide backbone of residues Gly-108 and Thr-109 (Scheme 10.11). The half-reaction whereby aspartic acid is converted into oxosuccinic acid involves formation of carbinolamine from the binary complex (the pyridoxal–lysine imine adduct) by attack of the aspartic acid amino function (Scheme 10.12).

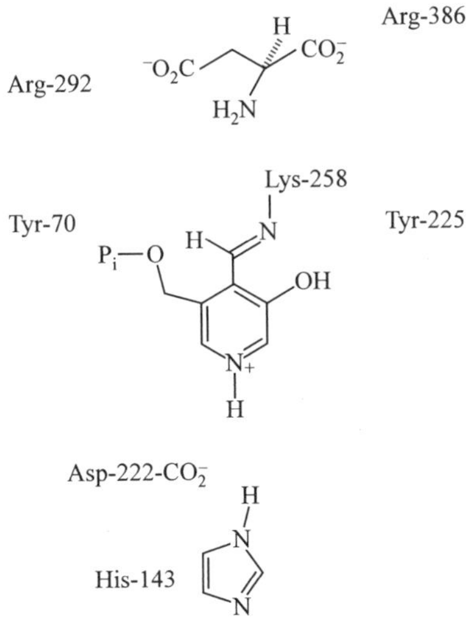

Scheme 10.11 Possible interactions between amino acid residues and the bound coenzyme-aspartic acid complex.

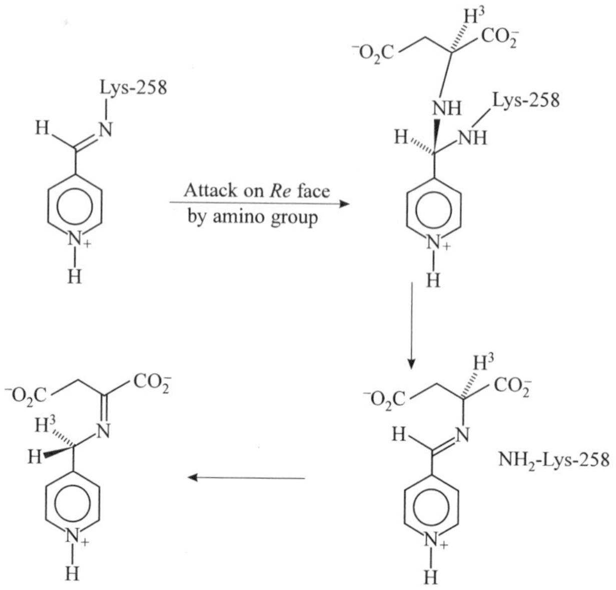

Scheme 10.12 Stereochemistry of formation of the ternary complex between aspartic acid, coenzyme and aspartate aminotransferase.

10.4 Carbon–carbon bond formation and fission

Introduction

Carbon–carbon bond formation and fission are reactions central to both primary and secondary metabolic processes. They most often occur via attack of a nucleophilic carbon (usually carbanionic) at an electrophilic carbon centre by aldol-type condensation reactions. (Hupe, 1987; Kluger, 1987, 1990; Gamblin *et al.*, 1991.)

Aldolases

Aldolases can be divided into two main groups: class I (from higher plants and mammals) possessing an essential lysine residue; and class II (fungi and bacteria), where a metal ion is essential for activity. Both classes of enzyme employ an electrophilic species to activate a ketonic or aldehydic group enabling it to be attacked by a carbon nucleophile.

The class I aldolase catalyses the aldol condensation reaction between dihydroxyacetone DHAP and glyceraldehyde-3-phosphate G3P to form FDP (fructose-1,6-diphosphate) (Scheme 10.13); only the pro-(S) hydrogen of DHAP is replaced by the C1 of G3P, and no label is incorporated at C3 of FDP when the synthesis is carried out in D_2O solvent. The reaction can be considered to occur in two stages, involving removal of the proton from C1 of DHAP followed by addition of the putative carbanion to the G3P. Aldolase also catalyses the stereospecific exchange of the 1-[(S)-^{3}H] proton of DHAP in the absence of G3P. The exchange rate can be reduced by adding G3P (Rose, 1958); the pH dependence of the kinetics is similar for both exchange and condensation reactions. The stereochemistry of the exchange is opposite to that catalysed by triose phosphate isomerase, where the (1-[(R)-^{3}H] proton is displaced, but is the same as that in the full aldolase reaction. The DHAP–enzyme complex exhibits increased absorbance between 240 and 270 nm consistent with formation of an imino function (Mehler & Bloom, 1963) and isolation of the enzyme–DHAP complex inhibited by labelled $NaBH_4$ reagent, followed by amino acid sequence analysis, indicates that the link is between the DHAP and Lys-229 of the enzyme.

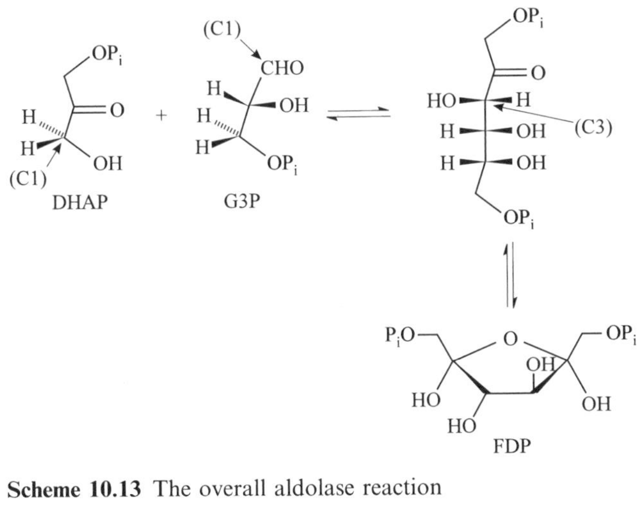

Scheme 10.13 The overall aldolase reaction

The simplest mechanism for catalysis of exchange based on the above results and utilizing the known chemistry of aldol condensations is shown in Scheme 10.14. This mechanism is confirmed by the inactivation by CN$^-$, which intervenes to yield a non-reactive aminonitrile at the imino function (Cash & Wilson, 1986). Reduction of the imine intermediate by borohydride is stereospecific; observation of the stereochemical configuration of the product (Scheme 10.15) indicates the side of the imino group which is open to solvent (Di Iasio *et al.*,1977). Thus the *Si* face of the DHAP attached to the enzyme by the imino function is facing out into the bulk solvent. Modelling of the open-chain FDP substrate with an imino link at Lys-229 provides a picture of the possible interactions at the active site (Gamblin *et al.*, 1991).

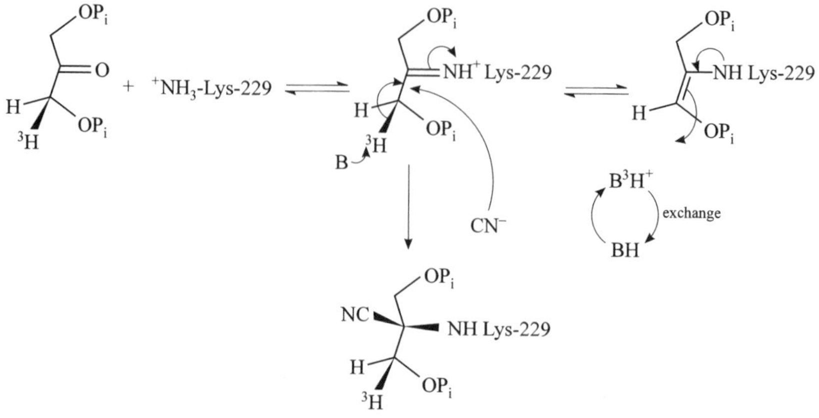

Scheme 10.14 The mechanism of proton exchange and CN⁻ inactivation in dihydroxyacetone phosphate catalysed by aldolase.

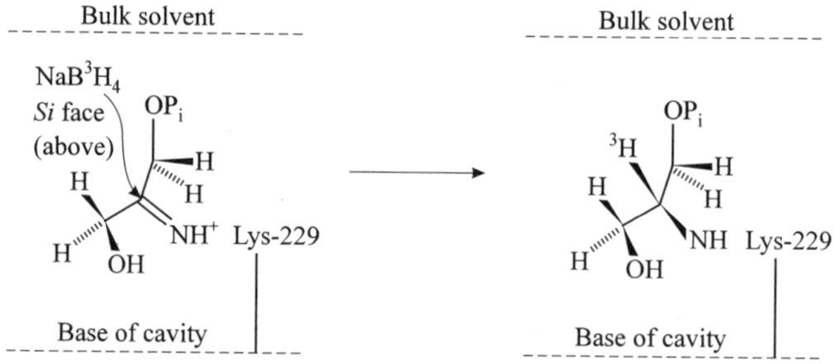

Scheme 10.15 Stereochemistry of the reduction of the DHAP–aldolase adduct by sodium borohydride.

The imine-forming step is pictured in Scheme 10.16: the water which departs from the carbinolamine intermediate presumably departs from the face (*Si*) which is attacked by NaBH₄. The side-chain amino group of Lysine-229 points up from the floor of a cavity towards the centre of a β-barrel type structure. The model, which presents the *Si* face of the imino group to the solvent, has been interpreted to involve the groups Lys-146/ Arg-148 and Lys-41/Arg-42/Arg-303 respectively in interaction with the C1 and C6 phosphate ions. Lysine-146 appears to be close enough to be involved in proton transfer at the carbinolamine nitrogen in lysine-229 and at the oxygen in the dehydration step (Scheme 10.16). The hydroxyl group of tyrosine-363 appears to be involved in addition of a proton to C3. Scheme 10.17 illustrates the interactions of the amino acid residues with the bound FDP (Littlechild & Watson, 1993; Morris & Tolan, 1994; Gefflant *et al.*, 1995).

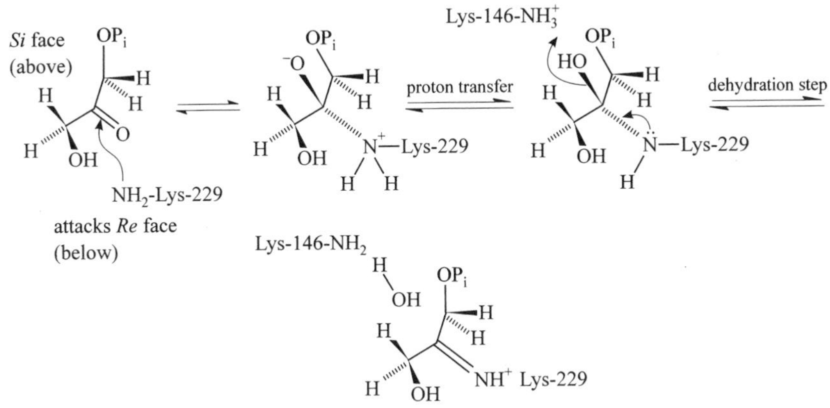

Lys-146-NH$_3^+$

Si face   OP$_i$ OP$_i$ OP$_i$

(above)

H   H

H   H

H   OH

NH$_2$-Lys-229

attacks *Re* face
(below)

proton transfer   dehydration step

HO

$^-$O

N$^+$—Lys-229   N—Lys-229

Lys-146-NH$_2$

H

OH

OP$_i$

H

H

NH$^+$ Lys-229

H   OH

Scheme 10.16 Stereochemistry of the imine-forming step in the aldolase-catalysed reaction (the lysine-146 appears to have a role in proton transfer at both nitrogen and oxygen of the carbinolamine). The *Re* face of the dihydroxyacetone phosphate is the same face as the *Si* face of fructose 1,6-diphosphate shown by Di Iasio *et al.* (1977) to be the site of attack by the terminal amino group of lysine-229.

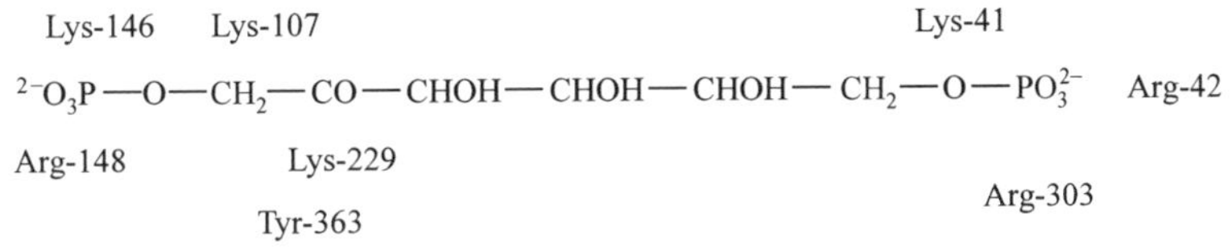

Lys-146   Lys-107   Lys-41

$^{2-}$O$_3$P—O—CH$_2$—CO—CHOH—CHOH—CHOH—CH$_2$—O—PO$_3^{2-}$ Arg-42

Arg-148   Lys-229

             Tyr-363   Arg-303

Scheme 10.17 Diagrammatic locations of amino acid residues relative to the substrate in aldolase-catalysed reactions.

10.5   Transfer of the hydride ion

Introduction

Hydride ion transfer plays an important part in primary metabolism and is involved in many redox reactions with coenzymes. Reactions involving the transfer of the hydrogen with its pair of electrons occur through mechanisms where the hydride ion is not expelled into bulk solution as a separate entity; such mechanisms are illustrated in this section with examples taken from the dehydrogenase family of enzymes.

    The classical Cannizzaro reaction involves direct transfer of the hydrogen from the tetrahedral adduct to the unsaturated acceptor (Section 2.16). It is possible to write a mechanism involving either a hydride ion transfer or a proton transfer for the glyoxalase reaction (Chapter 2); similar schemes could be written for the triose phosphate isomerase reaction. Isotope exchange with solvent, which in the case of glyoxalase is very small and for triose phosphate isomerase is quite large, indicates transfer of a proton in both reactions. A proton transfer mechanism where the conjugate acid of the base delivers its proton to the substrate faster than to solvent (Douglas, 1987) is consistent with little or no proton exchange. Stereochemical integrity of the transfer does not require any particular mechanistic type.

Pyridine–dihydropyridine systems

Liver alcohol dehydrogenase (LADH) is a representative of redox enzymes which catalyse the interconversion of alcohol and aldehyde or aldehyde and carboxylic acid. The alcohol group is oxidized by hydrogen transfer to the pyridine ring of a nicotinamide coenzyme residue (such as $NADP^+$, nicotinamide–adenine dinucleotide phosphate) to form a reduced pyridine nucleus; *in vivo*, the reduced pyridine is oxidized back to the nicotinamide group via an electron-transport pathway linked to oxygen (Adams, 1987). The C4 position of the nicotinamide nucleus is prochiral and the hydrogen can be transferred to the A (or *Re*) face or to the B (or *Si*) face (Karabatsos *et al.*, 1966); this is used to classify the enzymes which catalyse either of the stereochemistries (Scheme 10.18). LADH involves transfer to and from the *Re* face of the aldehyde function. The reaction also specifies the stereochemistry of the hydrogen transfer from the alcohol: LADH catalyses transfer of the pro-(R) hydrogen as shown in Scheme 10.18. The stereochemistry of the LADH enzyme system requires that the coenzyme and substrate occupy the active site of the enzyme in such a way that only the H_R hydrogen is transferred between the *Re* face of the pyridinium ion and the *Re* face of the aldehyde.

The redox reaction of LADH with an aldehyde/alcohol pair is essentially half of a catalysed redox reaction; the other half could be a conjugate oxidation–reduction of alcohol to aldehyde. The relatively large primary isotope effect for the transfer of the hydrogen indicates that this step is rate-limiting.

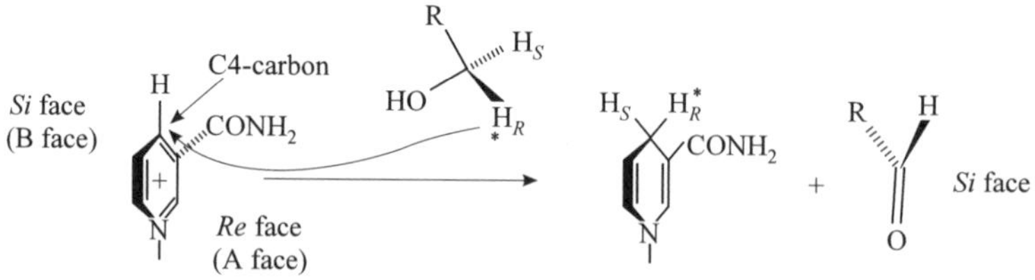

Scheme 10.18 Stereochemistry of hydrogen transfer between pyridinium ion nucleotide and the primary alcohol in alcohol dehydrogenases.

The structure of LADH has been determined by X-ray crystallography for a variety of systems including complexes with coenzyme and substrate analogues. The ternary complex of LADH with $NADP^+$ and the substrate 4-bromobenzyl alcohol includes the alcohol in a non-productive binding mode (Eklund *et al.*, 1982). The $NADP^+$ coenzyme lies at the bottom of a cavity lined with 'hydrophobic' residues (a 'hydrophobic barrel'). The *Re* face of the

$NADP^+$ species is exposed for hydrogen transfer, which is consistent with the observed stereochemistry. It is necessary to 'model-build' the alcohol for the 'productive' mode, as the structure of the non-productive binding site presents the wrong hydrogen for transfer and moreover it is not possible for X-ray crystallography to map a molecule in its reactive conformation.

Residues implicated in the active site include histidine-67, cysteine-46 and cysteine-174 which ligate one of the zinc(II) ions at the active site, and in the free enzyme there is a water molecule which completes a distorted tetrahedral complex of the zinc. It is possible that the coordinated water, with a reduced pK_a, acts as an acid catalyst according to Scheme 10.19.

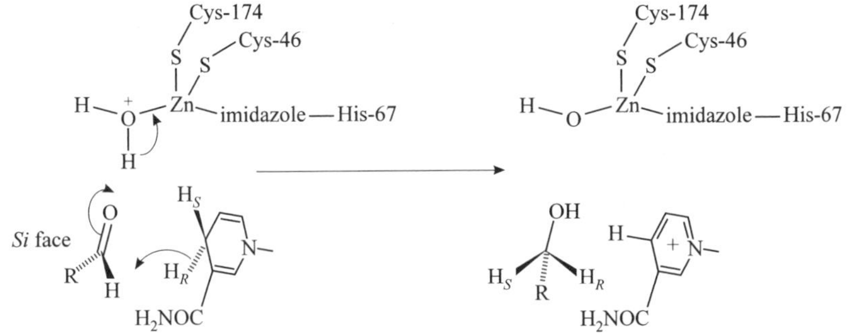

Scheme 10.19 A mechanism for the reduction of the aldehyde involving water coordinated to the zinc(II) in liver alcohol dehydrogenase.

The 'hydrophobic barrel' becomes 'filled' when the alcohol is arranged in the structure so that the pro-(R) hydrogen can transfer to the nicotinamide. If Scheme 10.20 operates, whereby the alcohol oxygen ligates directly with the zinc, then there is a requirement for processes by which the water is expelled and by which a proton can be transferred to and from solvent.

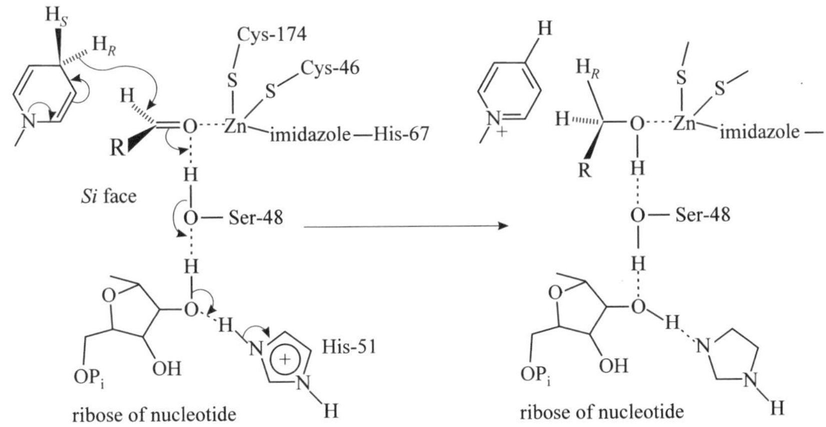

Scheme 10.20 A mechanism for the reduction of the aldehyde with carbonyl directly coordinated to zinc(II) in liver alcohol dehydrogenase catalysis.

The network of possible hydrogen bonds seen in the structure for serine-48, histidine-51 and the ribose of $NADP^+$ conveniently provides a relay system (cf. Section 10.2) to shuttle a proton between the solvent and the substrate atom ligated to the zinc. It is difficult to discuss other proposed mechanisms at this stage because of the lack of diagnostic information. The space at the active site close to the zinc suggests little room for a five-coordinate intermediate or transition state; this would give problems for the ease of substitution unless the alcohol displaces the water via a $D_N + A_N$ process or the ligated water acts as the general acid/base, as in Scheme 10.19. A radical mechanism is excluded because the product of oxidation of bicyclo[4.1.0]heptan-2-ol is the ketone wherein the cyclopropyl ring remains intact (MacInnes *et al.*, 1983).

Becker & Roberts (1984) studied a pyrazole adduct (represented by structure 3) with the enzyme by ^{15}N-NMR, and showed that $^{15}N2$ and $^{15}N1$ chemical shifts of the pyrazole group are consistent with covalent attachment of N1 with the pyridine ring (at C4); the NMR signal of N2 is highly shielded, consistent with direct complexation with the metal ion.

3

Intrinsic ^2H and ^{13}C isotope effects for oxidation of benzyl alcohol catalysed by the liver alcohol dehydrogenase enzyme indicate an aldehyde-like transition state (Scharschmidt *et al.*, 1984); the absence of normal solvent isotope effects (Schmidt *et al.*, 1979) is consistent with proton transfer steps which are not rate-limiting and not concerted with the hydride ion transfer.

The role of zinc(II) as an electrophile is replaced in lactate dehydrogenase by a combination of an imidazolium ion (His-195) and a guanidinium ion (Arg-109); the aspartate–histidine diad (Asp-168/His-195) probably functions to position the imidazolium ion for interaction with the oxygen of the lactate moiety (Scheme 10.21) (see Section 10.2).

The glyceraldehyde phosphate dehydrogenase system (Scheme 10.22) involves reduction from acid to aldehyde level; this includes the imidazolium ion of histidine-176 acting as the electrophilic activator of the carbonyl function in the substrate.

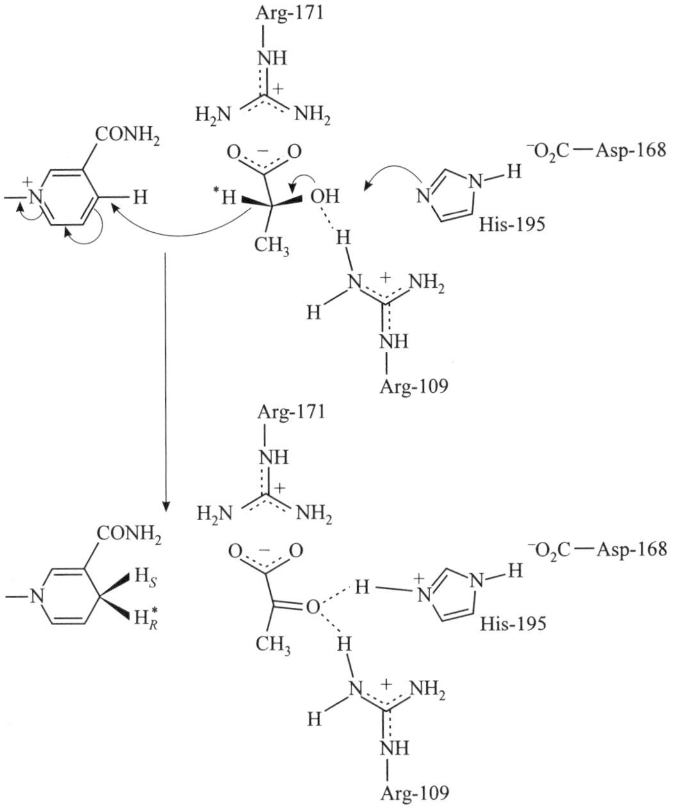

Scheme 10.21 Schematic diagram of the mechanism for lactate dehydrogenase.

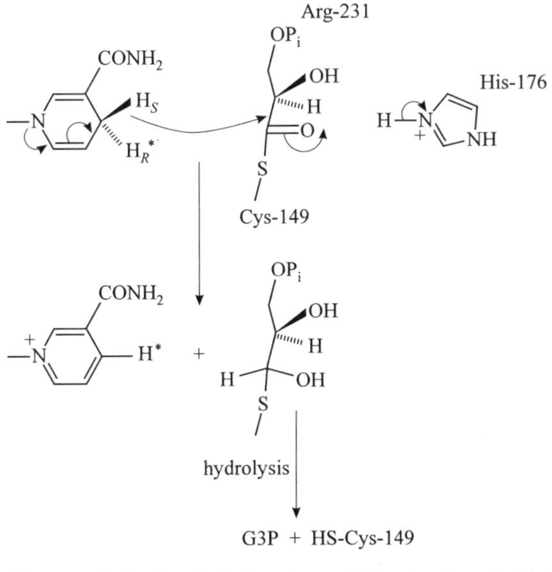

Scheme 10.22 Catalysis by glyceraldehyde phosphate dehydrogenase.

10.6   Alkyl group transfer

Introduction

Transfer of the alkyl group by nucleophilic substitution processes (A_N, D_N; see Appendix A.1), classically described as S_N2 and S_N1, is exemplified in biology by the wealth of reactions involving the C1 position of carbohydrate rings. Other transfer systems exist—examples are that of the methyl group from *S*-adenosylmethionine to a nucleophilic acceptor and reaction of dUMP with N^5,N^{10}-methylene tetrahydrofolate to give dTMP—but we shall consider in this section the carbohydrases lysozyme and β-galactosidase. Lysozyme from hen's egg white was the first enzyme to have its structure solved by X-ray crystallography and today the structures of many lysozymes have been determined, including mutant species and enzyme–inhibitor complexes. Structural data on β-galactosidase have been very difficult to obtain possibly because of the extraordinary long primary sequence (Fowler & Zabin, 1977) and only recently has the X-ray crstyallographic structure been achieved (Jacobson *et al.*, 1994); the enzyme, however, has a large amount of associated kinetic and mechanistic data, unlike the lysozymes, which have proved difficult to study as catalysts.

Alkyl group transfer of the A_N, D_N type is a well-known class of organic reaction mechanism and was largely responsible in forming the basis of mechanistic organic chemistry which originated from the Ingold school at University College London (see Chapters 2, 3 and 4); in recent years the emphasis in mechanism has been on the relationship between the lifetime of a carbenium ion and the concerted process. Indeed the methoxycarbenium ion ($CH_3OCH_2^+$), which is related to the action of carbohydrases, has been estimated to have a lifetime of about 10^{-15} s in water and its components are therefore involved in transfer reactions in an open or 'exploded' A_ND_N transition state (4) (Knier & Jencks, 1980; Richard, 1995).

Carbinolamine derivatives (5) are in equilibrium with the protonated imine (6) and suitably hydroxylated derivatives are potent inhibitors of β-mannosidases. Protonated secondary amines (7) are also potential transition state analogues and hydroxyl-substituted derivatives are weak to potent inhibitors (Page, 1990).

4

5   **6**   **7**

Glycosyl group transfer

The enzyme lysozyme isolated from hen's egg white catalyses the fission of the glycoside link in oligosaccharides of *N*-acetylglucosamine and its derivatives. The original X-ray crystallographic studies of the protein in the presence of oligosaccharide inhibitors indicate a cleft which could accommodate six hexose residues (A to F) located at 'sites' similarly defined. Two amino acid residues, aspartate-52 and glutamate-35, are located in the cleft on opposite sides of a position between the D and E sugar residues where fission occurs in catalysis. The mechanism was postulated to involve general acid catalysis by the glutamate-35 residue and stabilization of an oxocarbenium ion intermediate by the anion of the aspartic acid-52 residue (Scheme 10.23). The enzyme catalyses transfer of the glycosyl group to acceptors other than water and the configuration at C1 is retained in the products, providing good evidence for an intermediate. A frontside $A_N D_N$ non-enzyme catalysed process is unlikely as this mechanism has no substantial precedent in non-enzyme catalysed substitution reactions at carbon (see Chapter 1). Although the site of the catalytic process is in little doubt, it is chastening to note that model-building studies provide the only evidence that the carboxylic residues 35 and 52 are indeed involved. The glutamic acid-35 residue is located in a hydrophobic region, which explains the high observed acid pK_a of between 6 and 7 for lysozyme catalysis; the aspartate-52 is in a polar region and is almost certainly fully ionized in the pH region of catalysis. The glutamic acid is proposed to act in the undissociated form and the aspartate in its base form, but the problem of kinetic ambiguity (Section 2.13) has not been solved. In the models of the enzyme substrate complex, the glutamate carboxylic acid is close to the ring oxygen and the aspartate carboxylate anion to the C1 carbon. To what extent these models are conjectural, based on subjective opinion, is an open question; replacement of the glutamic acid by glutamine destroys activity but replacement of Asp-52 by an asparagine residue gives a mutant enzyme with a small amount of residual activity (Sinnott, 1987, 1990).

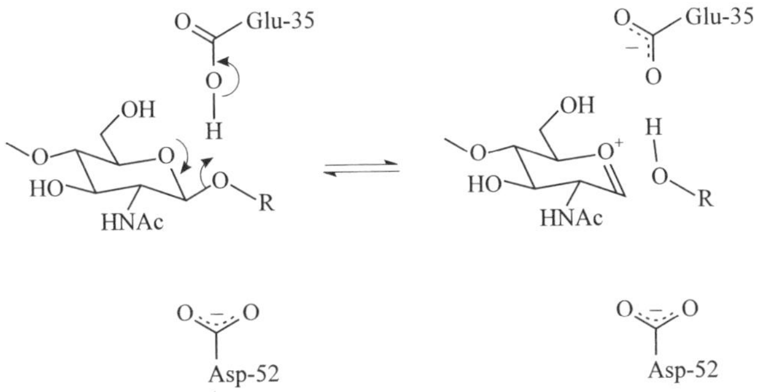

Scheme 10.23 A mechanism for the first step in lysozyme-catalysed hydrolysis.

The imidazolyl group (so often involved in enzyme catalytic mechanisms) is absent in some species variants of lysozyme and therefore appears to have no catalytic involvement; moreover, it is remote from the proposed active site in the enzyme from hen's egg white. The amino acid residues of the cleft region of lysozyme are illustrated in Scheme 10.24.

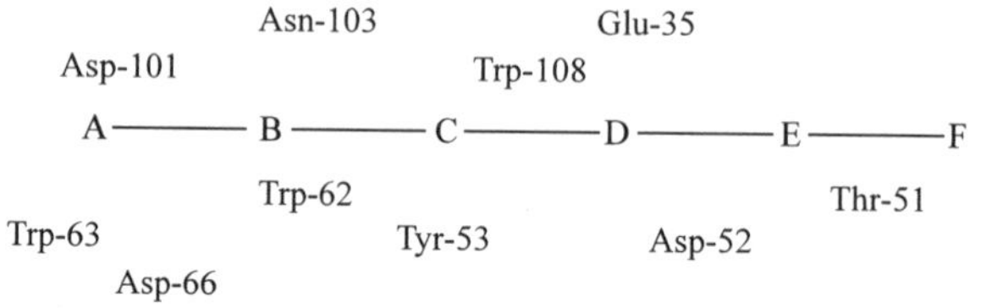

Scheme 10.24 Illustration of amino acid residues in the cleft region of lysozyme.

The gluconolactone inhibitor (**8**) binds much more strongly to lysozyme than regular tetrasaccharides. This was proposed to be due to the structural similarity of the C1 position of the gluconolactone to that of the putative oxocarbenium ion intermediate (Secemski *et al.*, 1972). The gluconolactone tetrasaccharide is a transition-state analogue for the interaction of the tetrasaccharide substrate with lysozyme.

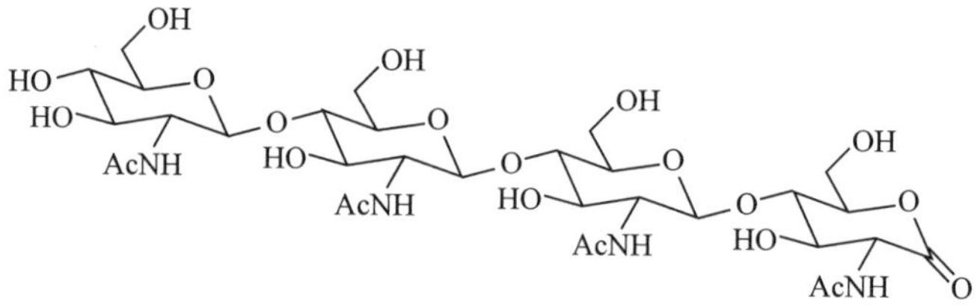

8

Glycosyl enzyme intermediates

The identity of the intermediate in the lysozyme mechanism has been postulated to be an acylal species whereby the C1 is attached covalently to the oxygen of the aspartate-52 group. All the data point to an open transition state for formation of the intermediate. Studies of the effect of the leaving group structure on the reactivity are consistent with substantial bond fission. The ^{18}O isotope effect on the leaving oxygen (1.047) compares with the value (1.025) for model reactions where the oxygen receives a proton. The ^{2}H-isotope effect at C(1) of between 1.11 and 1.19 from various substrates indicates a reduced bond order at C1 consistent with $sp^3 \rightarrow sp^2$ rehybridization.

The nature of an enzyme intermediate could be difficult to identify if its formation (k_2) were slower than its decomposition (k_3); such is the case with the glycosyl lysozyme, where all the kinetic data refer to its formation. In lysozyme it is therefore difficult to distinguish between a transition state on the way to either an sp^2 or an sp^3 intermediate. For β-galactosidase $k_2 > k_3$ and the observed $k_3^{H/D}$ values of 1.1 and 1.21 for the enzyme and a number of isozymes are consistent with an $sp^3 \rightarrow sp^2$ rehybridization as expected for a covalent intermediate. An ionic intermediate requires an inverse $k_3^{H/D}$ isotope effect.

Early structural studies considered that the putative oxocarbenium ion would interact covalently with aspartate-52 to form an acylal only with difficulty, due to changes elsewhere in the oligosaccharide. In oligosaccharide–enzyme complexes the distance between C1 of the putative fission site and the nearest oxygen of the carboxylate ion is of the order of 3 Å. It seems inconceivable that the known, very high, reactivity of such a species does not force it to collapse by conjugation with any weak adjacent nucleophile. It is well known that oxocarbenium ions are *very* reactive in water. In the *lysozyme* case, where there is retention of configuration at C1, one face of the C1 in the glycosyl cation must be protected so that the lifetime of the cation would have to exceed that of a diffusion process, enabling the leaving group to depart and be replaced by an acceptor nucleophile. The demonstration of acylal intermediates in other glycosidases involving retention is difficult to ignore; this, together with the absence of data to the contrary, forces us to accept the simplest explanation, namely the formation of an acylal intermediate in the lysozyme mechanism despite the bond distance problems (which might be *apparent* only—see Section 10.1) (Wang & Withers, 1995).

Primary alcohols increase the k_{cat} value for the solvolysis of aryl β-D-glucosides catalysed by β-glucosidase, consistent with breakdown of the glucosyl enzyme being slower than the acylalation step (Gopalan *et al.*, 1992). The rate of the β-glucosidase-catalysed hydrolysis of substituted aryl β-glucosides generates a non-linear Brønsted plot which is also compatible with a two-step mechanism involving a glucosyl enzyme intermediate. The Brønsted β_{lg} value of -0.7 indicates a large degree of bond fission and secondary kinetic isotope effects suggest an A_ND_N-like transition state for glycosylation (Kempton & Withers, 1992).

The site of acylalation in β-galactosidase has been identified as the carboxyl group of glutamic acid-537 by use of an altered substrate where the deacylalation step is slowed down sufficiently to enable even isolation and NMR studies of the intermediate. Gebler *et al.* (1992) showed that glycoside derivatives with fluorine in the axial 2-position (9, 10, 11) inhibits β-galactosidase; the ^{19}F-NMR signals of the intermediate from the glycoside with fluorine in the equatorial 2-position indicated that the stereochemistry of the attachment to the carboxyl group of the glutamic acid-358 residue of the β-glucosidase from *Agrobacterium* sp. has the α-configuration at C1 (Withers & Street, 1988; Street *et al.*, 1992). The glycosyl enzyme has a half-life of 11.5 h at neutral pH and tritiated inhibitor enables the site of acylalation to be identified as the glutamine-537 residue. The use of the fluoroglycoside provides further evidence of an intermediate because 'titration' kinetics are

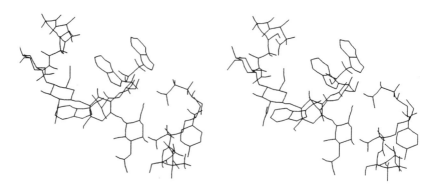

Fig. 10.4 Structure of the complex between tetra (*N*-acetyl)glucosamine and lyso-zyme, showing the main residues in the active-site region. The diagram is drawn from atomic coordinates deposited at the Brookhaven Protein Database (Bernstein *et al.*, 1977). For assistance with viewing see the note below the Contents list (p. vii).

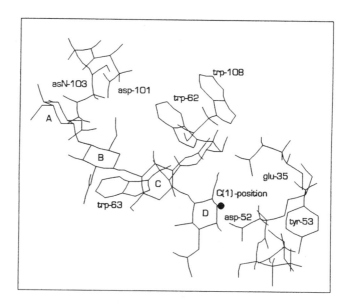

Key for stero-diagram

observed for the liberation of the C1 fluoride, and moreover there is a slow reactivation process. The source of the decreased reactivity is undoubtedly the electron-withdrawing power of the fluoride adjacent to C1 which destabilizes the formation of positive charge. Thus the low reactivity is consistent with an exploded $A_N D_N$ transition state.

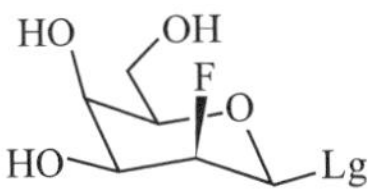

9 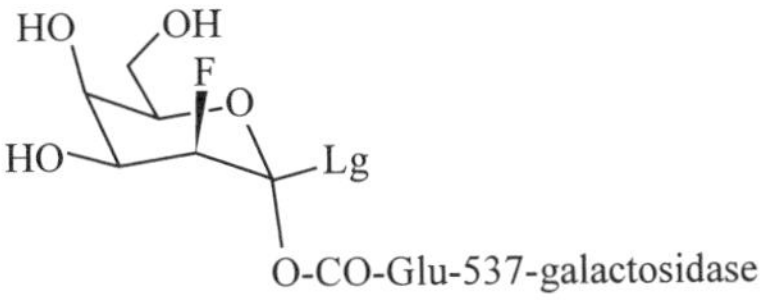   10

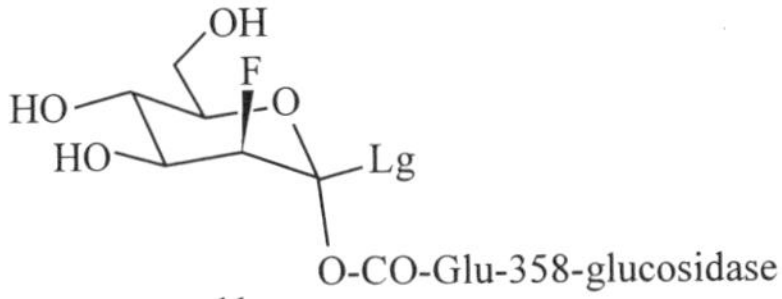

O-CO-Glu-537-galactosidase

O-CO-Glu-358-glucosidase

11

The nature of the transition state may be deduced from the isotope effects described above to involve considerable bond fission. This is borne out by the relatively large negative β_{lg} effects for leaving-group variation. Both aryloxy and substituted pyridinium ion leaving groups have been studied, and the effect of substituents on the oxygen leaving group could involve an acid component; the fact that the pyridinium species is a good substrate for galactosidase indicates that the acid component of the catalysis cannot be of vital importance. Figure 10.4 illustrates the structure of the binding site of a tetra(*N*-acetyl)glucosamine oligosaccharide bound to a lysozyme and clearly demonstrates the proximity of Glu-35 and Asp-52 to the C1 position of the D-residue.

Further reading

Auld, D. S. (1987) Acyl group transfer—metalloproteinases, in Page & Williams, (1987), p. 240.

Bender, M. L., Kezdy, F. J. & Gunter, C. R. (1964) The anatomy of an enzymatic catalysis. α-Chymotrypsin, *J. Am. Chem. Soc.*, **86**, 3714.

Blow, D. M. (1975) Structure and mechanism of chymotrypsin, *Acc. Chem. Res.*, **9**, 145.

Brocklehurst, K. (1987) Acyl-group transfer—cysteine proteases, in Page & Williams, (1987), p. 140.

Christen, P. & Metzler, D. E. (eds) (1985) *Transaminases*, Vol. 2, John Wiley, New York.

Dugas, H. (1989) *Bio-organic Chemistry*, 2nd edn, Springer, New York, p. 563–577.

Dunathan, H. C. (1971) Stereochemical aspects of pyridoxal phosphate catalysis, *Adv. Enzymol.*, **35**, 79.

Emery, V. C. & Akhtar, M. (1987) Pyridoxal phosphate dependent enzymes, in Page & Williams (1987), p. 345.

Fersht, A. R. (1984) *Enzyme Structure and Mechanism*, 2nd edn, W. H. Freeman, New York.

Fink, A. L. (1987) Acyl group transfer—the serine proteases, in Page & Williams (1987), p. 159.

Fischer, G. (1987) Acyl group transfer—aspartic proteases, in Page & Williams (1987), p. 229.

Kirsch, J. F. (1972) Linear free energy relationships in enzymology, in *Advances in Linear Free Energy Relationships*, Chapman, N. B. & Shorter, J. (eds), Plenum, New York, p. 369.

Knowles, J. R. (1991) Enzyme catalysis: not different, just better, *Nature (London)*, **350**, 121.

Lowe, G. (1976) The cysteine proteinases, *Tetrahedron*, **32**, 291.

Page, M. I. (1984) *The Chemistry of Enzyme Action*, Elsevier, Amsterdam.

Page, M. I. (1990) Enzyme catalysis, in *Comprehensive Medicinal Chemistry*, Vol. 2, Sammes, P. G. (ed.), Pergamon, Oxford, p. 45.

Page, M. I. & Williams, A. (eds) (1987) *Enzyme Mechanisms*, Royal Society of Chemistry, London.

Richard, J. P. (1995) A consideration of the barrier for carbocation–nucleophile combination reactions, *Tetrahedron*, **51**, 1535.

Rieder, S. V. & Rose, I. A. (1959) The mechanism of the triose phosphate isomerase reaction, *J. Biol. Chem.*, **234**, 1007.

Rose, I. A. (1958) The absolute configuration of dihydroxyacetone phosphate tritriated by aldolase reaction, *J. Am. Chem. Soc.*, **80**, 5835.

Sinnott, M. L. (1987) Glycosyl group transfer, in Page & Williams (1987), p. 259.

Sinnott, M. L. (1990) Catalytic mechanisms of enzymatic glycosyl transfer, *Chem. Rev.*, **91**, 1171.

Wells, G. B., Mustafi, D. & Makinen, M. W. (1994) Structure at the active-site of an acyl enzyme of α-chymotrypsin and implications for the catalytic mechanism, *J.Biol.Chem.*, **269**, 4577.

Structural information

Atomic coordinate data of enzymes and enzyme–substrate/inhibitor complexes may be obtained from the Brookhaven Data Bank (Bernstein *et al.*, 1977). Information on this can be found on the home page of the PDB (http://www.pdb.bnl.gov/). In the United Kingdom the database can be accessed directly via the Daresbury Laboratory Chemical Database Service (Daresbury Laboratory, Daresbury, Warrington WA4 4AD, UK (uig@ daresbury.ac.uk) home page (http://www.dl.ac.uk/)). (Fletcher *et al.*, 1996). The data may be visualized on personal computers by a variety of software packages such as Hyperchem, RasMol or Mage programs, which are readily available (http://www.prosci.uci.edu).

Appendices

A.1 IUPAC system for symbolic representation of reaction mechanisms

Since the 1950s organic chemists have employed a scheme of symbolic representation proposed by Ingold (1953) for reaction mechanisms. The benefits of a unified system have become apparent in recent years and a new system (Guthrie, 1975; Guthrie & Jencks, 1989; Littler, 1989), under scrutiny by the IUPAC, incorporates the direction of electron movement during bond fission or formation in the symbolism. Although the new system has vigorous opponents it is undoubtedly being used, and we therefore include a brief survey. As an example, the Ingold scheme distinguishes between nucleophilic aromatic substitution and ester hydrolysis; although the mechanisms are similar, each being bimolecular and involving nucleophilic substitution, they are given different symbols, namely S_NAr2 and $B_{Ac}2$. Nucleophilic displacement mechanisms at phosphorous and sulphur are given the symbols $S_N2(P)$ and $S_N2(S)$ respectively, to distinguish them from nucleophilic aliphatic substitution. There is indeed a mechanistic similarity between all the reactions enumerated above which can be exploited in their further consideration and demonstration. A unified symbolic system of representation should greatly assist this effort.

The basis of the symbolic system involves the recognition of *core* atoms joined by bonds wherein the main electronic reorganization occurs during reaction. A bond fission step is denoted by D and bond formation by A. These processes are called *primitive* changes and give an overall framework for the symbolic representation; for example, a bimolecular nucleophilic aliphatic substitution process (S_N2) is an 'AD' mechanism. Ester hydrolysis ($B_{Ac}2$) becomes $A + D$ and an $E1_{cb}$ process for elimination mechanisms becomes $AD + D$; the '+' term in the last code indicates that the dissociative step is separate from the addition step. It is necessary to indicate the direction of flow of the electrons, and if the bond to the core atom breaks to yield a nucleofuge or an electrofuge the symbol of the *primitive* step is given the subscript N or E respectively. Thus the S_N2 mechanism is fully symbolized as A_ND_N and the $B_{Ac}2$ as $A_N + D_N$, the S_N1 mechanism becomes $D_N + A_N$ and the $E1_{cb}$ mechanism is $A_ND_E + D_N$; the subscript n is given to 'A' because the bond is being formed to a *non-core* atom (the atom in the base accepting the proton), D_E because bond fission is to an atom (H) leaving as an electrophile and D_N because the subsequent step is fission of a bond between a *core* atom and a nucleofuge. The nucleophilic aromatic substitution mechanism ($S_N Ar2$) becomes $A_N + D_N$. These symbols are perfectly general as they distinguish only the direction of bond fission and not the type of atom involved: they refer to *all core* atoms, for example P, C, S, Si, B, etc.

A.2   Supplementary tables

Included here are tables which are useful adjuncts to enable calculations and manipulations to be carried out using equations quoted in Chapters 3, 4 and 5.

Table 3.a1 Some Hammett substituent constants.[a]

Substituent	σ_m	σ_p	Substituent	σ_m	σ_p
H	0.00	0.00	Me	−0.07	−0.17
Et	−0.07	−0.15	Pr^n	−0.05	−0.13
Pr^i	−0.07	−0.15	Bu^n	−0.07	−0.06
Bu^i	−0.08	−0.12	CH(Me)Et	−0.08	−0.12
Bu^t	−0.10	−0.20	CH_2Bu^i		−0.23
$C(Me_2)Et$	−0.06	−0.18	$3,4-(CH_2)_3$ (fused)		−0.26
$3,4-(CH_2)_4$ (fused)		−0.48	Ph	0.06	0.01
CH=CH−Ph	0.14		CH_2OH	0.08	0.08
CH_2Cl	0.11	0.12	CH_2CN	0.16	0.18
CF_3	0.47	0.54	CH_2CH_2COOH	−0.003	−0.07
CHO	0.36	0.22	COOH	0.37	0.41
COO^-	−0.10	0.00	COPh	0.34	0.46
$COCF_3$	0.63	0.80	COOMe	0.32	0.39
COOEt	0.37	0.45	$CONH_2$	0.28	0.36
$COCH_3$	0.38	0.50	CN	0.61	0.66
NH_2	−0.04	−0.66	NHMe	−0.30	−0.84
NHEt	−0.24	−0.61	$NHBu^n$	−0.34	−0.51
NMe_2	−0.05	−0.83	NH_3^+	1.13	1.70
NH_2Me^+	0.96		NMe_3^+	0.86	0.82
NH_2Et^+	0.96		$NHCOCH_3$	0.21	0.00
NHCOPh	0.22	0.08	$NHNH_2$	−0.02	−0.55
NHOH	−0.04	−0.34	N=NPh	0.32	0.39
NO_2	0.71	0.78	NO	0.62	0.91
OH	0.10	−0.37	O^-	−0.71	−0.52
OMe	0.08	−0.27	OEt	0.07	−0.24
OPr^n	0.10	−0.25	OPr^i	0.10	−0.45
OBu^n	0.10	−0.32	OC_5H_{11}	0.10	−0.34
OCH_2Ph		−0.42	OPh	0.25	−0.32
$3,4-(OCH_2O)$ (fused)		−0.16	OCF_3	0.36	0.32
$OCOCH_3$	0.39	0.31	SH	0.25	0.15
SMe	0.15	0.00	SEt	0.18	0.03
SiPr	0.23	0.07	SMe_2^+	1.00	0.90
SOMe	0.52	0.49	SO_2Me	0.68	0.72
SO_2NH_2	0.46	0.57	SO_3^-	0.05	0.09
SCN	0.51	0.51	SCOMe	0.39	0.44
SCF_3	0.35	0.38	F	0.34	0.06
Cl	0.37	0.23	Br	0.39	0.27
I	0.35	0.30	IO_2	0.70	0.76
AsO_3H^-	0.00	0.00	$B(OH)_2$	0.01	0.45
PO_3H^-	0.20	0.26			

[a] The data are taken from Perrin *et al.* (1981) and Hansch *et al.* (1991).

Table 3.a2 Transmission factors for Hammett substituent

Interposed group	Attenuation factor (f)
$-CH_2-$	0.47
$-(CH_2)_2-$	0.22
$-CH=CH-$	0.54
$-C\equiv C-$	0.39
$-$Phenylene$-$	0.24
	0.39
	0.28

Table 3.a3 Hammett's σ^--values for *para* substituents.[a]

Substituent	σ^-	Substituent	σ^-
CH=CHPh	0.13	SCF$_3$	0.64
CONH$_2$	0.63	SMe$_2$	0.83
COOMe	0.75	SOMe	0.73
COOEt	0.75	N$_3$	0.11
SO$_2$Me	1.13	COOH	0.77
SO$_2$CF$_3$	1.63	COMe	0.84
SO$_2$NH$_2$	0.94	CHO	1.03
CN	1.00	NO$_2$	1.24
NO	1.63	N$_2^+$	3.43

[a] Data from Perrin *et al.* (1981) and Hansch *et al.* (1991).

Table 3.a4 Hammett's σ^+-values for *para* substituents.[a]

Substituent	σ^+	Substituent	σ^+
NMe$_2$	-0.17	F	-0.07
NHPh	-1.40	Cl	0.11
NH$_2$	-1.30	Br	0.15
NHCOMe	-0.60	I	0.14
NHCOPh	-0.60	NMe$_3^+$	0.41
Ph	-0.18	OH	-0.92
2-Naphthyl	-0.14	OMe	-0.78
CH$_2$COOEt	-0.16	OPh	-0.50
SMe	-0.60		

[a] Data taken largely from Perrin *et al.* (1981) and Hansch *et al.* (1991).

Table 3.a5 Some applications of the Yukawa–Tsuno equation.

Reaction	ρ	r
Hydrolysis of ArCMe$_2$Cl	-4.54	1.00
Brominolysis of ArB(OH)$_2$	-3.84	2.29
Methanolysis of Ar$_2$CHCl	-4.02	1.23
Bromination of ArH	-5.28	1.15
Nitration of ArH	-6.38	0.90
Ethanolysis of Ar$_3$CCl	-2.52	0.88
Basicity of ArN=NPh	-2.29	0.85
Decomposition of ArCOCHN$_2$	-0.82	0.56
Beckmann rearrangement of ArC(Me)=NOH	-1.98	0.43
Semicarbazone formation from ArCHO	1.35	0.40
Rearrangement of ArCH(OH)CH=CHMe to ArCH=CHCH(OH)Me	-4.06	0.40
Decomposition of Ar$_2$CN$_2$ by benzoic acids	-1.57	0.19
Water + ArOSiEt$_3$	2.84	0.95
Hydroxide ion + ArOSiEt$_3$	3.52	0.46
Hydroxide ion + PhCOOAr	1.20	0.10

Table 3.a6 Some values of Charton's σ_I parameters.[a]

Substituent	σ_I	Substituent	σ_I
		CH_2 — Heteroatom	
CH_2Cl	0.17	CH_2I	0.17
CH_2Br	0.20	CH_2F	0.15
CH_2SCH_2Ph	0.06	CH_2SH	0.12
$CH_2NH_3^+$	0.36	$CH_2NHCOMe$	0.07
CH_2OEt	0.11	CH_2SCH_3	0.12
CH_2OCOMe	0.15	CH_2OMe	0.11
CH_2OH	0.11	CH_2NH_2	0.08
CH_2OPh	0.12	$CH_2Si(CH_3)_3$	−0.05
		CH_2CR_3	
CH_2CCl_3	0.12	CH_2Ph	0.03
CH_2CCH	0.14	$CH_2CHMe(OH)$	−0.01
CH_2Bu^t	−0.02	CH_2 — (2-thienyl)	0.05
$CH_2Me_2C(OH)$	−0.04	$CH_2CH=CH_2$	0.02
$CH_2CH=CHCH_3$	0.00	CH_2 — (1-naphthyl)	0.07
CH_2CN	0.20	$CH_2C_3F_7$	0.14
CH_2COOCH_3	0.19	CH_2COOEt	0.15
CH_2cyclohexyl	−0.03	$CH_2CONHPh$	0.00
CH_2COO^-	0.01	CH_2COMe	0.19
CH_2n-hexyl	−0.04	Et	−0.01
CH_2Pr^i	−0.03	Me	−0.01
CH_2CONH_2	0.08	$CH_2CONHMe$	0.28
		CH_2CH_2R	
CH_2CH_2OMe	0.00	CH_2CH_2OH	0.06
$CH_2CH_2SiMe_3$	−0.04	$CH_2CH_2CONH_2$	0.03
$(CH_2)_3CONH_2$	0.02	CH_2Pr^n	−0.04
$CH_2CH_2Bu^n$	0.04	$(CH_2)_3Bu^n$	−0.06
CH_2Et	−0.02	$(CH_2)_3SiMe_3$	−0.04
$(CH_2)_4$-cyclohexyl	−0.06	$CH_2CH_2NH_2$	0.04
$CH_2CH_2Pr^n$	−0.04	CH_2CH_2Ph	0.01
$CH_2CH_2CHCl_2$	0.03	$CH_2CH_2CCl_3$	0.04
$(CH_2)_4Ph$	−0.02		
		CHR_2	
$CH(OH)Me$	0.02	$CH(OH)Ph$	0.08
$CH(CH_3)Ph$	0.06	Bu^s	−0.01
Pr^i	0.01	CHF_2	0.32
$CHCl_2$	0.31	Cyclohexyl	−0.02
		CR_3	
$C{\equiv}CH$	0.29	CF_3	0.40
Bu^t	−0.01	2-Naphthyl	0.13
CN	0.57	3-Indolyl	−0.01
1-Naphthyl	0.14	2-Thienyl	0.19
2-Furyl	0.17	Phenyl	0.12
CCl_3	0.38		

Table 3.a6 (*contd*)

Substituent	σ_I	Substituent	σ_I
		$CH{=}CR_2$	
$CH{=}CCl_2$	0.18	$CH{=}CH_2$	0.11
$CH{=}CHEt$	0.07	$CH{=}CMe_2$	0.05
$CH{=}CHMe$	0.07		
		$CO{-}R$	
COO^-	−0.17	$CONH_2$	0.28
COOH	0.30	COMe	0.30
COOEt	0.30	COOMe	0.32
CONHPh	0.26		
		Nitrogen	
NHCOMe	0.31	$NHCONH_2$	0.21
NHCHO	0.32	$NHC(CH_3){=}NPh$	0.38
NO_2	0.76	N_2^+	0.42
NHCOPh	0.28	$NHMe_2{}^+$	0.70
NH_3^+	0.60	$NH_2(Pr^n)^+$	0.60
$NH_2(Bu^i)^+$	0.60	NCS	0.58
$N(O){=}CMe_2$	0.29	$NHSO_2Ph$	0.32
NH_2Me^+	0.60	NH_2Et^+	0.60
NH_2Bu^+	0.60	NHCOMe	0.28
N(COMe)(2-naphthyl)	0.27	N(COMe)(1-naphthyl)	0.26
NH_2	0.17	$NHNH_2$	0.22
NHPh	0.30	N(Me)Ph	0.15
NHMe	0.13	NMe_2	0.17
1-Pyrazolyl	0.30	1-Imidazolyl	0.51
		$N{=}CR_2$	
$N{=}C(CF_3)_2$	0.32	$N{=}CCl_2$	0.26
$N{=}CH_2$	0.20	$N{=}CHPh$	0.13
$N{=}C(Me)NHPh$	<0.15	$N{=}C(NH_2)_2$	<0.10
		Oxygen	
OPh	0.40	OMe	0.30
OEt	0.28	OH	0.24
OPr^i	0.27	OPr^n	0.28
OBu^n	0.28	$OCH_2cyclohexyl$	0.22
ONH_3^+	0.47	Ocyclohexyl	0.31
ONO_2	0.66	OCOMe	0.38
		Sulphur	
SH	0.27	SPh	0.31
SMe	0.30	SCH_2CH_2Ph	0.25
SO_3^-	0.15		
SEt	0.26	SPr^n	0.25
SBu^n	0.26	SCH_2Ph	0.26
$SCPh_3$	0.12	SO_2Pr^i	0.57
SOPh	0.51	SO_2Ph	0.56
SO_2Me	0.59	SO_2Pr^n	0.57
$SCONH_2$	0.33	SO_2Et	0.59

(*continued on p. 254*)

Table 3.a6 (*contd*)

Substituent	σ_I	Substituent	σ_I
	Silicon		
SiMe$_3$	−0.11	SiMe$_2$Ph	−0.12
Si(Me$_2$)OSiMe$_3$	−0.11	SiPr^n	0.25
	Heteroatoms		
AsO$_3$H$^-$	0.13	F	0.54
Cl	0.47	Br	0.40
I	0.40	H	0.00

[a] Parameters taken from Charton (1964, 1981).

Table 3.a7 Some Taft σ^* parameters.[a]

Substituent	σ^*	Substituent	σ^*
	Heteroatom		
H	0.49		
	CH$_2$Heteroatom		
CH$_2$NMe$_3^+$	1.90	CH$_2$SO$_2$Me	1.32
CH$_2$CN	1.30	CH$_2$I	0.85
CH$_2$OH	0.56	CH$_2$OMe	0.52
CH$_2$F	1.10	CH$_2$OPh	0.85
CH$_2$Cl	1.05	CH$_2$Br	1.00
CH$_2$SiMe$_3$	−0.26	CH$_3$	0.00
	CH$_2$CR$_2$		
CH$_2$ cyclohexyl	−0.06	CH$_2$COOH	1.05
CH$_2$Bu^t	−0.165	CH$_2$CF$_3$	0.92
CH$_2$Ph	0.22	CH$_2$CH=CHMe	0.13
Bu^i	−0.125	CH$_2$COMe	0.60
	CH$_2$CH$_2$R		
CH$_2$CH$_2$Ph	0.08	(CH$_2$)$_3$Ph	0.02
(CH$_2$)$_3$CF$_3$	0.12	CH$_2$CH$_2$Cl	0.39
Et	−1.00	Pr^n	−0.115
Bu^n	−0.13	CH$_2$CH$_2$CF$_3$	0.32
CH$_2$CH$_2$NO$_2$	0.50		
	CHR$_2$		
Pr^i	−0.19	Cyclohexyl	−0.15
Bu^s	−0.21	CH(OH)Ph	0.77
CHPh$_2$	0.41	CHF$_2$	2.05
CHCl$_2$	1.94	CH(Et)Ph	0.04
CH(CH$_3$)Ph	0.11	CHEt$_2$	−0.225
CH(Bu^t)Me	−0.28		
	CH=CR$_3$		
trans-CH=CHNO$_2$	1.7	CH=CH$_2$	0.65
CH=CHPh	0.41	*trans*-CH=CHCCl$_3$	1.19
trans-CH=CHCOOH	1.01	*trans*-CH=CHCl	0.90
CH=CCl$_2$	0.88	CH=CHMe	0.36
	COR		
COMe	1.65	COOMe	2.00
	CR$_3$		
Ph	0.60	Bu^t	−0.30
C≡CPh	1.35	CCl$_3$	2.65

[a] Data taken from Taft (1956); Hine (1962); Shorter (1972); Isaacs (1995).

Table 3.a8 Some Taft–Pavelich steric parameters E_s.[a]

Substituent	E_s	Substituent	E_s
CH_2Me	−0.19	CH_2Cl	−0.24
CH_2F	−0.24	CH_2Br	−0.27
CH_2SMe	−0.34	CH_2I	−0.37
$CH_2CH_2CMe_3$	−0.34	CH_2OPh	−0.33
CH_2Ph	−0.38	CH_2CH_2Ph	−0.38
$(CH_2)_3Ph$	−0.45	CH_2CH_2OMe	−0.77
CH_2CH_2Cl	−0.90	CH_2Bu^t	−1.74
$CH(Me)Et$	−1.13	$CH(Me)Ph$	−1.19
$CH(Et)Ph$	−1.50	$CHCl_2$	−1.54
$CHPh_2$	−1.76	$CH(Me)(neopentyl)$	−1.85
$CHBr_2$	−1.86	$CH(Bu^i)_2$	−2.47
$CH(neopentyl)_2$	−3.18	$CH(Me)Bu^t$	−3.33
CF_3	−1.16	$CMe(Bu^t)(neopentyl)$	−4.00
CCl_3	−2.06	CBr_3	−2.43
$CMe_2(neopentyl)$	−2.57	CMe_2Bu^t	−3.90
CEt_3	−3.80	Cyclobutyl	−0.06
Cyclopentyl	−0.51	Cyclohexyl	−0.79
Cyclohexylmethyl	−0.98	Cycloheptyl	−1.10
Et	−0.07	H	1.24
Bu^i	−0.93	Isopentyl	−0.35
Pr^i	−0.47	Me	(0.00)
Bu^n	−0.39	n-Octyl	−0.33
n-Pentyl	−0.40		

[a] Parameters taken from Shorter (1972); Taft (1956); Isaacs (1996). The reader should be aware that E_s is occasionally referred to as $E_s(H) = 0$; in that case subtract 1.24 from the figures quoted above.

Table 4.a1 Fractionation factors for hydrogen in some species, in water solvent.[a]

Species	ϕ	Species	ϕ
$R_3CC\underline{H}CR_2$	0.87	$RC\underline{H}(NH_3R^+)R$	1.17
$R_2CC\underline{H}_2$	0.81	$RC\underline{H}(OPO_3^{2-})R$	1.19
$RC\underline{H}O$	0.83	$RC\underline{H}(OR)OH$	1.24
$RCOC\underline{H}_3$	0.84	$R_2CC\underline{H}_2$	0.81
$R_3CC\underline{H}_3$	0.88	$R_2CC\underline{H}R$	0.87
$RCOC\underline{H}_2CR_3$	0.93	$RC\underline{H}O$	0.83
$RCOC\underline{H}_2OH$	0.99	$RCOC\underline{H}_3$	0.84
$R_3CC\underline{H}(R)SR$	1.01	$R_3CC\underline{H}_3$	0.88
$RC\underline{H}_2OH$	1.04	$RCOC\underline{H}_2R$	0.93
$RCOC\underline{H}(OH)R$	1.10	$RC\underline{H}_2R$	0.98
$RC\underline{H}(NH_3^+)R$	1.12	$RCOC\underline{H}_2OH$	0.99
$RC\underline{H}(OH)_2$	1.14	$RC\underline{H}R(SR)$	1.01
$RC\underline{H}(OH)R$	1.16	$\underline{H}O^-$	0.43
$RCOO\underline{H}$	0.92	$\underline{H}_3O^+$	0.69
$RO\underline{H}$	1.04	$RN\underline{H}_2$	1.13
$CH_3CON\underline{H}CH_3$	1.12	$RN\underline{H}_3^+$	1.08

[a] Parameters taken from Quinn & Sutton (1991).

Table 5.a1 Grunwald–Winstein Y-values from some solvents and solvent mixtures.[a]

Solvent	Y	Solvent	Y
EtOH/H_2O		Pr^{i}OH	−2.73
100% EtOH	−2.03	Bu^{i}OH	−3.26
95% EtOH	−1.29	HCOOH	2.05
90% EtOH	−0.75	50% HCOOH/H_2O	2.64
80% EtOH	0.00	MeCOOH/H_2O	
70% EtOH	0.60	100% MeCOOH	−1.64
50% EtOH	1.66	50% MeCOOH	1.94
30% EtOH	2.72	50% MeCOOH/HCOOH	0.76
0% EtOH (H_2O)	3.49	Dioxan/H_2O	
MeOH/H_2O		90% Dioxan	−2.03
100% MeOH	−1.09	50% Dioxan	1.36
90% MeOH	−0.30	Acetone/H_2O	
70% MeOH	0.96	90% Acetone	−1.86
50% MeOH	1.97	50% Acetone	1.40
30% MeOH	2.75	HCONH$_2$	0.60
		n-C$_3$F$_7$COOH	1.70

[a] Parameters from Leffler & Grunwald (1963); Hine (1966).

A.3   Estimation of ionization constants

Linear free-energy relationships are particularly useful if it is necessary to have accurate knowledge of the pK_a of an intermediate which is unstable or for which other experimental difficulties such as solubility mitigate against explicit determination (see, for example, Section 8.3). For example, an indirect experimental method gives 13.7 as the pK_a of the unstable hydrate of formaldehyde, which is a model of many putative intermediates in biosynthetic pathways. The pK_a of primary alcohols (RCH_2OH) is governed by the equation $pK_a = 15.9 - 1.42\sigma^*$. Allowing for a statistical factor of 0.3 (there are two identical hydroxyl functions in the hydrate), substitution of σ^* for R = H gives a pK_a of 13.3, in reasonably good agreement with the observed value.

Tables a.1 and a.2 collect representative linear free-energy equations for the ionization constants of acids (and of the conjugate acids of bases). Entering the appropriate substituent parameters will yield the pK_a of the acid to a reasonable degree of accuracy (Perrin et al., 1981).

Sometimes the simple linear free-energy relationship for the pK_a of the unknown acid is not available and recourse must be taken to less direct routes of calculation with assumptions being made about the Charton, Taft, Hammett or Brønsted parameters. Fox & Jencks (1974) and Taylor (1993) showed how the pK_a of an acid can be calculated using a standard acid of known pK_a starting from a structure which is closely related to that of the unknown. The method has been extensively applied to alcohols and ammonium ions of structure X–C–Y where Y is the acid and X is the substituent separated by a carbon unit; it assumes that the effect of change of X on the property of Y has a ρ_1-value of -9.1 and when the structure is X–C–C–Y the ρ_1 is -4.4 (see Taylor, 1993). When the substituent is directly attached to Y (X–Y) the ρ_1-value is -19.1. The method is exemplified by the calculation of the pK_a of the ammonio group of $HOCH_2NH_2^+(OCH_3)$ which is representative of a common intermediate type in acyl group transfer (see Section 8.3). The standard known ionization is that of $HCH_2NH_2^+(OCH_3)$ (4.75) and the calculations are best tabulated as follows:

Calculation 1

$HCH_2NH_2^+(OCH_3)$ $\xrightarrow{\hspace{3cm}}$ $HOCH_2NH_2^+(OCH_3)$

4.75 $\quad\quad -9.1 \times (0.1 - 0.0) \quad\quad$ 3.75

Calculation 2

$HCH_2NH_2^+CH_3 \xrightarrow{\hspace{1.5cm}} HCH_2NH_2^+(OCH_3) \xrightarrow{\hspace{1.5cm}} HOCH_2NH_2^+(OCH_3)$

10.7 $\quad -19.1 \times (0.30 - 0.01) \quad$ 4.78 (calc.) $\quad -9.1 \times (0.11 - 0.0) \quad$ 3.78

Calculation 3 shows that it is important to use a standard as close as possible in structure to that of the unknown acid. It is also important, for confidence in the final result, that several independent routes are employed with different starting points and that any deviant results can be explained. Calculation 3 includes statistical terms due to the different numbers of

Table a.1 Some Hammett relationships for ionization constants.[a]

Acid	ρ^g	$pK_{intercept}$ [h.]
ArCOOH (water)[b]	1.00	4.20
ArCOOH (40% EtOH/water)	1.67	4.87
ArCOOH (70% EtOH/water)	1.74	6.17
ArCOOH (EtOH)	1.96	7.21
ArCOOH (CH$_3$OH)	1.54	6.51
ArCOOH (50% butylcellosolve/H$_2$O)	1.42	5.63
2-Hydroxybenzoic acids	1.10	4.00
ArCH$_2$COOH	0.47	4.30
ArCH$_2$CH$_2$COOH	0.22	4.55
trans-ArCH=CHCOOH	0.47	4.45
4-ArC$_6$H$_4$COOH (50% butylcellosolve/H$_2$O)	0.48	5.64
ArC≡CCOOH (35% dioxan/H$_2$O)	0.80	3.26
ArOCH$_2$COOH	0.30	3.17
ArSCH$_2$COOH	0.32	3.38
ArSeCH$_2$COOH	0.35	3.75
ArSOCH$_2$COOH	0.17	2.73
ArSO$_2$CH$_2$COOH	0.25	2.51
ArB(OH)$_2$ (25% EtOH/H$_2$O)	2.15	9.70
ArPO(OH)$_2$	0.76	1.84
ArPO$_2$(OH)$^-$	0.95	6.96
ArAsO$_2$(OH)$^-$	0.87	8.49
ArAsO(OH)$_2$	1.05	3.54
ArSeO$_2$H	0.90	4.74
anti-ArCH=NOH	0.86	10.70
ArOH	2.23	9.92
ArCOCH$_2$PPh$_3^+$ (80% EtOH/H$_2$O)	2.40	6.00
ArCOCH$_2$CONHPh (water/dioxan)	0.79	9.40
ArCOCH$_2$COCH$_3$	1.72	8.53
ArSH (48.9% EtOH/H$_2$O)	2.24	7.67
ArNH$_3^+$	2.89	4.58
ArCH$_2$NH$_3^+$	1.06	9.39
ArNHNH$_3^+$	1.17	5.19
ArNHPh	4.07	22.40
trans-ArC(OH)=CHOMe	1.10	8.24
ArCOCH$_2$(Me)Ph$^+$	2.00	7.32
ArS(Me)CH$_2$COPh$^+$	1.40	7.32
ArCH$_2$NO$_2$	0.83	6.88
ArCH$_2$NO$_2$ (50% MeOH)	1.22	7.93
ArCH(Me)NO$_2$	1.03	7.39
ArCH(Me)NO$_2$ (50% dioxan/H$_2$O)	1.62	10.30
ArCH(NO$_2$)$_2$	1.47	3.89
ArCH$_2$CH(Me)NO$_2$ (50% MeOH/H$_2$O)	0.40	9.13
Substituted fluorenes (water/DMSO)[c]	6.30	22.10
2-Nitrophenols	2.16	6.89
ArSO$_2$NH$_2$	0.88	10.02

(*continued on p. 266*)

Table a.1 (*contd*)

Acid	ρ^a	$pK_{intercept}{}^h$
ArSO$_2$NHPh	1.16	8.31
PhSO$_2$NHAr	1.74	8.31
ArC$^+$(OH)Me	2.60	−6.00
ArC(OH)$_2^+$	1.20	−7.26
ArNHMe$_2^+$	3.46	5.13
Conjugate acids of the bases		
Pyridines	5.90	5.25
Pyrimidines	5.90	1.23
Quinolines^d	5.90	4.90
Isoquinolines^e	5.90	5.40
1-Naphthylamines^d	2.81	3.90
2-Naphthylamines^f	2.81	4.35

[a] Parameters taken from Perrin *et al.* (1981).
[b] Solvent in parentheses.
[c] DMSO, Dimethyl sulphoxide.
[d] 2-, 3- and 4-substituted.
[e] 1-, 3- and 4-substituted.
[f] 3- and 4-substituted.
[g] In the case of multi-substitution $\Sigma\sigma$ is employed.
[h] Calculated value at $\sigma = 0$.

protons in the various acids. It fails, however, because chemical intuition tells us that the value of 0.73 is much too low for a hydroxylamine derivative with only an alkyl substituent.

Calculation 3

$$HNH_2^+H \xrightarrow{\hspace{3cm}} HOCH_2NH_2^+H \xrightarrow{\hspace{3cm}} HOCH_2NH_2^+(OCH_3)$$

9.25 $-19.1 \times (0.11 + 0.01) + \log(4/3)$ 6.83 (calc.) $-19.1 \times (0.30 + 0.01) + \log(3/2)$ 0.73 (calc.)

As can be seen from the above example, cross-checking of the resultant calculated pK_a gives confidence in the result; however, the technique is time-consuming and requires considerable background skill in obtaining and choosing standards. If the pK_a values of a large number of structures are needed for a *guide*, then recourse to software is possible: a program (CAMEO) has been developed by Jorgensen *et al.* (1990) for pK_a calculation for acids in dimethyl sulphoxide solvent (see also Gushurst & Jorgensen, 1986) which starts with a structure-building routine assuming that the pK_a is an additive property of the effects of functional groups. A further program, p*Kalc* (version 3.1), that may be employed to calculate pK_a values has been described by Csizmadia *et al.* (1993); this software appears to be a computational version of the approaches described in the monograph of Perrin *et al.*

Table a.2 Some Taft and Charton relationships for ionization constants.[a]

Acid	ρ^* or ρ_1[b]	$pK_{intercept}$
Taft relationships		
RNH_3^+	-3.14	10.15
$RR_1NH_2^+$	-3.33	10.59
$RR_1R_2NH^+$	-3.30	9.61
RPH_3^+	-2.64	2.46
$R_2PH_2^+$	-2.61	3.59
R_3PH^+	-2.67	7.85
RCOOH	-1.62	4.66
RCH_2COOH	-0.67	4.76
RCH_2OH	-1.42	15.90
RSH	-3.50	10.22
RCH_2SH	-1.47	10.54
ROH_2^+	-2.36	-2.18
$RCH(OH)_2$	-1.42	14.40
$CH_3COCHRCOOEt$	-3.44	12.59
$RCH(NO_2)_2$	-2.23	5.35
$RNHC(NH_2)_2^+$	-3.60	14.00
$RNHNH_3^+$	-2.80	7.80
Charton relationships		
$XCH_2NH_3^+$	-10.02	10.39
$XCH_2NH_2CH_3^+$	-10.53	10.93
$XCH_2NH(CH_3)_2^+$	-11.01	10.03
$XCH_2CH_2NH_3^+$	-4.71	10.54
$X_1X_2X_3COH$	-8.23	15.88

[a] Parameters for ρ^* taken from Perrin *et al.* (1981) and for ρ_1 from Taylor (1993).
[b] In the case of multi-substitution $\Sigma\sigma^*$ or $\Sigma\sigma_1$ is employed.
[c] Calculated value at σ^* or $\sigma_1 = 0$.

(1981) which utilizes free-energy correlations such as the Hammett, Brønsted and Taft equations.

Linear free-energy relationships also enable the calculation of regular equilibrium constants. There have been no extensive compilations of these although reference should be made to the article by Jaffé (1953) and the books of Hine (1975) and Leffler & Grunwald (1963); techniques similar to those elaborated above enable equilibrium constants to be calculated with reasonable accuracy.

A.4 Estimation of partition coefficients

Partition coefficients are extremely important in medicinal chemistry as they may be related to drug transfer across membranes; they are also of importance in studies of catalysis because they can relate to binding at active centres. The parameter 'log P' has been employed to model various kinds of transport between aqueous and lipid phases. The value P is defined as the partition coefficient of a substance between water and n-octanol. The free energy of the transport process is often linearly related to standard 'hydrophobic' parameters (π) for substituents (x) determined from P_x values for substituted phenoxyacetic acids and that for the unsubstituted acid P_H (Eqn [a.1])

$$\pi = \log P_x - \log P_H \tag{a.1}$$

Linear free-energy equations for a variety of transport processes are collected in Table a.3; these include both polar and hydrophobic parameters. Table a.4 collects π-values for a number of substituents; values of $\log P$ may therefore be calculated for selected compounds, simply by substitution. Partition coefficients expressed on a logarithmic scale are an additive constitutive molecular property (like, for example, molecular volume) and Hansch & Leo (1979) describe the calculation of $\log P$ values by a 'fragmental hydrophobicity' system. The overall value of $\log P$ is given by Eqn [a.2].

$$\log P = \sum_i a_i f_i \tag{a.2}$$

where a_i indicates the population of a given fragment i in a structure and f_i is the absolute contribution of its substituents and substructures to the total lipophilicity. Corrections are necessary and are usually incorporated via

Table a.3 Linear free-energy relationships for some transport and complexation processes.

Reaction	Equation
Hydrolysis of 4-nitrophenyl esters by serum albumin	$\log k = 0.95 \log P + 3.5 E_s - 0.47$
Inhibition of chymotrypsin	
With aromatic acids	$\log K_i = 0.94 \log P - 0.96 pK + 3.66$
With PhCOR	$\log K_i = -0.31\rho - 2.53$
With ArOH	$\log K_i = -0.95 \log P + 1.88$
With hydrocarbons	$\log K_i = -1.47 \log P + 1.2$
Reaction of chymotrypsin	
With PhCONHCH$_2$COOR	$\log K_m = -0.41\rho - 0.4 E_s + 0.71$
With Ph(CH$_2$)$_2$COOR	$\log K_m = -0.21\rho - 3.16$
Binding of serum albumin	
With barbiturates	$-\log C = 0.58 \log P + 2.39$
With phenols	$-\log C = 0.68\rho + 3.48$
Partition coefficient between n-octanol and water	
Aryloxyacetic acids	$\log P = \pi^a + 1.21$
Phenols	$\log P = \pi^b + 1.46$

[a] Taking π-values from the phenoxyacetic acid in Table a.4.
[b] Taking π-values from the phenol in Table a.4.

Table a.4 Some Leo–Hansch π parameters for use in calculating $\log P$ values.

Substituent	$ArOCH_2CO_2H$	ArOH	Substituent	$ArOCH_2CO_2H$	ArOH
H	0.00	0.00	2-F	0.01	0.25
3-F	0.13	0.47	4-F	0.15	0.31
2-Cl	0.59	0.69	3-Cl	0.76	1.04
2-COMe	0.01		3-COMe	−0.28	−0.07
4-COMe	−0.37	−0.11	3-CN	−0.30	−0.24
4-CN	0.32	0.14	2-Br	0.75	0.89
3-Br	0.94	1.17	4-Br	1.02	1.13
2-I	0.92	1.19	3-I	1.15	1.47
4-I	1.26	1.45	2-Me	0.68	
3-Me	0.51	0.56	4-Me	0.52	0.48
2-Et	1.22		3-Et	0.97	0.94
3-Pr^n	1.43		3-Pr^i	1.30	
4-Pr^i	1.40		3-Bu^n	1.90	
4-Bu^s	1.82		3-Bu^t	1.68	
4-Cyclopentyl	2.14		4-Cyclohexyl	2.51	
3-Ph	1.89		3,4-$(CH_2)_3$	1.04	
3,4-$(CH_2)_4$	1.39		3,4-$(CH)_4$	1.24	
3-CF$_3$	1.07	1.49	3-CH$_2$OH		−1.02
4-COOH		0.12	3-OH	−0.49	−0.66
4-OH	−0.61	−0.87	2-OMe	−0.33	
3-OMe	0.12	0.12	4-OMe	−0.04	−0.12
3-OCF$_3$	1.21		3-OCH$_2$COOH		−0.70
4-OCH$_2$COOH		−0.81	3-NH$_2$		−1.29
4-NH$_2$		−1.63	3-NMe$_2$		0.10
2-NO$_2$	−0.23	0.33	3-NO$_2$	0.11	0.54
4-NO$_2$	0.24	0.50	3-NHCOMe	−0.79	
3-NHCOPh	0.72		4N≡NPh	1.71	
3-NHCONH$_2$	−1.01		3-SMe	0.62	
3-SCF$_3$	1.58		3-SO$_2$CF$_3$	0.93	
4-CH$_2$OH		−1.26	3-CH$_2$COOH		−0.61
3-COOH	−0.15	0.04			

additive values of f_i (Leo *et al.*, 1975; Hansch & Leo, 1979; Kubinyi, 1979). A simple example calculates the partition coefficient of $PhCH_2CH_2CH_2Cl$ from f-values:

$$
\begin{aligned}
\log P &= f_{Ph} + 3f_{CH_2} + 3f_b + f_{aliph.Cl} \\
&= 1.90 + 1.98 - 0.36 + 0.06 \\
&= 3.58 \ (\text{experimental} \log P = 3.55)
\end{aligned}
\tag{a.3}
$$

Table a.5 collects additivity values (f) which relate to fragments such as groups, functions and bonds making up the structure of a molecule.

The calculations can also be carried out with software possessing suitable standard values. This is useful for the large numbers of substances often requiring processing in the pharmaceutical industry. A standard software

Table a.5 Fragmental constants for calculating log P.[a]

Calculation step	f	Calculation step	f
CH_2	0.66	Single bond in chain	−0.12 (fb)
H	0.23	CH	0.65
Single carbon (C)	0.20	Branching in C chain	−0.13 (fcbr)
Single bond in ring	−0.09(fb)		
Phenyl group	1.90	Me	0.80
Aliphatic groups		Aromatic groups	
COOH	−1.09	COOH	−0.03
CN	−1.28	CN	−0.34
Cl	0.06	Cl	0.94
NH	−2.11	NH	−1.03
CO	−1.90	CO	−0.32
F	−0.38	F	0.37
NO_2	−1.26	NO_2	−0.02
NH_2	−1.54	NH_2	−1.00
$CONH_2$	−2.18	$CONH_2$	−1.26
I	0.60	I	1.35
OH	−1.64	OH	−0.40
Oxyanion	−1.81	Oxyanion	−0.57
Br	0.20	Br	1.09
-S-	−0.79	-S-	0.03

[a] From Leo *et al.* (1975).

package 'MEDCHEM' is available, of which the algorithm 'CLOGP' is employed to calculate the log P values. The confidence in a particular calculated value will derive from comparison of a number of estimates from experimental log P values of different standard compounds. The manual approach will give the most reliable calculations starting with a known log P value for a standard compound, and is preferred for the non-specialist mechanistic laboratory.

A.5   Fitting data to theoretical equations

The aim in kinetic studies is to design the experimental protocol to yield data which in its simplest form is fitted to linear equations. The theoretical equation is often not linear and the data may need further processing to obtain it in linear form; such processing often magnifies errors and introduces significant 'weighting' problems; it should not be employed to determine the parameters of an equation (such as C and K in Eqn [a.4]).

The non-linear equation [a.4], where y is a dependent variable such as a rate and x is an independent variable such as a concentration, is the general form for a large number of processes including enzyme kinetics (Michaelis–Menten equation), ionization, 1:1 complexation and the rate law for a two-step reaction involving a change in rate-limiting step. Methods (such as the Lineweaver–Burk plot in enzyme kinetics) have been developed for obtaining linear forms of Eqn [a.4] but these are not satisfactory (Cornish-Bowden, 1979) and special fitting methods are available to fit data by a least-squares criterion to a hyperbolic function (Wilkinson, 1961).

$$y = \frac{C.[x]}{[x] + K} \qquad\qquad [a.4]$$

When the theoretical law is neither Eqn [a.4] nor linear, the fitting process is often not exact and the most satisfactory approach employs *guessed* values of the parameters which are tested by determining the sum of the squares of the differences between experimental data and values calculated from the *guesses*. The *guessed* parameters which give the least sum are taken as the fit. Software programs (such as ENZFITTER or PFIT) carry out such calculations, each with its own technique for arriving efficiently at a set of guesses which best fit the data. The type of equation which can be fitted varies according to the software, which is designed for general equations, and can sometimes tackle equations with more than one independent variable. Such programs are now used almost exclusively to determine first-order rate constants because they are relatively cheap and work on modest desk-top microcomputers.

The graphical method of fitting parameters to a linear equation using ruler and graph paper retains the advantage of giving the observer a good 'feel' for the reliability of the derived parameters. Maximum and minimum possible slope and intercept may be readily obtained by inspection. The only other equation that can be fitted graphically in a similar fashion is the non-linear equation [a.4]. The data are reduced to a normalized form on double logarithmic paper (log parameter versus log x); it is overlaid with a normalized plot of the theoretical equation (to the same scale) drawn on hard tracing paper. Vertical and horizontal adjustment of the theoretical line is made to fit the experimental data (in the same way that a ruler is used to fit a linear plot) and the parameters are then read from the transparent overlay. Computational techniques using software have the advantage that they enable graphical comparison of the fitted line with the experimental data even when this is not possible via a manual approach; good laboratory practice includes at least a graphical comparison of fit, even if this is only on the screen of the microcomputer.

One of the advantages of the modern high-speed desk computer is its ability to find the best fit of the data to an equation according to a pre-defined condition such as a minimum in the sum of the squares of the differences between the experimental values and the values calculated from a large range of guessed parameters. Thus the Jaffé equation may be fitted without recourse to algebraic manipulation, simply by making sufficient guesses of the parameters over reasonable ranges. The original method of analysis involves a linearization procedure which introduces weighting problems referred to above.

$$\log k = \rho_1 \sigma_{x(1)} + \rho_2 \sigma_{x(2)} + \text{constant} \qquad [a.5]$$

Another form of fitting which is made possible by high-speed computers is that of two sets of data to two equations, each with some common parameters. Thus the dependent of Brønsted β-values on the pK_a of leaving and nucleophilic groups (cf., for example, Jencks & Jencks, 1977) is governed by Eqns [a.6] and [a.7]

$$\beta_{nuc} = p_{xy}.pK_a^{lg} + C_1 \qquad [a.6]$$

$$\beta_{lg} = p_{xy}.pK_a^{nuc} + C_2 \qquad [a.7]$$

Without a global fit of β_{nuc} and β_{lg} to them, the separate equations often give differing values of p_{xy} because of the lack of large numbers of data points associated with this type of work. Global fitting, of course, gives a single best fit for p_{xy} and this technique is exemplified in a recent study of nucleophilic substitution (Renfrew *et al.*, 1995).

References

Abraham, M. H. (1974) Solvent effects on transition states and reaction rates, *Prog. Phys. Org. Chem.*, **11**, 1.

Abraham, M. H., Dodd, D., Johnson, M. D., Lewis, E. S. & More O'Ferrall, R. A. (1971) Microscopic reversibility and the symmetrical isotopic exchange reactions of organometallic compounds, *J. Chem. Soc. Sect. B*, 762.

Abraham, M. H., Grellier, P. L., Abboud, J. M., Doherty, R. M. & Taft, R. W. (1988) Solvent effects in organic chemistry—recent developments, *Can. J. Chem.*, **66**, 2673.

Adams, M. J. (1987) Oxido-reductases—pyridine nucleotide-dependent enzymes, in Page & Williams (1987), p. 477.

Akhtar, M., Emery, V. C. & Robinson, J. A. (1984) Pyridoxal phosphate dependent enzyme reactions: mechanism and stereochemistry, in Page (1984c), p. 302.

Al-Awadi, N. & Williams, A. (1990) Effective charge development in ester hydrolysis catalysed by cationic micelles, *J. Org. Chem.*, **55**, 2001.

Albery, W. J. (1975) Solvent isotope effects, in *Proton Transfer Reactions*, Caldin, E. F. & Gold, V. (eds), Chapman and Hall, London, p. 263.

Albery, W. J. (1993) Transition state theory revisited, *Adv. Phys. Org. Chem.*, **28**, 139.

Albery, W. J. & Davies, M. H. (1972) Mechanistic conclusions for the nature of solvent isotope effects, *J. Chem. Soc., Faraday Trans.*, **1**, 167.

Albery, W. J. & Knowles, J. R. (1976a) Deuterium and tritium exchange in enzyme kinetics, *Biochemistry*, **15**, 5588.

Albery, W. J. & Knowles, J. R. (1976b) Free energy profile from the reaction catalysed by triose phosphate isomerase, *Biochemistry*, **15**, 5627.

Allison, M. F. L., Bamford, C. & Ridd, J. (1958) Correlation of reaction rates with the *H*-function in concentrated solutions of sodium methoxide, *Chem. Ind. (London)*, 718.

Al-Rawi, H. & Williams, A. (1977) Elimination–addition mechanisms of acyl group transfer: the hydrolysis and synthesis of carbamate esters, *J. Am. Chem. Soc.*, **99**, 2671.

do Amaral, L., Bastos, M. P., Bull, H. G., Otis, J. J. & Cordes, E. H. (1989) Secondary deuterium isotope effects for certain acyl group transfer reactions of phenyl formate, *J. Am. Chem. Soc.*, **101**, 169.

Amyes, T. L. & Jencks, W. P. (1989) Lifetimes of oxocarbenium ions in aqueous solutions from common inhibition of the solvolysis of α-azido ethers by added ozide ion, *J. Am. Chem. Soc.*, **111**, 7888.

Amyes, T. L. & Richard, J. P. (1992) Generation and stability of a simple thiol ester enolate in aqueous solution, *J. Am. Chem. Soc.*, **114**, 10 297.

Anderson, B. M. & Jencks, W. P. (1960) The effect of structure on reactivity in semicarbazone formation, *J. Am. Chem. Soc.*, **82**, 1773.

Anh, N. T. & Minot, C. (1980) Conditions favouring retention of configuration in S_N2 reactions. A perturbational study, *J. Am. Chem. Soc.*, **102**, 103.

Anet, F. A. & Kopelwick, G. (1989) Detection and assignments of diasterotopic chemical shifts in partially deuteriated methyl groups of a chiral molecule, *J. Am. Chem. Soc.*, **111**, 3429.

Antonov, V. K.. Ginodman, L. M., Rumsh, L. D., Kapitamikov, Y. V., Barschevs-kaya, T. N., Yavashev, L. P., Gurova, A. G. & Volkova, L. I. (1981) Studies on the mechanism of action of proteolytic enzymes using heavy oxygen exchange, *Eur. J. Biochem.*, **117**, 195.

Applequist, D. E. & Kaplan, L. (1965) The decarbonylation of aliphatic and bridge-head aldehydes, *J. Am. Chem. Soc.*, **87**, 2194.

Applequist, D. E. & Klug, J. H. (1978) Energies of the cycloalkyl and 1-methyl-cycloalkyl free radicals by the decarbonylation method, *J. Org. Chem.*, **43**, 1729.

Arnett, E. M. & Reich, R. (1980) Electronic effects in the Menschutkin reaction. A complete kinetic and thermodynamic dissection of alkyl transfer to 3- and 4-substituted pyridines, *J. Am. Chem. Soc.*, **102**, 5892.

Arora, M., Cox, B. G. & Sorensen, P. E. (1979) A kinetic and thermodynamic study of the addition of methoxide ion to substituted benzaldehydes, *J. Chem. Soc., Perkin Trans. 2*, 103.

Asboth, A. & Polgar, L. (1983) Transition-state stabilisation at the oxyanion binding sites of serine and thiol proteinases: hydrolysis of thiono and oxygen esters, *Biochemistry*, **22**, 117.

Ashby, E. C. & Pham, T. N. (1987) Evidence for electron transfer in reactions of nucleophiles with optically active alkyl halides. A challenge to the S_N2 transition state, *Tetrahedron Lett.*, **28**, 3183.

Auld, D. S. (1987) Acyl group transfer—metalloproteinases, in Page & Williams (1987), p. 240.

Auld, D. S. & Bruice, T. C. (1967) Catalytic reactions involving azomethines (IX). General base catalysis of the transamination of 3-hydroxypyridine-4-aldehyde by alanine, *J. Am. Chem. Soc.*, **89**, 2098.

Bacaloglu, R., Blasko, A., Bunton, C. A. & Ortega, F. (1990) Single electron transfer in deacylation of ethyl dinitrobenzoates, *J. Am. Chem. Soc.*, **112**, 9336.

Baggott, J. (1989) Molecules caught in the act, *New Scientist*, 17 June, p. 1958.

Bagno, A., Scorrano, G. & More O'Ferrall, R. A. (1987) Linear free energy relation-ships for acidic media, *Rev. Chem. Ind.*, **7**, 313.

Bagno, A. Lorato. G. & Scorrano, G. (1993) Thermodynamics of protonation and hydration of aliphatic amides, *J. Chem. Soc., Perkin Trans. 2*, 1091.

Bagno, A., Boso, R. L., Ferrari, N. & Scorrano, G. (1995) Steric effects on the solvation of protonated ketones, *J. Chem. Soc., Perkin Trans. 2*, 2053.

Barton, P., Laws, A. P. & Page, M. I. (1994) Structure–activity relationships in the esterase-catalysed hydrolysis and transesterification of esters, *J. Chem. Soc., Perkin Trans 2*, 2021.

Ba-Saif, S. A., Luthra, A. K. & Williams, A. (1987) Concertedness in acyl group transfer: a single transition state in acetyl transfer between phenolate ion nucleo-philes, *J. Am. Chem. Soc.*, **109**, 6362.

Ba-Saif, S. A., Waring, M. A. & Williams, A. (1990) Single transition state in the transfer of a neutral phosphoryl group between phenoxide ion nucleophiles in aqueous solution, *J. Am. Chem. Soc.*, **112**, 8115.

Ba-Saif, S. A., Colthurst, M., Waring, M. A. & Williams, A. (1991) An open transition state in carbonyl acyl group transfer in aqueous solution, *J. Chem. Soc., Perkin Trans. 2*, 1901.

Beak, P. (1992) Determinations of transition-state geometries by the endocyclic restriction test—mechanisms of substitution at non-stereogenic atoms, *Acc. Chem. Res.*, **25**, 215.

Becker, N. N. & Roberts, J. D. (1984) Structure of the liver alcohol dehydrogenase–NAD$^+$-pyrazole complex as determined by ^{15}N-nmr spectroscopy, *Biochemistry*, **23**, 3336.

Belasco, W. J. G. & Knowles, J. R. (1980) Direct observation of substrate distortion by triose phosphate isomerase using Fourier transform infra-red spectroscopy, *Biochemistry*, **19**, 472.

Belasco, J. G., Albery, W. J. & Knowles, J. R. (1983) Double isotope fractionation: test for concertedness and for transition state dominance, *J. Am. Chem. Soc.*, **105**, 247.

Bell, R. P. (1959) *The Proton in Chemistry*, 1st edn, Methuen, London.

Bell, R. P. (1980) *The Tunnel Effect in Chemistry*, Chapman and Hall, New York.

Bell, R. P. & Sorensen, P. E. (1976) Kinetics of addition of hydroxide ions to substituted benzaldehydes, *J. Chem. Soc., Perkin Trans. 2*, 1594.

Bell, R. P., Critchlow, H. E., Page, M. I. (1974) Ground state and transition state effects in the acylation of α-chymotrypsin in organic-solvent mixtures, *J. Chem. Soc., Perkin Trans. 2*, 66.

Bender, M. L. (1951) Oxygen changes as evidence for the existence of an intermediate in ester hydrolysis, *J. Am. Chem. Soc.*, **73**, 1626.

Bender, M. L. & Chen, M. C. (1963) Acylium ion formation in the reaction of carboxylic acid derivatives, *J. Am. Chem. Soc.*, **85**, 37.

Bender, M. L. & Glasson, W. A. (1960) Kinetics of α-chymotrypsin-catalyzed hydrolysis and methanolysis of acetyl-L-phenylalanine methylester—evidence for the specific binding of water on the enzyme surface, *J. Am. Chem. Soc.*, **82**, 3336.

Bender, M. L. & Kezdy, F. J. (1965) Mechanism of action of proteolytic enzymes, *Annu. Rev. Biochem.*, **34**, 49.

Bender, M. L. & Williams, A. (1996) Ketimine intermediate in amine-catalysed enolization of acetone, *J. Am. Chem. Soc.*, **88**, 2502.

Bender, M. L., Chow, Y. L. & Chloupek, F. (1958) Intramolecular catalysis of hydrolytic reactions II. The hydrolysis of phthalamic acid, *J. Am. Chem. Soc.*, **80**, 5380.

Bender, K. L., Kezdy, F. J. & Gunter, C. R. (1964) The anatomy of an enzymatic catalysis. α-Chymotrypsin, *J. Am. Chem. Soc.*, **86**, 3714.

Bender, M. L., Reinstein, J. A., Silver, M. S. & Mikulak, R. (1965) Kinetics and mechanism of hydroxide ion and morpholine-catalysed hydrolysis of methyl *o*-formyl benzoate. Participation of the neighbouring aldehyde group, *J. Am. Chem. Soc.*, **87**, 4545.

Bender, M. L., Begue-Canton, M. L., Blakely, R. L., Brubacher, L. J., Feder, J., Gunter, C. R., Kezdy, F. J., Killheffer, J. V., Marshall, T. H., Miller, C. G., Roseke, R. W. & Stoops, J. K. (1966) The determination of the concentration of hydrolytic enzyme solutions: α-chymotrypsin, trypsin, papain, elastase, subtilisin and acetylcholinesterase, *J. Am. Chem. Soc.*, **88**, 5890.

Bentley, T. W. & Llewellyn, G. (1990) Y_x scales of solvent ionising power, *Prog. Phys. Org. Chem.*, **17**, 121.

Bergeron, R. J. & Burton, P. S. (1982) Role of cyclohexaamylose C-3 hydroxyls in catalytic hydrolysis, *J. Am. Chem. Soc.*, **104**, 3664.

Bergman, N. A., Chiang, Y. & Kresge, A. J. (1978) An isotope effect maximum for proton transfer between normal acids and bases, *J. Am. Chem. Soc.*, **100**, 5954.

Bernasconi, C. F. (1992a) The principle of non-perfect synchronisation, *Adv. Phys. Org. Chem.*, **27**, 119.

Bernasconi, C. F. (1992b) The principle of non-perfect synchronisation: more than a qualitative concept? *Acc. Chem. Res.*, **25**, 9.

Bernstein, F. C., Koetzle, T. F., Williams, G. J. B., Meyer, E. F. Jr, Brice, M. D., Rodgers, J. R., Kennard, O., Shimanouchi, T. & Tasumi, M. (1977) The protein data bank. A computer-based archival file for macromolecular structures, *J. Mol. Biol.*, **112**, 535.

Bevington, P. R. (1969) *Data Reduction and Error Analysis for the Physical Sciences*, McGraw-Hill, New York.

Billups, W. E., Houk, K. E. & Stevens, R. V. (1986) Stereochemistry and reaction mechanism, in *Investigation of Rates and Mechanisms of Reactions*, 4th edn, Part I, Bernasconi, C. F. (ed.), Wiley–Interscience, New York, p. 663.

Birktoft, J. J., Kraut, J. & Freer, S. T. (1976) A detailed structural comparison between the charge relay system in chymotrypsinogen and in α-chymotrypsin, *Biochemistry*, **15**, 4481.

Blokzijl, W. & Engberts, J. B. F. N. (1993) Hydrophobic effects. Opinions and facts, *Angew. Chem., Int. Ed. Engl.*, **32**, 1545.

Blow, D. M. (1976) Structure and mechanism of chymotrypsin, *Acc. Chem. Res.*, **9**, 145.

Boger, D. L. and Mathvink, R. J. (1992) Acyl radicals: intermolecular and intra-molecular alkene addition reactions, *J. Org. Chem.*, **57**, 1429.

Bordwell, F. G. (1970) Are nucleophilic bimolecular concerted reactions involving four or more bonds a myth? *Acc. Chem. Res.*, **3**, 281.

Bordwell, F. G. & Boyle, W. C. (1971) Brønsted coefficients and ρ values as guides to transition state structures in deprotonation reactions, *J. Am. Chem. Soc.*, **93**, 511.

Bordwell, F. G. & Boyle, W. C. (1972) Acidities, Brønsted coefficients and transition state structure for 1-arylnitroalkanes, *J. Am. Chem. Soc.*, **94**, 3907.

Bordwell, F. G. & Wilson, C. A. (1987) Distinguishing between polar and electron transfer mechanisms for reactions of anions and alkyl halides, *J. Am. Chem. Soc.*, **109**, 5470.

Bordwell, F. G., Boyle, W. C., Hautala, J. A. & Lee, K. C. (1969) Brønsted coefficients larger than 1 and less than 0 for proton removal from carbon acids, *J. Am. Chem. Soc.*, **91**, 4002.

Born, M. and Oppenheimer, R. (1927) Quantum theory of the molecules, *Ann. Phys. (Leipzig)*, **84**, 457.

Botting, N. P. & Gani, D. (1989) Probing the mechanism of methylaspartase, in *Molecular Recognition: Chemical and Biocheical Problems*, Roberts, S. M. (ed.) Royal Society of Chemistry, Cambridge, RSC Special Publication 78, p. 134.

Bourne, N. & Williams, A. (1984) Evidence for a single transition state in the transfer of the phosphoryl group ($-PO_3^{2-}$) to nitrogen nucleophiles from pyridine-*N*-phosphonates, *J. Am. Chem. Soc.*, **106**, 7591.

Bourne, N., Hopkins, A. R. & Williams, A. (1983) A preassociation concerted mechanism in the transfer of the sulphate group between isoquinoline-*N*-sulphon-ate and pyridines, *J. Am. Chem. Soc.*, **105**, 3358.

Bourne, N., Chrystiuk, E., Davis, A. M. & Williams, A. (1988) A single transition state in the reaction of aryl diphenylphosphinate esters with phenolate ions, *J. Am. Chem. Soc.*, **110**, 1840.

Brayer, G. D., Delbaere, L. T. J., James, M. N. G., Bauer, C.-A. & Thompson, R. C. (1979) Crystallographic and kinetic investigations of the covalent complex formed by a specific tetrapeptide aldehyde and serine protease from *Streptomyces griseus*, *Proc. Nat. Acad. Sci. U.S.A.*, **76**, 96.

Breslow, R. (1995) Biomimetic enzymes and artificial enzymes, *Acc. Chem. Res.*, **28**, 146.

Breslow, R. & Campbell, P. (1969) Selective aromatic substitution within a cyclodextrin mixed complex, *J. Am. Chem. Soc.*, **91**, 3085.

Breslow, R. & Campbell, P. (1971) Selective aromatic substitution by hydrophobic binding of a substrate to a simple cyclodextrin catalyst, *Bioorg. Chem.*, **1**, 100.

Breslow, R., Doherty, J. B., Guillot, G. & Lipsey, C. (1978) β-Cyclodextrin-bis-imidazole—a model for ribonuclease, *J. Am. Chem. Soc.*, **100**, 3227.

Brønsted, J. N. (1928) Acid and base catalysis, *Chem. Rev.*, **5**, 23.

Buchwald, S. L., Friedman, J. M. & Knowles, J. R. (1984) Stereochemistry of nucleophilic displacement on two phosphoric monoesters and a phosphoguanidine: the role of metaphosphate, *J. Am. Chem. Soc.*, **106**, 4911.

Buncel, E., Wilson, H. & Chuaqui, C. (1982) Reactivity–selectivity correlations 4. The α-effect in S_N2 reactions at sp^3 carbon. The reactions of hydrogen peroxide anion with methyl phenyl sulphate, *J. Am. Chem. Soc.*, **104**, 4896.

Bunnett, J. F. (1961) Kinetics of reaction in moderately concentrated acids—an empirical criterion of mechanism, *J. Am. Chem. Soc.*, **83**, 4968.

Bunton, C. A. (1984) Reaction in micelles and similar self organised aggregates, in Page (1984c), p. 401.

Bunton, C. A. & Savelli, G. (1987) Organic reactivity in aqueous micelles and similar assemblies, *Adv. Phys. Org. Chem.*, **22**, 213.

Burgi, H. B. (1975) Stereochemistry of reaction paths as determined from crystallographic structure data—a relationship between structure and energy, *Angew. Chem., Int. Ed. Engl.*, **14**, 460.

Burgi, H. B. & Dunitz, J. D. (1983) From crystal statics to chemical-dynamics, *Acc. Chem. Res.*, **16**, 153.

Burkhardt, G. N., Ford, W. G. K. and Singleton, E. (1936) The hydrolysis of arylsulphuric acids, *J. Chem. Soc.*, 17.

Burwell, R. L. & Pearson, R. G. (1966) The principle of microscopic reversibility, *J. Phys. Chem.*, **70**, 300.

Caldwell, G., Magnera, T. F. & Kebarle, P. (1984) S_N2 reactions in the gas phase. Temperature dependence of the rate constants and energies of the transition states. Comparison with solution, *J. Am. Chem. Soc.*, **106**, 6959.

Caldwell, S. R., Raushel, F. M., Weiss, P. M. & Cleland, W. W. (1991) Transition state structures for enzymatic and alkaline phosphotriester hydrolysis, *Biochemistry*, **30**, 7444.

Calvo, K. C. (1985) The Conant–Swan fragmentation reaction—stereochemistry of phosphate ester formation, *J. Am. Chem. Soc.*, **107**, 3690.

Campbell, P., Nashed, N. T., Lapinskas, B. A. & Gurrieri, J. (1983) Thion esters as a probe for electrophilic catalysis in the serine protease mechanism, *J. Biol. Chem.*, **258**, 59.

Capon, B. & Page, M. I. (1971) The kinetics and mechanism of the hydrolysis of esters of *cis* and *trans* 2-hydroxycyclopentane carboxylic acids, *J. Chem. Soc. (B)*, 741.

Capon, B., Ghosh, A. K. & Grieve, D. McL. A. (1981) Direct observation of simple tetrahedral intermediates, *Acc. Chem. Res.*, **14**, 306.

Cash, D. & Wilson, I. B. (1986) The cyanide adduct of the aldolase dihydroxyacetone phosphate imine, *J. Biol. Chem.*, **241**, 4290.

Cevasco, G., Guanti, G., Thea, S. & Williams, A. (1984) Control of the dissociative mechanism in the hydrolysis of aryl 4-hydroxybenzoates, *J. Chem. Soc., Chem. Commun.*, 783.

Cevasco, G., Guanti, G., Hopkins, A. R., Thea, S. and Williams, A. (1985) A novel dissociative mechanism in acyl transfer from aryl 4-hydroxybenzoate esters in aqueous solution, *J. Org. Chem.*, **50**, 479.

Chandrasekhar, J. & Jorgensen, W. L. (1985) Energy profile for a non-concerted S_N2 reaction in solution, *J. Am. Chem. Soc.*, **107**, 2974.

Chang, S. and le Noble, W. J. (1983) Study of ion pair return in 2-norbornyl brosylate by means of ^{17}O-nmr, *J. Am. Chem. Soc.*, **105**, 3708.

Chang, T. K., Chaing, Y., Guo, H. -X., Kresge, A. J., Matthew, L., Powell, M. F. & Wells, J. A. (1996) Solvent isotope effects in H_2O–D_2O mixtures on serine-protase-catalysed hydrolysis reactions, *J. Am. Chem. Soc.*, **118**, 8802.

Chao, Y. & Cram, D. J. (1976) Catalysis and chiral recognition through designed complexation of transition states in transacylations of amino ester salts, *J. Am. Chem. Soc.*, **98**, 1015.

Charton, M. (1964) Definition of inductive substituent constants, *J. Org. Chem.*, **29**, 1222.

Charton, M. (1981) Electrical substituent constants for correlation analysis, *Prog. Phys. Org. Chem.*, **13**, 119.

Chatfield, D. C. & Brooks, B. R. (1995) HIV-1 protease cleavage mechanism elucidated with molecular dynamics simulation, *J. Am. Chem. Soc.*, **117**, 5561.

Chattaway, F. D. and Chapman, D. L. (1912) The transformation of ammonium cyanate into carbamide, *Trans. Chem. Soc.*, **101**, 170.

Chiang, Y., Kresge, A. J., Chang, T. K., Powell, M. F. & Wells, J. A. (1995) Solvent isotope effects on a hydrolysis reaction catalysed by subtilisin, *J. Chem. Soc., Chem. Commun.*, 1587.

Christen, P. & Metzler, D. E. (eds) (1965) *Transaminases*, Vol. 2, John Wiley, New York.

Chrystiuk, E. & Williams, A. (1987) A single transition state in the transfer of methoxycarbonyl group between isoquinoline and substituted pyridines, *J. Am. Chem. Soc.*, **109**, 3040.

Clark, J. D., O'Keefe, S. J. & Knowles, J. R. (1988) Malate synthase: proof of a stepwise Claisen condensation using the double-isotope fractionation test, *Biochemistry*, **27**, 5961.

Cohen, S. G., Vaidja, V. M. & Schulz, R. M. (1970) Active site of α-chymotrypsin. Activation by association–desolvation, *Proc. Nat. Acad. Sci. U.S.A.*, **66**, 249.

Colter, A. K., Wang, S. S., Megerle, G. H. & Ossip, P. S. (1964) Chemical behaviour of charge-transfer complexes II phenanthrene catalysis in acetolysis of 2,4,7-trinitro-9-fluorenyl *p*-toluenesulfonate, *J. Am. Chem. Soc.*, **86**, 3106.

Connors, K. A. (1987) *Binding Constants: The Measurement of Molecular Complex Stability*, Wiley–Interscience, New York.

Cordes, E. H. (1970) Secondary deuterium isotope effects for hydrolysis of acetals and ortho formates, *J. Chem. Soc., Chem. Commun.*, 527.

Cornforth, J. W. (1969) Exploration of enzyme mechanisms by asymmetric labelling, *Q. Rev. Chem. Soc.*, **23**, 125.

Cornforth, J. W., Redmond, J. W., Eggerer, H., Buckel, W. & Gutschow, C. (1969) Asymmetric methyl groups and the mechanism of malate synthase, *Nature (London)*, **221**, 1212.

Cornish-Bowden, A. (1979) *Fundamentals of Enzyme Kinetics*, Butterworths, London.

Cox, B. G. & Gibson, A. (1975) Kinetics and thermodynamic basicities of anions in mixed solvents, *Symp. Far. Soc.*, **10**, 107.

Cox, M. M. & Jencks, W. P. (1978) General acid catalysis of the aminolysis of phenyl acetate by a preassociation mechanism, *J. Am. Chem. Soc.*, **100**, 5956.

Crich, D. & Fortt, S. M. (1989) Acyl radical cyclisations in synthesis Part I. Substituent effects on the mode and efficiency of cyclisation of 7-heptenoyl radicals, *Tetrahedron*, **45**, 6581.

Csizmadia, F., Szegezdi, J. & Darvas, F. (1993) Expert system approach for predicting pK_a, in *Trends in QSAR and Modelling*, Wermuth, C. G. (ed.), ESCOM, Leiden, p. 507.

Cullis, P. M. & Nicholls, D. (1987) The existence of monomeric metaphosphate in hydroxylic solvent: a positional isotope exchange study, *J. Chem. Soc., Chem. Commun.*, 783.

Curran, T. P., Farrar, C. R., Niazy, O. and Williams, A. (1980) Structure activity studies on the equilibrium reaction between phenolate ions and 2-aryloxazolin-5-ones—data consistent with a concerted acyl group transfer mechanism, *J. Am. Chem. Soc.*, **102**, 6828.

Davis, A. M., Hall, A. D. & Williams, A. (1988a) Charge description of base catalysed alcoholysis of aryl phosphodiesters: a ribonuclease model, *J. Am. Chem. Soc.*, **110**, 5105.

Davis, A. M., Regan, A. C. & Williams, A. (1988b) Experimental charge measurement at leaving oxygen in the bovine ribonuclease-A catalysed cyclisation of uridine 3'-phosphate aryl esters, *Biochemistry*, **27**, 9042.

del la Mare, S., Coulson, A. F. W., Knowles, J. R., Priddle, J. D. & Offord, R. E. (1972) Active site labelling of triose phosphate isomerase, *Biochem. J.*, **129**, 321.

Delbaere, L. T. J. & Brayer, G. D. (1980) Structure of the complex formed between the bacterial-produced inhibitor chymostatin and the serine enzyme *Streptomyces griseus* protease A, *J. Mol. Biol.*, **139**, 45.

Deslongchamps, P. (1983) *Stereoelectronic Effects in Organic Chemistry*, Pergamon, Oxford.

Dewar, M. J. S. (1948) *The Electronic Theory of Organic Chemistry*, Oxford University Press, Oxford, p. 117.

Diederich, F., Schurmann, G. & Chao, I. (1988) Designed water-soluble macrocyclic esterases: from non-productive to productive binding, *J. Org. Chem.*, **53**, 2744.

Diederich, F., Smithrud, D. B., Sandford, E. M., Wyman, T. B., Ferguson, S. B., Carcanague, D., Chao, I. & Houk, K. N. (1992) Solvent effects in molecular recognition, *Acta Chem. Scand.*, **46**, 205.

Dietze, P. E. & Jencks, W. P. (1986) Swain–Scott correlations for reactions of nucleophilic reagents and solvents with secondary substrates, *J. Am. Chem. Soc.*, **108**, 4549.

Di Iasio, A., Trombetta, G. & Grazi, E. (1977) Fructose-1,6-bisphosphate aldolase from liver: the absolute configuration of the intermediate carbinolamine, *FEBS Lett.*, **73**, 244.

Douglas, K. T. (1987) Glutathione-dependent enzymes—chemistry, in Page & Williams (1987), p. 442.

Douzou, P. & Petsko, G. A. (1984) Proteins at work: 'stop action' pictures at subzero temperatures, *Adv. Prot. Chem, 36*, 245.

Dunathan, H. C. (1971) Stereochemical aspects of pyridoxal phosphate catalysis, *Adv. Enzymol., 35*, 79.

Eckstein, F. (1983) Phosphorothioate analogues of nucleotides—tools for the investigation of biochemical processes, *Angew. Chem., Int. Ed. Engl., 22*, 423.

Eigen, M. (1964) Proton transfer, acid base catalysis and enzymatic hydrolysis, *Angew. Chem., Int. Ed. Engl., 3*, 1.

Eklund, H., Plapp, B. V., Samama, J.-P. & Branden, C.-I. (1982) Binding of substrate in a ternary complex of horse liver alcohol dehydrogenase, *J. Biol. Chem, 257*, 14349.

Elrod, J. P., Gandour, R. D., Hogg, J. L., Kise, M., Maggiore, G. M., Schowen, R. L. & Venkatasubban, K. S. (1975) Proton bridges in enzyme catalysis, *Faraday Soc. Symp., 10*, 145.

Emery, V. C. & Akhtar, M. (1987) Pyridoxal phosphate dependent enzymes, in Page & Williams (1987), p. 345.

Epand, R. M. & Wilson, I. B. (1963) Evidence for the formation of hippuryl-chymotrypsin during the hydrolysis of hippuric acid esters, *J. Biol. Chem., 238*, 1718.

Faroz, J. F. & Cordes, E. H. (1979) Kinetic and secondary deuterium isotope effects for *O*-ethyl-*S*-phenylbenzaldehyde acetal hydrolysis, *J. Am. Chem. Soc., 101*, 1488.

Fersht, A. R. & Kirby, A. J. (1967) Structure and mechanism in intramolecular catalysis. The hydrolysis of substituted aspirins, *J. Am. Chem. Soc., 89*, 4853.

Fersht, A. R. & Sperling, J. (1973) The charge relay system in chymotrypsin and chymotrypsinogen, *J. Mol. Biol., 74*, 137.

Fersht, A. R., Shi, J. P., Wilkinson, A. J., Blow, D. M., Carter, P., Waye, M. M. Y. & Winter, G. P. (1984) Analysis of enzyme structure and activity by protein engineering, *Angew. Chem., 23*, 467.

Fersht, A. R., Leatherbarrow, R. J. & Wells, T. N. C. (1986) Quantitative analysis of structure–activity relationships in engineered proteins by linear free energy relationships, *Nature (London), 322*, 284.

Fersht, A. R., Leatherbarrow, R. J. & Wells, T. N. C. (1987) Structure–activity relationships in engineered proteins: analysis of use of binding energy by linear free energy relationships, *Biochemistry, 26*, 6030.

Fischer, G. (1987) Acyl group transfer—aspartic proteases, in Page & Williams (1987), p. 229. Also Fink, A. L. (1987) in Page & Williams (1987), p.159.

Fishbein, J. C. & McClelland, R. A. (1987) Azide ion trapping of the intermediate in the Bauberger rearrangement—lifetime of a free nitrenium ion in aqueous solution, *J. Am. Chem. Soc., 109*, 2824.

Fleming, I. (1976) *Frontier Orbitals and Organic Chemical Reactions*, John Wiley, London.

Fletcher, D. A., McMeeking, R. F. & Perkin, D. (1996) The United Kingdom Chemical Database Service, *J. Chem. Inf. Comput. Sci., 36*, 746.

Foundling, S. I., Cooper, J., Watson, F. E., Cleasby, A., Pearl, L. H., Sibanda, B. L., Hemmings, A., Wood, S. P., Blundell, T. L., Valler, M. J., Norey, C. G., Kay, J., Boger, J., Dunn, B. M., Leckie, B. J., Jones, D. M., Atrash, B., Hallett, A. & Szelke, M. (1987) High resolution x-ray analysis of renin inhibitor-aspartic proteinase complexes, *Nature (London), 327*, 349.

Fowler, A. V. & Zabin, I. (1977) The amino acid sequence of β-galactosidase of *Escherica coli, Proc. Nat. Acad. Sci. U.S.A., 74*, 1507.

Fox, J. P. & Jencks, W. P. (1974) General acid base catalysis of the methoxyamino-lysis of 1-acetyl-1,2,4-triazole, *J. Am. Chem. Soc.*, **96**, 1436.

Frey, P. A. (1982) Stereochemistry of enzymatic reactions of phosphates, *Tetrahedron*, **38**, 1541.

Fyfe, C. A. (1973) Molecular motion in molecular complexes, adducts and related species in the solid state, in *Molecular Complexes*, Foster, R. (ed.), Elek Science, London, Vol. 1, p. 209.

Gamblin, S. J., Davies, G. J., Grimes, J. M., Jackson, R. M., Littlehead, J. A. & Watson, H. C. (1991) Activity and specificity of human aldolases, *J. Mol. Biol.*, **219**, 573.

Gebler, J. C., Aebersold, R. & Withers, S. G. (1992) Glu-537 and not Glu-461 is the nucleophile in the active site of (*lac Z*) β-galactosidase from *Escherichia coli*, *J. Biol. Chem.*, **267**, 11 126.

Gefflant, T., Blonski, C., Perie, J. & Willson, M. (1995) Class I aldolases: substrate specificity, mechanism, inhibition and structure, *Progr. Biophys. Mol. Biol.*, **63**, 301.

Gensmantel, N. P. & Page, M. I. (1982) The micelle catalysed hydrolysis of penicillin derivatives (Part I) and the effect of increasing the hydrophobicity of penicillin on its micellar-catalysed hydrolysis (Part II), *J. Chem. Soc., Perkin Trans. 2*, **147**, 155.

Gensmantel, N. P., Proctor, P. & Page, M. I. (1980) Metal-ion catalysed hydrolysis of some β-lactam antibiotics, *J. Chem. Soc., Perkin Trans. 2*, 1725.

Gerrard, A. F. & Hamer, N. K. (1968) Evidence for a planar intermediate in alkaline solvolysis of methyl *N*-cyclohexylphosphoramidothioic chloride, *J. Chem. Soc. (B)*, 539.

Giese, B. (1986) *Radicals in Organic Synthesis: Formation of Carbon Carbon Bonds*, Pergamon, Oxford.

Goering, H. L. & Levy, J. F. (1961) The solvolysis of benzhydryl 4-nitrobenzoate carbonyl-^{18}O. A new method for detecting ion pair return, *Tetrahedron Lett.*, 644.

Goering, H. L., Briody, R. G. & Levy, J. F. (1963) The stereochemistry of ion pair return associated with solvolysis of 4-chlorobenzhydryl 4-nitrobenzoate, *J. Am. Chem. Soc.*, **85**, 3059.

Gopalan, V., Vanderjagt, D. J., Libell, D. P. & Glen, R. H. (1992) Transglycosyla-tion as a probe of the mechanism of action of mammalian cytosolic β-galactos-idase, *J. Biol. Chem.*, **267**, 9629.

Goto, T., Kishi, Y., Takahasi, S. & Hirata, Y. (1965) Tetrodotoxin, *Tetrahedron*, **21**, 2059.

Grob, C. A. (1976) Derivatives of inductive substituent constants from pK_a values of 4-substituted quinuclidines, *Helv. Chim. Acta*, **59**, 264.

Grunwald, E. (1985) Structure–energy relations, reaction mechanism, and disparity of progress of concerted reaction events, *J. Am. Chem. Soc.*, **107**, 125.

Grunwald, E. & Winstein, S. (1948) The correlation of solvolysis rates, *J. Am. Chem. Soc.*, **70**, 846.

Gushurst, A. J. & Jorgensen, W. L. (1986) Computer-assisted mechanistic evalua-tions of organic reactions 12. pK_a predictions for organic compounds in Me$_2$SO, *J. Org. Chem.*, **51**, 3513.

Guthrie, J. P. (1990) Concertedness and E2 elimination reactions. Prediction of transition state position and reaction rates using two-dimensional reaction sur-faces based on quadratic and quartic approximations, *Can. J. Chem.*, **68**, 1643.

Guthrie, J. P. (1991) Concerted mechanism for alcoholysis of esters: an examination of the requirements, *J. Am. Chem. Soc.*, **113**, 3941.

Guthrie, R. D. (1975) A suggestion for the revision of mechanistic designations, *J. Org. Chem.*, **40**, 402.

Guthrie, R. D. & Jencks, W. P. (1989) IUPAC recommendations for the representation of reaction mechanisms, *Acc. Chem. Res.*, **22**, 343.

Gutte, B. & Merrifield, R. B. (1969) The synthesis of ribonuclease-A, *J. Am. Chem. Soc.*, **91**, 501.

Gutte, B. & Merrifield, R. B. (1971) The synthesis of ribonuclease-A, *J. Biol. Chem.*, **246**, 1922.

Haake, P. (1987) Thiamine-dependent enzymes, in Page & Williams (1987), p. 390.

Hall, S. & Inch, T. D. (1980) Phosphorus stereochemistry, *Tetrahedron*, **36**, 2059.

Hall, S., Doweyko, A. & Jordan, F. (1976) Glyoxalase I enzyme studies. Nuclear magnetic resonance evidence for an enediol–proton transfer mechanism, *J. Am. Chem. Soc.*, **98**, 7460.

Hammett, L. P. (1937) The effect of structure upon the reactivity of organic compounds. Benzene derivatives, *J. Am. Chem. Soc.*, **59**, 96.

Hammett, L. P. & Deyrup, A. J. (1932) A series of simple basic indicators (I). The acidity functions of mixtures of sulphonic and perchloric acids in water, *J. Am. Chem. Soc.*, **54**, 2721.

Hammett, L. P. & Pfluger, H. L. (1933) The rate of addition of methyl esters to trimethylamine, *J. Am. Chem. Soc.*, **55**, 4079.

Hammond, G. S. (1955) A correlation of reaction rates, *J. Am. Chem. Soc.*, **77**, 334.

Hansch, C. & Leo, A. (1979) *Substituent Constants for Correlation Analysis in Chemistry and Biology*, Wiley–Interscience, New York.

Harder, S., Streitwieser, A., Petty, J. T. & Schleyer, P. V. R. (1995) Ion pair S_N2 reactions. Theoretical study of inversion and retention mechanisms, *J. Am. Chem. Soc.*, **117**, 3253.

Hartley, B. S. & Kilby, B. A. (1954) The reaction of *p*-nitrophenylesters with chymotrypsin and insulin, *Biochem. J.*, **56**, 288.

Henchman, M., Viggiano, A. A., Paulson, F., Freedman, A. & Wormhoudt, J. (1985) Thermodynamic and kinetic properties of the metaphosphate anion, PO_3^-, in the gas phase, *J. Am. Chem. Soc.*, **107**, 1453.

Hengge, A. C. (1992) Can acyl transfer occur via a concerted mechanism? Direct evidence from heavy-atom isotope effects, *J. Am. Chem. Soc.*, **114**, 6575.

Hengge, A. C. & Cleland, W. W. (1990) Direct measurement of transition state bond cleavage in hydrolysis of phosphate esters of *p*-nitrophenol, *J. Am. Chem. Soc.*, **112**, 7421.

Herlihy, J. M., Maister, S. G., Albery, W. J. & Knowles, J. R. (1976) Energetics of triose phosphate isomerase: the fate of the $1(R)$-^{3}H label of tritiated dihydroxyacetone phosphate in the isomerase reaction, *Biochemistry*, **15**, 5601.

Herschlag, D. & Jencks, W. P. (1989) Evidence that metaphosphate monoanion is not an intermediate in solvolysis reactions in aqueous solution, *J. Am. Chem. Soc.*, **111**, 7579.

Hill, S. V., Thea, S. & Williams, A. (1982) A test of Leffler's assumption for a simple addition reaction in aqueous solution, *J. Chem. Soc., Chem. Commun.*, 547.

Hine, J. (1960) Polar effects on rates and equilibria, *J. Am. Chem. Soc.*, **82**, 4877.

Hine, J. (1962) *Physical Organic Chemistry*, McGraw-Hill, New York.

Hine, J. (1966) The principle of least motion. Application to reactions of resonance stabilised species, *J. Org. Chem.*, **31**, 1236.

Hine, J. (1975) *Structural Effects on Equilibria in Organic Chemistry*, John Wiley, New York.

Hine, J. (1977) The principle of least nuclear motion, *Adv. Phys. Org. Chem.*, **15**, 1.

Hirschman, R., Nutt, R. F., Veber, D. F., Vitali, R. A., Vargas, S. L., Jacob, T. A., Holly, F. W. & Denkewalter, R. G. (1969) Studies on the total synthesis of an enzyme V. The preparation of enzymatically active material, *J. Am. Chem. Soc.*, **91**, 507.

Hogg, J. L., Phillips, M. K. & Jergers, D. E. (1977) Catalytic proton bridge in acetyl-imidazolium ion hydrolysis implicated by a proton inventory, *J. Org. Chem.*, **42**, 2459.

Hogg, J. L., Rodgers, J., Kovach, I. & Schowen, R. L. (1982) Kinetic isotope effect probes of transition state structure. Vibrational analysis of model transition state structures for carbonyl addition, *J. Am. Chem. Soc.*, **102**, 79.

Holmes, R. R. (1990) The stereochemistry of nucleophilic substitution at tetra-coordinated silicon, *Chem. Rev.*, **90**, 17.

Hopkins, A. R. and Williams, A. (1982a) Carboxy group participation in sulphate and sulphamate group transfer reactions, *J. Org. Chem.*, **47**, 1745.

Hopkins, A. R. and Williams, A. (1982b) The question of free sulphur trioxide as an intermediate in sulphonate group transfers, *J. Chem. Soc., Chem. Commun.*, 37.

Hosseini, M. W., Lehn, J. M. & Mertes, M. P. (1983) Efficient molecular catalysis of ATP hydrolysis by protonated macrocyclic ligands, *Helv. Chim. Acta.*, **66**, 2454.

Hosseini, M. W., Lehn, J. M., Jones, K. C., Plute, K. E., Mertes, K. B. & Mertes, M. P. (1989) Supramolecular catalysis—polyammonium macrocycles as enzyme mimics for phosphoryl transfer in ATP hydrolysis, *J. Am. Chem. Soc.*, **111**, 6330.

Hudson, R. F. & Brown, C. (1972) Reactivity of heterocyclic phosphorus compounds, *Acc. Chem. Res.*, **5**, 204.

Hudson, R. F. & Green, M. (1963) Stereochemistry of displacement reactions at phosphorus atoms, *Angnew. Chem. Int. Ed. Engl.*, **2**, 11.

Hudson, R. F. & Klopman, G. (1964) Nucleophilic reactivity IV. Competing bimolecular substitution and β-elimination, *J. Chem. Soc.*, 5.

Hughes, E. D., Juliusberger, F., Masterman, S., Topley, B. & Weiss, J. (1935) Aliphatic substitution and the Walden inversion, *J. Chem. Soc.*, 1525.

Humffray, A. A. & Ryan, J. J. (1967) Rate correlations involving the linear combinations of substituent parameters Part II. Hydrolysis of aryl benzoates, *J. Chem. Soc. (B)*, 468.

Hupe, D. J. (1987) Imino formation in enzymatic reactions, in Page & Williams (1987), p. 317.

Huskey, W. P. (1991) Origins and interpretation of heavy atom isotope effects, in *Enzyme Mechanism from Isotope Effects*, Cook. P. F. (ed.), CRC Press, Boca Raton.

Illuminati, G. & Mandolini, L. (1981) Ring closure reactions of bifunctional chain molecules, *Acc. Chem. Res.*, **14**, 95.

Ingold, C. K. (1953) *Structure and Mechanism in Organic Chemistry*, 1st edn, G. Bell, London (see also 2nd edn (1969), Cornell University Press, Ithaca).

Isaacs, N. S. (1981) *Liquid Phase High Pressure Chemistry*, John Wiley, New York.

Isaacs, N. S. (1995) *Physical Organic Chemistry*, 2nd edn, Longman Scientific, Harlow.

Israelachvili, J. (1987) Solvation forces and liquid structures, as probed by direct force measurements, *Acc. Chem. Res.*, **20**, 415.

Jacobson, R. H., Zhang, X. J., Dubose, R. F. & Matthews, B. W. (1994) 3-Dimensional structure of β-galactosidase from *Esherichia coli, Nature*, **369**, 761.

Jaffé, H. H. (1953a) Some extensions of Hammett's equation, *Science*, **118**, 246.

Jaffé, H. H. (1953b) A re-examination of the Hammett equation, *Chem. Rev.*, **53**, 191.

278 *References*

Jaffé, H. H. (1954) Application of the Hammett equation to fused ring systems, *J. Am. Chem. Soc.*, **76**, 4261.

Jaskolski, M., Tomasselli, A. G., Sawyer, T. K., Staples, D. G., Heinrikson, R. L., Schneider, J., Kent, S. B. H. & Wlodawer, A. (1991) Structure at 2.5 Å resolution of chemically synthesised human immunodeficiency virus type-1 protease complexed with a hydroxyethylene-based inhibitor, *Biochemistry*, **30**, 1600.

Jencks, D. A. & Jencks, W. P. (1977) On the characterisation of transition states by structure–reactivity coefficients, *J. Am. Chem. Soc.*, **99**, 7948.

Jencks, W. P. (1959) Studies on the mechanism of oxime and semicarbazone formation, *J. Am. Chem. Soc.*, **81**, 475.

Jencks, W. P. (1972) General acid–base catalysis of complex reactions in water, *Chem. Rev.*, **72**, 705.

Jencks, W. P. (1976) Enforced general acid–base catalysis of complex reactions and its limitations, *Acc. Chem. Res.*, **9**, 425.

Jencks, W. P. (1980) When is an intermediate not an intermediate? Enforced mechanisms of general acid–base catalysed, carbocation, carbanion, and ligand exchange reactions, *Acc. Chem. Res.*, **13**, 161.

Jencks, W. P. (1981) How does a reaction choose its mechanism? *Chem. Soc. Rev.*, **10**, 345.

Jencks, W. P. (1985) A primer for the Bema Hapothle. An empirical approach to the characterisation of changing transition states, *Chem. Rev.*, **85**, 51.

Jencks, W. P. (1988) Are structure-activity correlations useful? *Bull. Soc. Chim. Fr.*, 218.

Jencks, W. P. & Gilbert, H. F. (1979) Mechanism of the base catalysed cleavage of imido esters to nitriles, *J. Am. Chem. Soc.*, **101**, 5774.

Jencks, W. P., Brant, S. R., Gandler, J. R., Fendrich, G. & Nakamura, C. (1982) Non-linear Brønsted correlations. The roles of resonance, solvation, and changes in transition state structure, *J. Am. Chem. Soc.*, **104**, 7045.

Johnson, S. L. (1967) Nucleophilic catalysis of ester hydrolysis and related reactions, *Adv. Phys. Org. Chem.*, **5**, 237.

Jones, P. G. & Kirby, A. J. (1984) Simple correlation between bond length and reactivity. Combined use of crystallographic and kinetic data to explore a reaction coordinate, *J. Am. Chem. Soc.*, **89**, 5964.

Jorgensen, W. L. (1989) Free energy calculations: a breakthrough for modelling organic chemistry in solution, *Acc. Chem. Res.*, **22**, 184.

Jorgensen, W. L., Laird, E. R., Gushurst, A. J., Fleischer, J. M., Gothe, S. A., Helson, H. E., Paderes, G. D. & Sinclair, S. (1990) CAMEO: A program for the logical prediction of the products of organic reactions, *Pure Appl. Chem.*, **10**, 1921.

Julin, D. A. & Kirsch, J. F. (1989) Kinetic isotope effect studies on aspartate aminotransferase: evidence for a concerted 1,3 prototropic shift mechanism for the cytoplasmic isozyme and L-asparate and dichotomy of mechanism, *Biochemistry*, **28**, 3825.

Kallos, J. & Avatis, K. (1966) Study of the polarity of the active site of chymotrypsin, *Biochemistry*, **5**, 1979.

Kamlet, M. & Taft, R. W. (1985) Linear solvation free energy relationships. Local empirical rules or fundamental laws of chemistry? A reply to the chemometricians, *Acta Chem. Scand., Ser. B*, **39**, 611.

Karabatsos, G. J., Flemming, J. S. & Hsi, N. (1966) Comments on the factors controlling product specificity in the reduction of carbonyl compounds with alcohol dehydrogenase and diphosphopyridine nucleotide, *J. Am. Chem. Soc.*, **88**, 849.

Kempton, J. B. & Withers, S. G. (1992) Mechanism of *Agrobacterium* β-glucosidase—kinetic studies, *Biochemistry*, **31**, 9961.

Kice, J. C. (1980) Mechanisms and reactivity of organic oxyacids of sulphur and their anhydrides, *Adv. Phys. Org. Chem.*, **17**, 65.

Kim, H. & Hynes, J. T. (1992) A theoretical model for S_N1 ionic dissociation in solution. 1, Activation free energies and transition state structure, *J. Am. Chem. Soc.*, **114**, 10 508.

Kim, J. K. & Bunnett, J. F. (1970) Evidence for a radical mechanism of aromatic 'nucleophilic' substitution, *J. Am. Chem. Soc.*, **92**, 7463.

Kim, J. K. & Caserio, M. C. (1981) Acyl-transfer reactions in the gas phase. The question of tetrahedral intermediates, *J. Am. Chem. Soc.*, **103**, 2124.

Kim, J. K. & Caserio, M. C. (1982) Acyl-transfer reactions in the gas phase. Ion-molecule chemistry of vinyl acetate, *J. Am. Chem. Soc.*, **104**, 4624.

King, J. F. (1975) Return of sulfenes, *Acc. Chem. Res.*, **8**, 10.

Kintzinger, J. P., Kotzyba-Hibert, F., Lehn, J. M., Pagelot, A. & Saigo, K. (1981) Dynamic properties of molecular complexes and receptor–substrate complementarity. Molecular dynamics of macrotricyclic diammonium cryptates, *J. Chem. Soc., Chem. Commun.*, 833.

Kirby, A. J. (1980) Effective molarities for intramolecular reactions, *Adv. Phys. Org. Chem.*, **17**, 183.

Kirby, A. J. (1983) The anomeric effect and related stereoelectronic effects at oxygen, Springer, Berlin.

Kirby, A. J. (1994) Crystallographic approaches to transition state structures, *Adv. Phys. Org. Chem.*, **29**, 87.

Kirby, A. J. (1996) *Stereoelectronics Effects*, Oxford Science Publications, Oxford.

Kirsch, J. F. (1972) Linear free energy relationships in enzymology, in *Advances in Linear Free Energy Relationships*, Chapman, N. B. and Shorter, J. J. (eds), Plenum, London, p. 369.

Kirsch, J. F., Eicheles, G., Ford, G. C., Vincent, M. G., Jansonius, J. N., Gehring, H. & Christen, P. (1984) Mechanism of action of aspartate aminotransferase proposed on the basis of its spatial structure, *J. Mol. Biol.*, **174**, 497.

Klibanov, A. M. (1990) Asymmetric transformation catalysed by enzymes in organic solvents, *Acc. Chem. Res.*, **23**, 114.

Klinman, J. P. (1991) Hydrogen tunnelling and coupled motions in enzyme reactions, in *Enzyme Mechanism from Isotope Effects*, Cook, P. F. (ed.), CRC Press, Boca Raton, p. 127.

Klotz, I. M. (1987) Enzyme models—synthetic polymers, in Page & Williams (1987), p. 14.

Kluger, R. (1987) Thiamine diphosphate: a mechanistic update on enzymic and non-enzymic catalysis of decarboxylation, *Chem. Rev.*, **87**, 863.

Kluger, R. (1990) Ionic intermediates in enzyme-catalysed carbon–carbon bond formation: patterns, prototypes and proposals, *Chem. Rev.*, **90**, 1151.

Knier, B. L. & Jencks, W. P. (1980) Mechanism of reactions of *N*-(methoxymethyl)-*N*,*N*-dimethylanilinium ions with nucleophilic reagents, *J. Am. Chem. Soc.*, **102**, 6789.

Knowles, J. R. (1980) Enzyme-catalysed phosphoryl transfer reactions, *Annu. Rev. Biochem.*, **49**, 877.

Knowles, J. R. (1990) A feast for chemists, in 'Mechanistic Enzymology', *Chem. Rev.*, **90**, 1077.

Knowles, J. R. (1991) Enzyme catalysis: not different, just better, *Nature (London)*, **350**, 121.

Knowles, J. R., Norman, R. O. C. & Prosser, J. H. (1961) The transition state in nucleophilic aromatic substitution, *Proc. Chem. Soc.*, 341.

Komives, E. A., Chang, L. C., Lolis, E., Tilton, R. F., Petsko, G. A. & Knowles, J. R. (1991) Electrophilic catalysis in triose phosphate isomerase: the role of histidine-95, *Biochemistry*, **30**, 3011.

Koskinen, A. M. P. & Klibanov, A. M. (eds) (1996) *Enzymatic Reactions in Organic Media*, Blackie, London.

Kosower, E. M. (1958) A new measure of solvent polarity: Z-values, *J. Am. Chem. Soc.*, **80**, 3253.

Kreevoy, M. M. & Lee, I. S. H. (1984) Marcus theory of perpendicular effect on α for hydride transfer between NAD^+ analogues, *J. Am. Chem. Soc.*, **106**, 2550.

Kresge, A. J. (1973) The Brønsted relationship—recent developments, *Chem. Soc. Rev.*, **2**, 475.

Kresge, A. J. (1975a) What makes proton transfer fast? *Acc. Chem. Res.*, **8**, 354.

Kresge, A. J. (1975b) The Brønsted relationship: significance of the exponent, in *Proton Transfer Reactions*, Caldin, E. F. & Gold, V. (eds), Chapman and Hall, London, p. 179.

Kresge, A. J., More O'Ferrall, R. A. & Powell, M. F. (1987) Solvent isotope effects. Fractionation factors and mechanisms of proton transfer, in *Isotopes in Organic Chemistry*, Buncel, E. & Lee, C. C. (eds), Elsevier, Amsterdam, p. 177.

Kreutter, K., Steinmetz, A. C. U., Liang, T. C., Prorok, M., Abeles, R. H. & Ringe, D. (1994) The mechanism of inactivation of serine proteases by chloroketones, *Biochemistry*, **33**, 13792.

Krupka, R. M., Kaplan, H. & Laidler, K. J. (1966) Kinetic consequences of the principle of microscopic reversibility, *Trans. Faraday Soc.*, **62**, 2754.

Kubinyi, H. (1979) Lipophilicity and drug research design, *Prog. Drug Res.*, **23**, 97.

Kunitake, T. & Okahata, Y. (1976) Multifunctional hydrolytic catalysis Part 7. Co-operative catalysis of the hydrolysis of phenyl esters by a copolymer of N-methacrylohydroxamic acid and 4-vinylimidazole, *J. Am. Chem. Soc.*, **98**, 7793.

Kunitake, T., Okahata, Y. & Sakamoto, T. (1976) Multifunctional hydrolytic catalysis Part 8. Remarkable acceleration of the hydrolysis of p-nitrophenyl acetate by micellar bifunctional catalysts, *J. Am. Chem. Soc.*, **98**, 7799.

Kurz, J. L. (1963) Transition state characterisation for catalysed reactions, *J. Am. Chem. Soc.*, **85**, 987.

Kurz, J. L. (1972) Transition states as acids and bases, *Acc. Chem. Res.*, **5**, 1.

Kurz, J. L. and Lee, Y. N. (1975) The acidity of water in the transition state for methyl tosylate hydrolysis, *J. Am. Chem. Soc.*, **97**, 3841.

Lapworth, A. (1903) Reactions involving the addition of hydrogen cyanide to carbonyl compounds, *Trans. Chem. Soc.*, **83**, 995.

Lapworth, A. (1904) The action of halogens on compounds containing the carbonyl group, *Trans. Chem. Soc.*, **85**, 30.

Lapworth, A. (1907) Oxime formation and decomposition in the presence of mineral acids, *Trans. Chem. Soc.*, **91**, 1133.

Leffler, J. E. (1953) Parameters for the description of transition states, *Science*, **117**, 340.

Leffler, J. E. & Grunwald, E. (1963) *Rates and Equilibria of Organic Reactions*, John Wiley, New York.

Lehn, J-M. & Sirlin, C. (1978) Molecular catalysis: enhanced rates of thiolysis with high structural and chiral recognition in complexes of a reactive macrocyclic receptor molecule, *J, Chem. Soc. Chem. Commum.*, 949.

Leo, A., Hansch, C. & Elkins, D. (1971) Partition coefficients and their uses, *Chem. Rev.*, **71**, 525.

Leo, A., Jow, P. Y. C., Silipo, C. & Hansch, C. (1975) Calculation of hydrophobic constant ($\log P$) from π and f constants, *J. Med. Chem.*, **18**, 865.

Lewis, E. S. (1982) Nitroactivated carbon acids, in *The Chemistry of Amino, Nitroso and Nitro Compounds and Their Derivatives*, Patai, S. (ed.), John Wiley, Chichester, Supplement F, Part 2, p. 715.

Lewis, E. S. (1986) Linear free energy relationships, in *Investigations of Rates and Mechanisms of Reactions*, Bernasconi, C. F. (ed.), Wiley–Interscience, New York, Part I, p. 871.

Lewis, E. S. & Hu, D. D. (1984) Methyl transfers 8. The Marcus equation and transfers between arenesulphonates, *J. Am. Chem. Soc.*, **106**, 3292.

Lexa, D., Savéant, J.-M., Su, K. B. & Wang, D. L. (1987) Chemical versus redox catalysis of electrochemical reactions. Reduction of *trans*-1,2-dibromocyclohexane in electro-generated aromatic anion radicals, *J. Am. Chem. Soc.*, **109**, 6464.

Littlechild, J. A. & Watson, H. C. (1993) A data based reaction mechanism for type I fructose biphosphate aldolase, *Trends Biochem. Sci.*, **18**, 36.

Littler, J. S. (1989) The detailed linear representation of reaction mechanisms, *Pure Appl. Chem.*, **61**, 57.

Lodi, P. J. & Knowles, J. R. (1993) Direct evidence for the exploitation of an α-helix in the catalytic mechanism of triose phosphate isomerase, *Biochemistry*, **32**, 4338.

Lowe, G. (1976) The cysteine proteases, *Tetrahedron,* **32**, 291.

Lowe, G. (1983) Chiral [^{16}O,^{17}O,^{18}O]phosphate esters, *Acc. Chem. Res.*, **16**, 244.

Lowe, G. (1991a) Mechanisms of sulphate activation and transfer, *Philos. Trans. R. Soc. London, Ser. B*, **332**, 141.

Lowe, G. (1991b) The stereochemical course of sulphuryl transfer reactions, *Phosphorus, Sulfur Silicon*, **59**, 63.

Lowry, T. H. & Richardson, K. S. (1987) *Mechanism and Theory in Organic Chemistry*, 3rd edn, Harper and Row, New York, p. 596.

Lutly, J., Retey, J. & Arigoni, D. (1969) Preparation and detection of chiral methyl groups, *Nature (London)*, **221**, 1213.

McClelland, R. A. & Santry, L. J. (1983) Reactivity of tetrahedral intermediates, *Acc. Chem. Res.*, **16**, 394.

McInnes, I., Nonhebel, D. C., Orszulik, S. T. & Suckling, C. J. (1983) On the mechanism of hydrogen transfer by nicotinamide coenzymes and alcohol dehydrogenase, *J. Chem. Soc., Perkin Trans. 1*, 2777.

Mandolini, L. (1986) Intramolecular reactions of chain molecules, *Adv. Phys. Org. Chem.*, **22**, 1.

Marcus, R. A. (1992) Skiing the reaction rate slopes, *Science*, **256**, 1523.

Martell, A. E. (1989) Vitamin B_6-catalysed reactions of α-amino acids and α-keto acids, *Acc. Chem. Res.*, **22**, 115.

Martin, J. C. (1983) Frozen transition states—pentavalent carbon *et al.*, *Science*, **221**, 509.

Martin, J. C. & Scott, J. M. W. (1967) Thermodynamic parameters and solvent isotope effects as mechanistic criteria in the neutral hydrolysis of some alkyl trifluoracetates in water and deuterium oxide, *Chem. Ind. (London)*, 665.

Matthews, B. W., Sigler, P. B., Henderson, R. & Blow, D. M. (1967) Three-dimensional structure of tosyl-α-chymotrypsin, *Nature (London)*, **214**, 652.

Matthews, B. W., Alden, R. A., Birktoft, J. J., Freer, S. T. & Kraut, J. (1975) X-ray crystallographic study of boronic adducts with subtilisin BPN$'$ (Novo), *J. Biol. Chem.*, **250**, 7120.

Mehler, A. H. & Bloom, B. (1963) Interaction between rabbit muscle aldolase and dihydroxyacetone phosphate, *J. Biol. Chem.*, **238**, 105.

Melander, L. & Saunders, W. H. (1980) *Reaction Rates of Isotopic Molecules*, John Wiley, New York, p. 266.

Menger, F. M. (1979) On the structure of micelles, *Acc. Chem. Res.*, **12**, 111.

Menger, F. M. & Bender, M. L. (1966) The effect of charge transfer complexation on the hydrolysis of some carboxylic acid derivatives, *J. Am. Chem. Soc.*, **88**, 131.

Menger, F. M., Jerkunica, J. M. & Johnston, C. (1978) The water content of a micelle interior. The Fjord versus Reef models, *J. Am. Chem. Soc.*, **100**, 4676.

Meyerson, S., Kuhn, E. S., Ramirez, F., Maracek, J. F. & Okazaki, H. (1978) Electron impact and field ionisation mass spectrometry of α-ketol phosphate salts. Gas phase thermolysis of phosphodiester to monomeric alkyl metaphosphate, *J. Am. Chem. Soc.*, **100**, 4062.

Mills, I. M. (1977) Potential energy surfaces from vibrational rotational data, *Faraday Disc.*, **62**, 7.

Minor, S. S. & Schowen, R. L. (1973) One proton solvation bridge in intramolecular carboxylate catalysis of ester hydrolysis, *J. Am. Chem. Soc.*, **95**, 2279.

Mock, W. L. & Chua, D. L. Y. (1995) Exceptional active site H-bonding in enzymes? Significance of the 'oxyanion hole' in the serine proteases from a model study, *J. Chem. Soc., Perkin Trans. 2*, 2069.

Mock, W. L. & Shih, N.-Y. (1989) Dynamics of molecular recognition involving cucerbituril, *J. Am. Chem. Soc.*, **111**, 2697.

More O'Ferrall, R. A. (1975) Substrate isotope effects, in *Proton Transfer Reactions*, Caldin, E. F. & Gold, V. (eds), Chapman and Hall, London, p. 201.

More O'Ferrall, R. A. & Ridd, J. (1963) Kinetic dependence of some base-catalysed reactions on the concentration of sodium methoxide, *J. Chem. Soc.*, 5035.

Morris, A. J. & Tolan, D. R. (1994) Role of lysine-146 in class 1 aldolase, *Biochemistry*, **33**, 12291.

Morris, J. J. & Page, M. I. (1980a) Buffer catalysis in the hydrazinolysis of benzylpenicillin, *J. Chem. Soc., Perkin Trans. 2*, 220.

Morris, J. J. & Page, M. I. (1980b) Intramolecular nucleophilic and general acid catalysis in the hydrolysis of an amide. Some comments on the mechanism of catalysis of serine proteases, *J. Chem. Soc., Perkin Trans. 2*, 1131.

Morris, J. J. & Page, M. I. (1980c) Hydroxy group participation in the hydrolysis of amides and its effective concentration in the absence of strain effects, *J. Chem. Soc., Perkin Trans. 2*, 679.

Morris, J. J. & Page, M. I. (1980d) Structure–reactivity relationships and the mechanism of general base catalysis in the hydrolysis of a hydroxy amide, *J. Chem. Soc., Perkin Trans. 2*, 685.

Muller, N. (1990) Search for a realistic view of hydrophobic effects, *Acc. Chem. Res.*, **23**, 23.

Murray, C. J. & Webb, T. (1991) Characterization of transition states by isotopic mapping and structure–reactivity coefficients: vibrational analysis calculations for the general base catalysed addition of alcohols to acetaldehyde, *J. Am. Chem. Soc.*, **113**, 1684.

Nakagawa, Y., King Sun, L.-H. & Kaiser, E. T. (1976) Detection of covalent intermediates by nucleophilic trapping in the hydrolysis of phenyl tetrahydrofurfuryl sulfite catalysed by pepsin, *J. Am. Chem. Soc.*, **98**, 1616.

Newcomb, M. & Curran, D. P. (1988) A critical evaluation of studies employing alkyl halide 'mechanistic probes' as indicators of single electron transfer processes, *Acc. Chem. Res.*, **21**, 206.

Nickborg, E. B., Davenport, R. C., Petsko, G. A. & Knowles, J. R. (1988) Triose phosphate isomerase: removal of a putatively electrophilic histidine residue results in a subtle change in catalytic mechanism, *Biochemistry*, **27**, 5948.

Ohba, T., Tsuchiya, N., Nishimura, K., Ikeda, E., Wakagama, J. & Takei, H. (1996) Mechanism-based inactivation of α-chymotrypsin, *Bioorg. Med. Chem. Lett.*, **6**, 543.

Olah, G. A., O'Brien, D. H. & White, A. M. (1967) Stable carbonium ions LII. Protonated esters and their cleavage in fluorosulfonic acid–antimony pentafluoride solution, *J. Am. Chem. Soc.*, **89**, 5694.

O'Leary, M. H. (1988) Transition state structures in enzyme catalysed decarboxylation, *Acc. Chem. Res.*, **21**, 458.

Page, M. I. (1973) The energetics of neighbouring group participation, *Chem. Soc. Rev.*, **2**, 295.

Page, M. I. (1984a) The energetics and specificity of enzyme–substrate interactions, in Page (1984c), p. 1.

Page, M. I. (1984b) The mechanisms of chemical catalysis used in enzymes, in Page (1984c), p. 229.

Page, M. I. (ed.) (1984c) *The Chemistry of Enzyme Action*, Elsevier, Amsterdam.

Page, M. I. (1987) The mechanisms of reactions of β-lactam antibiotics, *Adv. Phys. Org. Chem.*, **23**, 165.

Page, M. I. (1990) Enzyme inhibition, in *Comprehensive Medicinal Chemistry*, Sammes, P. G. (ed.), Pergamon, Oxford, Vol. 2, p. 61.

Page, M. I. (1991) The energetics of intramolecular reactions and enzyme catalysis, *Philos. Trans. R. Soc. London, Ser. B*, **332**, 149.

Page, M. I. & Crombie, D. A. (1982) Reactions in macromolecular systems, *Macromol. Chem.*, **2**, 322.

Page, M. I. & Crombie, D. A. (1984) Reactions in macromolecular systems, *Macromol. Chem.*, **3**, 351.

Page, M. I. & Jencks, W. P. (1971) Entropic contributions to rate accelerations in enzymic and intramolecular reactions and the chelate effect, *Proc. Nat. Acad. Sci. U.S.A.*, **68**, 1678.

Page, M. I. & Williams, A. (eds) (1987) *Enzyme Mechanisms*, Royal Society of Chemistry, London.

Page, M. I., Casey, L. A. & Galt, R. (1993) The mechanism of hydrolysis of the γ-lactam isatin and its derivatives, *J. Chem. Soc., Perkin Trans. 2*, 23.

Patterson, J. F., Huskey, W. P., Venkatasubban, K. S. & Hogg, J. L. (1978) One proton catalysis in the intermolecular imidazole catalysed hydrolysis of esters, *J. Org. Chem.*, **43**, 4935.

Pau, J. K., Kim, J. K. & Caserio, M. C. (1978) Mechanisms of ionic reactions in the gas phase. Displacement reactions at carbonyl carbon, *J. Am. Chem. Soc.*, **100**, 3831.

Pellerite, M. J. & Brauman, J. I. (1980) Intrinsic barriers in nucleophilic displacements, *J. Am. Chem. Soc.*, **102**, 5993.

Perkins, C. W., Wilson, S. R. & Martin, J. C. (1985) Ground state analogues of transition states for attack of sulphonyl, sulphinyl and sulphenyl sulphur, *J. Am. Chem. Soc.*, **107**, 3209.

Perrin, D. D., Dempsey, B. & Serjeant, E. P. (1981) pK_a *Prediction for Organic Acids and Bases*, Chapman and Hall, London.

Petsko, G. A. & Ringe, D. (1984) Fluctuations in protein structure from X-ray diffraction, *Annu. Rev. Biophys. Bioeng.*, **13**, 331.

Pfenninger, J. P., Henberger, C. & Graf, W. (1980) The radical induced stannane reduction of seleno esters and seleno carbonates: a new method for the degradation of carboxylic acids to alkanes and for decarbonylation of alcohols to alkanes, *Helv. Chim. Acta*, **63**, 2328.

Pilling, M. J. & Seakins, P. W. (1995) *Reaction Kinetics*, Oxford Science Publications, Oxford.

Pilling, M. J. and Smith, I. W. M. (1987) *Modern Gas Kinetics: Theory, Experiment and Application*, Blackwell Scientific, Oxford.

Polanyi, J. C. & Zewail, A. H. (1995) Direct observation of the transition state, *Acc. Chem. Research*, **28**, 119.

Pollock, E., Hogg, J. L. & Schowen, R. L. (1973) One proton catalysis in the deacetylation of acetyl-α-chymotrypsin, *J. Am. Chem. Soc.*, **95**, 968.

Prince, R. H. & Wooley, P. R. (1972) Metal-ion assisted catalysis of nucleophilic attack, *J. Chem. Soc. Dalton*, 1548.

Pross, A. (1977) The reactivity–selectivity principle and its mechanistic applications, *Adv. Phys. Org. Chem.*, **14**, 69.

Pross, A. (1983) On the breakdown of rate equilibrium relationships, *Tetrahedron Lett.*, **24**, 835.

Pross, A. (1985) The single electron shift as a fundamental process in organic chemistry. The relationship between polar and electron transfer pathways, *Acc. Chem. Res.*, **18**, 212.

Pross, A. & Shaik, S. S. (1989) Brønsted coefficients. Do they measure transition state structure? *New J. Chem.*, **13**, 427.

Provalov, P. L. & Gill, S. J. (1989) The hydrophobic effect: a reappraisal, *Pure Appl. Chem*, **61**, 1097.

Quinn, D. M. & Sutton, L. D. (1991) Theoretical basis and mechanistic utility of solvent isotope effects, in *Enzyme Mechanism from Isotope Effects*, Cook, P F. (ed.), CRC Press, Boca Raton.

Quinn, D. M., Venkatasubban, K. S., Kise, M. & Schowen, R. L. (1980) Protonic reorganization and substrate structure in catalysis by amido hydrolase, *J. Am. Chem. Soc.*, **102**, 5365.

Rebek, J. (1990) Molecular recognition with model systems, *Angew. Chem., Int. Ed. Engl.*, **29**, 245.

Reichardt, C. (1965) Empirical parameters of the polarity of solvents, *Agnew. Chem., Int. Ed. Engl.*, **4**, 29.

Reichardt, C. (1979) *Solvent Effects in Organic Chemistry*, Verlag Chemie, Weinheim.

Reichardt, C. (1994) Solvatochromic dyes as solvent polarity indicators, *Chem. Rev.*, **94**, 2319.

Renfrew, A. H. M., Rettura, D., Taylor, J. A., Whitmore, J. M. J. & Williams, A. (1995) Stepwise versus concerted mechanisms at trigonal carbon, *J. Am. Chem. Soc.*, **117**, 5484.

Richard, J. P. (1995) A consideration of the barrier for carbocation-nucleophile combination reactions, *Tetrahedron*, **51**, 1535.

Richard, J. P. & Jencks, W. P. (1982) A simple relationship between carbocation lifetime and reactivity selectivity relationships for the solvolysis of ring-substituted 1-phenylethyl derivatives, *J. Am. Chem. Soc.*, **104**, 4689.

Rieder, S. C. & Rose, I. A. (1959) The mechanism of the triosephosphate isomerase reaction, *J. Biol. Chem.*, **234**, 1007.

Roberts, J. D. & Moreland, W. J. (1953) Electrical effects of substituent groups in saturated systems. Reactivities of 4-substituted bicyclo[2.2.2]octane-1-carboxylic acids, *J. Am. Chem. Soc.*, **75**, 2167.

Robertus, J. D., Kraut, J., Alden, R. A. & Birktoft, J. J. (1972) Subtilisin; a stereochemical mechanism involving transition state stabilisation, *Biochemistry*, **11**, 4293.

Rodgers, J., Femec, D. A. & Schowen, R. L. (1982) Isotopic mapping of transition-state structural features associated with enzymic catalysis of methyl transfer, *J. Am. Chem. Soc.*, **104**, 3263.

Rogers, G. A. & Bruice, G. A. (1974) Synthesis and evaluation of a model for the so-called 'charge-relay' system of the serine esterases, *J. Am. Chem. Soc.*, **96**, 2473.

Rose, I. A. (1958) Absolute configuration of dihydroxyacetone phosphate tritiated by aldolase reaction, *J. Am. Chem. Soc.*, **80**, 5835.

Rossi, R. A., Pierini, A. B. & Palacio, S. M. (1989) The $S_{RN}1$ mechanism as a route to nucleophilic substitution on alkyl halides, *J. Chem. Ed.*, **66**, 720.

Rynbrandt, J. D. and Rabinovitch, B. S. (1971) Intramolecular energy relaxation. Non-random decomposition of hexafluorobicyclopropyl, *J. Phys. Chem.*, **75**, 2164.

Salvadori, G. & Williams, A. (1971) Demonstration of intermediates in the hydrolysis of 4-ethoxy pyrylium salts, *J. Am. Chem. Soc.*, **93**, 2727.

Saunders, W. H. (1976) Distinguishing between concerted and nonconcerted eliminations, *Acc. Chem. Res.*, **8**, 19.

Saunders, W. H. (1986) Kinetic isotope effects, in *Investigation of Rates and Mechanisms of Reactions*, Bernasconi, C. F. (ed.), Wiley–Interscience, New York, Part I, p. 565.

Saunders, W. H., Cockerill, A. F., Asperger, S., Klasinc, L. & Stefanovic, D. (1966) The sulphur isotope effect in the E2 reaction of 2-phenylethyldimethylsulphonium bromide with hydroxide ion. A correction, *J. Am. Chem. Soc.*, **88**, 848.

Savéant, J.-M. (1990) Single electron transfer and nucleophilic substitution, *Adv. Phys. Org. Chem.*, **26**, 1.

Savéant, J.-M. (1993) Electron transfer, bond breaking and bond formation, *Acc. Chem. Res.*, **26**, 455.

Schadt, F. L., Bentley, T. W. & Schleyer, P. V. R. (1976) The S_N2–S_N1 spectrum. Quantitative treatments of nucleophilic solvent assistance. A scale of nucleophilicities, *J. Am. Chem. Soc.*, **98**, 7667.

Schaleger, L. L. & Long, F. A. (1963) Entropies of activation and mechanism of reactions in solution, *Adv. Phys. Org. Chem.*, **1**, 1.

Scharschmidt, M., Fisher, M. A. & Cleland, W. W. (1984) Variation of transition state structure as a function of the nucleophile in reactions catalysed by dehydrogenases: I. Liver alcohol dehydrogenase with benzyl alcohol and yeast alcohol dehydrogenase with benzaldehyde, *Biochemistry*, **23**, 547.

Schmidt, J., Chen, J., De Traglia, M., Muikel, D. & McFarland, J. J. (1979) Solvent isotope effect on the liver alcohol dehydrogenase reaction, *J. Am. Chem. Soc.*, **101**, 3634.

Schneider, J. & Kent, S. B. H. (1988) Enzymatic activity of a synthetic 99 residue protein corresponding to the putative HIV-1 protease, *Cell*, **54**, 363.

Schowen, K. B. J. (1978) Solvent hydrogen isotope effects, in *Transition States of Biochemical Processes*, Schowen, R. L. & Gandour, R. D. (eds), Plenum, New York, p. 225.

Schowen, R. L. (1972) Mechanistic deductions from solvent isotope effects, *Prog. Phys. Org. Chem.*, **9**, 275.

Scott, F. L., Flynn, E. J. & Fenton, D. F. (1971) Ambident neighbouring groups Part V. Effects of solvent on several O-5 ring closures, *J. Chem. Soc. B.*, 277.

Secemski, I. I., Lehrer, S. S. & Lienhard, G. E. (1972) Transition state analogue for lysozyme, *J. Biol. Chem.*, **247**, 4740.

Shaik, S. S. (1985) The collage of S_N2 reactivity patterns: a state correlation diagram model, *Progr. Phys. Org. Chem.*, **15**, 197.

Shaik, S. S. (1990) The S_N2 and single electron transfer concepts. A theoretical and experimental overview, *Acta Chem. Scand. Ser. B*, **44**, 205.

Shorter, J. (1972) The separation of polar, steric and resonance effects by the use of linear free energy relationships, in *Advances in Linear Free Energy Relationships*, Chapman, N. B. & Shorter, J. J. (eds), Plenum, London, Chapter 2, p. 71.

Shorter, J. (1978) The conversion of ammonium cyanate into urea—a saga in reaction mechanisms, *Chem. Soc. Rev.*, **7**, 1.

Sinnott, M. L. (1987) Glycosyl group transfer, in Page & Williams (1987).

Sinnott, M. L. (1988) The principle of least nuclear motion and the theory of stereoelectronic control, *Adv. Phys. Org. Chem.*, **24**, 113.

Sinnott, M. L. (1990) Catalytic mechanisms of enzymic glycosyl transfer, *Chem. Rev.*, **91**, 1171.

Skoog, M. T. & Jencks, W. P. (1984) Reactions of pyridines and primary amines with *N*-phosphorylated pyridines, *J. Am. Chem. Soc.*, **106**, 7597.

Smith, I. W. M. (1990) Femtosecond chemistry, *Nature (London)*, **343**, 691.

Smith, I. W. M. (1992) Probing the transition state, *Nature (London)*, **358**, 279.

Smith, P. J. & Bourns, A. N. (1970) Isotope effect studies on elimination reactions VI. The mechanism of the bimolecular elimination reaction of 2-arylethylammonium ions, *Can. J. Chem.*, **48**, 125.

Sneen, R. A. (1973) Organic ion pairs as intermediates in nucleophilic substitution and elimination reactions, *Accs. Chem. Res.*, **6**, 46.

Sommer, L. H. (1965) *Stereochemistry, Mechanism and Silicon*, McGraw-Hill, New York.

Spector, L. B. (1982) *Covalent Catalysis of Enzymes*, Springer, New York.

Stark, G. R. (1965) Reactions of cyanate with functional groups of proteins III. Reactions with amino and carboxyl groups, *Biochemistry*, **4**, 1030.

Stauffer, D. A., Barrass, R. E. & Dougherty, D. A. (1990) Biomimetic catalysis of an S_N2 reaction resulting from a novel form of transition state stabilisation, *Angew. Chem., Int. Ed. Engl.*, **29**, 915.

Stein, R. L., Elrod, J. P. & Schowen, R. L. (1983) Correlative variations in enzyme-derived and substrate-derived structures of catalytic transition states. Implications for the catalytic strategy of acyl-transfer enzymes, *J. Am. Chem. Soc.*, **105**, 2446.

Steitz, T. A., Henderson, R. & Blow, D. M. (1969) Structure of crystalline α-chymotrypsin III. Crystallographic studies of substrates and inhibitors bound to the active site of α-chymotrypsin, *J. Mol. Biol.*, **46**, 337.

Steitz, T. A. & Steitz, J. A. (1993) A general 2-metal ion mechanism for catalytic RNA, *Proc. Nat. Acad. Sci. U.S.A.*, **90**, 6498.

Street, I. P., Kempton, J. B. & Withers, S. G. (1992) Inactivation of a β-glucosidase through the accumulation of a stable 2-deoxy-2-fluoro-α-D-glucopyranosyl-enzyme intermediate—a detailed investigation, *Biochemistry*, **31**, 9970.

Streitwieser, A., Jagow, R. H., Fahey, R. C. & Suzuki, S. (1958) Kinetic isotope effects in the acetolysis of deuterated cyclopentyl tosylates, *J. Am. Chem. Soc.*, **80**, 2326.

Ta-Shma, R. & Rappoport, Z. (1991) Solvent induced changes in the selectivity of solvolyses in aqueous alcohols and related mixtures, *Adv. Phys. Org. Chem.*, **27**, 239.

Taylor, P. J. (1993) On the calculation of tetrahedral intermediate pK_a values, *J. Chem. Soc., Perkin Trans. 2*, 1423.

Tedder, J. M. & Nechvatal, A. (1988) *Pictorial Orbital Theory*, Pitman, London.

Tee, O. S. (1994) The stabilisation of transition states by cyclodextrins and other catalysts, *Adv. Phys. Org. Chem.*, **29**, 1.

Thatcher, G. R. J. & Kluger, R. (1989) Mechanism and catalysis of nucleophilic substitution in phosphate esters, *Adv. Phys. Org. Chem.*, **25**, 99.

Thea, S. & Williams, A. (1986) Measurement of effective charge in organic reactions in solution, *Chem. Soc. Rev.*, **15**, 125.

Thea, S., Harun, M. G. & Williams, A. (1979) Mechanism in sulphonyl group transfer: effective charge in ground-transition-, and product-states during base catalysed hydrolysis of aryl sulphonate esters possessing associative (AE) and dissociative (EA) mechanisms, *J. Chem. Soc., Chem. Commun.*, 717.

Thea, S., Guanti, G., Kashefi-Naini, N. & Williams, A. (1983) Steric and electronic control of the dissociative hydrolysis of 4-hydroxybenzoate esters, *J. Chem. Soc., Chem. Commun.*, 529.

Thea, S., Cevasco, G., Guanti, G., Kashefi-Naini, N. & Williams, A. (1985) Reactivity in the *para*-oxoketene route of ester hydrolysis: the effect of internal nucleophilicity and the irrelevance of 'B' strain, *J. Org. Chem.*, **50**, 1867.

Thornton, E. R. (1967) A simple theory for predicting the effects of substituent changes on transition state geometry, *J. Am. Chem. Soc.*, **89**, 2915.

Todd, A. R. (1959) Some aspects of phosphate chemistry, *Proc. Natl. Acad. Sci. U.S.A.*, **45**, 1389.

Toney, M. D. & Kirsch, J. F. (1989) Direct Brønsted analysis of the restoration of activity to a mutant enzyme by exogenous amines, *Science*, **243**, 1485.

Traylor, P. S. & Westheimer, F. H. (1965) Mechanisms in the hydrolysis of phosphorodiamidic chlorides, *J. Am. Chem. Soc.*, **87**, 553.

Tso, P. O. P., Melvin, I. S. & Olson, A. (1963) Interaction and association of bases and nucleosides in aqueous solution, *J. Am. Chem. Soc.*, **85**, 1289.

Ulmer, K. M. (1983) Protein engineering, *Science*, **219**, 666.

van Bekkum, H., Verkade, P. E. & Wepster, B. M. (1959) A simple re-evaluation of the Hammett ρ–σ relation, *Rec. Trav. Chim. Pays Bas*, **78**, 815.

Van Etten, R. L., Sebastian, J. F., Clowes, G. A. & Bender, M. L. (1967) Acceleration of phenyl ester cleavage by cycloamyloses. A model for enzymatic specificity, *J. Am. Chem. Soc.*, **89**, 3242.

Verbit, L. (1970) Optically active deuterium compounds, *Prog. Phys. Org. Chem.*, **7**, 51.

Waley, S. G., Mill, J. C., Rose, I. A. & O'Connell, E. L. (1970) Identification of site in triose phosphate isomerase labelled by glycidophosphate, *Nature (London)*, **227**, 181.

Walter, C. J., Anderson, H. L. & Sanders, J. K. M. (1993) Exo-selective acceleration of an intermolecular Diels–Alder reaction by a trimeric host, *J. Chem. Soc., Chem. Commun.*, 458.

Wang, Q. & Withers, S. G. (1995) Substrate-assisted catalysis in glycosidases, *J. Am. Chem. Soc.*, **117**, 10 137.

Weatherhead, R. H., Stacey, K. A. & Williams, A. (1980) Studies with receptor and microgels as matrices for molecular receptor and catalytic sites; the exceptional reactivity of hydroxamic groups in polymers with a high level of cross-linking monomer feed, *Makromol. Chem.*, **181**, 2529.

Weiner, H., White, W. N., Hoare, D. G. & Koshland, D. E. (1966) The formation of anhydrochymotrypsin by removing the elements of water from the serine at the active site, *J. Am. Chem. Soc.*, **88**, 3851.

Wescott, C. M. & Klibanov, A. M. (1993) Solvent variation inverts substrate specificity of an enzyme, *J. Am. Chem. Soc.*, **115**, 1629.

Westheimer, F. H. (1961) The magnitude of the primary kinetic isotope effect for compounds of hydrogen and deuterium, *Chem. Rev.*, **61**, 265.

Westheimer, F. H. (1968) Psuedo-rotation in the hydrolysis of phosphate esters, *Accs. Chem. Res.*, **1**, 70.

Westheimer, F. H. (1981) Monomeric metaphosphate, *Chem. Rev.*, **81**, 313.

Westheimer, F. H. (1987) Why nature chose phosphates, *Science*, **235**, 1173.

Westheimer, F. H. and Bender, M. L. (1962) Imidazole catalysis of the hydrolysis of δ-thiovalerolactone, *J. Am. Chem. Soc.*, **84**, 4908.

Whalley, E. (1964) Use of volumes of activation for determination of reaction mechanisms, *Adv. Phys. Org. Chem.*, **2**, 93.

Wilkinson, F. (1980) *Chemical Kinetics and Reaction Mechanism*, Van Nostrand–Reinhold, Wokingham, UK.

Wilkinson, G. N. (1961) Statistical estimation in enzyme kinetics, *Biochem. J.*, **80**, 324.

Williams, A. (1971) The hydrolysis of pyrylium salts: kinetic evidence for hemiacetal intermediates, *J. Am. Chem. Soc.*, **93**, 2733.

Williams, A. (1984) Effective charge and Leffler's index as mechanistic tools for reactions in solution, *Acc. Chem. Res.*, **17**, 425.

Williams, A. (1985) Bonding in phosphoryl ($-PO_3^{2-}$) and sulphuryl ($-SO_3^-$) group transfer between nitrogen nucleophiles as determined from rate constants for identity reactions, *J. Am. Chem. Soc.*, **107**, 6335.

Williams, A. (1989) Concerted mechanisms of acyl group transfer reactions in solution, *Acc. Chem. Res.*, **22**, 387.

Williams, A. (1991) Effective charge and transition state structure in solution, *Adv. Phys. Org. Chem.*, **27**, 1.

Williams, A. & Douglas, K. T. (1972) E1cB mechanisms Part II. Base hydrolysis of substituted phenyl phosphorodiamidates, *J. Chem. Soc., Perkin Trans. 2*, 1454.

Williams, A. & Douglas, K. T. (1973) E1cB mechanisms Part IV. Base hydrolysis of substituted phenyl phosphoro- and phosphorothio-diamidates, *J. Chem. Soc., Perkin Trans. 2*, 318.

Williams, A. & Douglas, K. T. (1975) Elimination–addition mechanisms of acyl transfer reactions, *Chem. Rev.*, **75**, 627.

Williams, A. & Jencks, W. P. (1974) Urea synthesis from amines and cyanic acid: kinetic evidence for a zwitterionic intermediate, *J. Chem. Soc., Perkin Trans. 2*, 1753.

Wilson, J. M., Bayer, R. J. & Hupe, D. J. (1977) Structure reactivity correlations for the thiol–disulphide interchange reaction, *J. Am. Chem. Soc.*, **97**, 7922.

Wintner, E. A., Conn, M. M. & Rebek, J. (1994) Studies in molecular replication, *Acc. Chem. Res.*, **27**, 198.

Withers, S. G. & Street, I. P. (1988) Identification of a covalent α-D-glucopyranosyl–enzyme intermediate formed on a β-glucosidase, *J. Am. Chem. Soc.*, **110**, 8551.

Yukawa, Y. & Tsuno, Y. (1959) Resonance effect in Hammett relationships III. The modified Hammett relationships for electrophilic reactions, *Bull. Chem. Soc. Jpn*, **32**, 971.

Zefirov, N. S. (1977) General equation of the relationship between product ratio and conformational equilibrium, *Tetrahedron*, **33**, 2719.

Zewail, A. H. (1988) Laser femtochemistry, *Science*, **242**, 1645.

Zucker, L. & Hammett, L. P. (1939) Kinetics of iodination of acetone in sulphuric and perchloric acid solutions, *J. Am. Chem. Soc.*, **61**, 2791.

Index